STRONTIUM ALUMINATE - CEMENT FUNDAMENTALS, MANUFACTURING, HYDRATION, SETTING BEHAVIOUR AND APPLICATIONS

Authored by **Petr Ptáček**

Strontium Aluminate - Cement Fundamentals, Manufacturing, Hydration, Setting Behaviour and Applications
http://dx.doi.org/10.5772/57363
Authored by Petr Ptáček

Contributors
Petr Ptáček

Published by InTech
Janeza Trdine 9, 51000 Rijeka, Croatia

Publishing Process Manager Danijela Duric
Technical Editor InTech DTP team
Cover InTech Design team

First published July, 2014
Printed in Croatia

Additional hard copies can be obtained from orders@intechopen.com

Strontium Aluminate - Cement Fundamentals, Manufacturing, Hydration, Setting Behaviour and Applications, Authored by Petr Ptáček
p. cm.
ISBN 978-953-51-1591-5

For EU product safety concerns:
IN TECH d.o.o., Prolaz Marije Krucifikse Kozulić 3, 51000 Rijeka, Croatia,
info@intechopen.com or visit our website at intechopen.com.

Meet the author

Petr Ptáček was born in 1978 in Kutná Hora, Czech Republic. He received his PhD in materials science and technology from Brno University of Technology, Faculty of Chemistry in 2005. Since 2005 up to now he has been working at the Institute of Materials Chemistry and at the Materials Research Centre of Brno University of Technology as an Assistant Professor and a Senior Researcher, respectively. Current teaching and research activities include chemistry, materials structure and thermodynamics, heat treatment of silicates, ceramics and advanced ceramics, refractory materials, special inorganic cements, geopolymers, colloidal chemistry and heterogeneous kinetics. He was awarded several times for his out-standing academic performance in the fields mentioned above. Ptáček has published over 75 scientific articles, including full-length articles, book chapters and contributions to international conferences; he registered 4 national patents and participated in the development of 4 applied industrial technologies.

Contents

Preface

The goal of this book, "Strontium Aluminate Cement: Fundamentals, Manufacturing, Hydration, Setting behaviour and Applications", is to provide the reader with basic overview of this non-traditional binder and to support the researcher`s interest in this topic. Despite the fact, that massive production of strontium aluminate or tri-strontium aluminate cements for industrial application is improbable, the binder shows a few favorable properties. Some of them are described in this book, other must be discovered first.

Really promising options are the control of reactivity and rate of heat release during the hydration of strontium aluminate cement, the control of processing of strontium aluminate based MDF, of preparation of ceramic foam stabilized by strontium aluminate cement and of dense or thermal insulating refractory castables with tailored thermal expansion for high temperature applications. Moreover, an immense number of combinations can be made for the preparation of blended and multicomponent cements, influence of substitutions on the properties of strontium aluminate. On behalf of all authors I hope that that this publication hold your interest.

Petr Ptáček
University of Technology,
Czech Republic

Special Thanks

The authors of this book are honored to spotlight the following co-workers for their immeasurable contributions in laboratory preparations and specimen analyses: Eva Bartoníčková (Raman spectroscopy), Jiří Másilko (X-ray diffraction analysis), Radoslav Novotný (calorimetry), Jaromír Wasserbauer (electron microscopy), Jiří Švec (thermal analysis and infrared spectroscopy) and Martin Boháč (rheology) and Lukáš Kalina (X-ray photoelectron spectroscopy). I would like to express my appreciation to Halina Szklorzová for language corrections.

Acknowledgements

This work was supported by the project "Popularization of Brno University of Technology (BUT) R&D results and the support of systematic collaboration with Czech students" No. CZ.1.07/2.3.00/35.0004.

INVESTICE DO ROZVOJE VZDĚLÁVÁNÍ

Chapter 1

Introduction

1. Discovery of strontium aluminate cements and current state of knowledge

The experimental study of the system CaO-Al_2O_3-H_2O have considerable practical value not only for its direct contribution to the setting of high-alumina cements (which are composed of mainly anhydrous calcium aluminates) and Portland cements, but also for possible applications in the fields of geology, soil equilibrium, water purification, and the extraction of purified hydrated alumina from crude bauxite [1-3,12]. Unfortunately, so far the study of the system CaO-Al_2O_3-H_2O has many difficulties, including [1]:

- The existence of numerous highly hydrated compounds, the determination of the degree of hydration of which is often a problem.
- Relatively low solubility of these compounds.
- The fact that most of the compounds are metastable, yet, once formed; persist over long periods of time.
- Close similarity in crystal habit and optical properties of compounds occurring as thin hexagonal (or pseudohexagonal) plates.
- The necessity for strict exclusion of carbon dioxide from air.

According to the results from numerous investigations of chemical reactions related to the hardening of hydraulic cements [4,12], lime and alumina in solution may combine to form isometric tri-calcium aluminate hexahydrate ($3CaO{\bullet}Al_2O_3{\bullet}6H_2O$, C_3AH_6). This compound is isomorphic to C_3FH_6 garnet as well as grossularite garnet ($3CaO{\bullet}Al_2O_3{\bullet}3SiO_2$, C_3AS_3) and andradite garnet ($3CaO{\bullet}Fe_2O_3{\bullet}3SiO_2$). Complete solid solutions exist between each of these compounds and the three others. The hydrous members of the series may be termed hydrogarnets [4,5,198]. The general formula may be written as $3CaO{\bullet}Al_2O_3{\bullet}mSiO_2{\bullet}(6\text{-}2m)H_2O$, where m may vary from 0 to 3 [6]. Hydrogarnets crystallize in various cubic forms, of which icositetrahedra are probably the most usual at ordinary temperatures. In recent mineralogical nomenclature, the phases in the C_3AH_6 – C_3AS_3 series are collectively called hydrogrossular [7].

Tri-calcium aluminate hexahydrate decomposes under hydrothermal conditions at about equilibrium temperature 220 – 226°C, yielding calcium hydroxide and less basic aluminate. The equation may be written as [4,6]:

$$3\,(3CaO\cdot Al_2O_3\cdot 6\,H_2O) \leftrightarrow 4CaO\cdot 3\,Al_2O_3\cdot 3\,H_2O + 5\,Ca(OH)_2 + 10\,H_2O \qquad (1)$$

Although strontium is rather minor constituent of commonly applied cements such as Portland cement (PC) or alumina cement (AC), the strontium compounds may be formed via the reaction with silica, alumina and water. Despite the fact that the ionic radius of Sr^{2+}(1.26 and 1.44 Å for the coordination number 8 and 12, respectively [92]) is larger than that of Ca^{2+}(1.06 Å), strontium can partially replace calcium in numerous metal oxides and numerous studies concerned to Ca/Sr mixed oxides can be found in literature [8]. The existence of tri-strontium aluminate hexahydrate (3SrO•Al2O3•6SiO2) has been reported by Brandenger [9] and Maekawa [10].

Carlson [4] recognizes tri-strontium aluminate hexahydrate as the only hydrated phase formed in the system strontium – alumina – water. This phase is isomorphous with corresponding calcium compound of tri-calcium aluminate hexahydrate and the complete solid solution exists between those two phases. This is proved by the X-ray diffraction patterns, which indicate the same reflecting planes with such changes in relative intensities as might be expected for isomorphous substitutions. There is also a shift corresponding to somewhat larger unit cell of strontium compounds. The refraction index of tri-strontium aluminate hexahydrate is 1.588 [4,10].

$3SrO\bullet Al_2O_3\bullet 6H_2O$ has higher temperature stability than $3CaO \bullet Al_2O_3\bullet 6H_2O$, which decomposes under hydrothermal conditions according to Eq.1.1. Tri-strontium aluminate hexahydrate decomposes under hydrothermal conditions at temperatures as low as 305 °C (2 atm.) and 454 °C (420 atm.). The product of the decomposition is dependent on both the temperature and the pressure. In most cases the decomposition is accompanied by the liberation of strontium hydroxide. The lower hydrate with 1.5 molecules of water ($3CaO \bullet Al_2O_3\bullet 1/2H_2O$, $C_3AH_{0.5}$) is probably formed. The hexagonal prism of crystalline phase $5SrO\bullet 4Al_2O_3$ was also identified as a product. It decomposes under further heating with the formation of $SrO\bullet Al_2O_3$ and $3SrO\bullet Al_2O_3$ [4].

Therefore, with SrO formally replacing CaO in calcium aluminate cement, the strontium aluminate binder belongs to the family of aluminous cements based on the system of CaO – SrO – BaO – Al_2O_3 – ZrO_2 – HfO_2 [11-14]. The differences and the similarities between the strontium – alumina and lime – alumina systems are briefly mentioned below [14]:

1. There are no compounds like Sr_5A and Sr_4A in the lime – alumina system.
2. No analog of $C_{12}A_7$ in the strontium – alumina system exists[1].
3. The similarities exist to some extent between Sr_3A and C_3A, SrA and CA, SrA_2 and CA_2 as well as between SrA_6 and CA_6, although the melting behaviour and the temperature differ significantly.

1 Phase has not own field in the equilibrium diagram of SrO-Al_2O_3 system, but the synthesis of $Sr_{12}A_7$ is described by Yamaguchi at al [41].

The discovery of calcium aluminates, the study of their crystal chemistry and phase equilibrium, subsequently developing the family of alumina cements and their ultimate field applications covered a span of one and half century, even though Bied received the French and the British patents for the method to produce calcium aluminate cement from bauxite and lime in a cupola furnace in 1908–1909, just about a century ago. The use of AC in basic refractories is rather limited as it leads to the formation of low-melting compounds like monticellite. In certain non-refractory applications AC shows comparatively poorer resistance to aggressive environments as well as to X-rays and gamma radiation [14].

Strontium aluminate cement can be prepared by solid-state sintering of approximately equimolar blend of $SrCO_3$ and Al_2O_3 at the temperature of 1500 °C. The main constituent of strontium aluminate cement is mono-strontium aluminate ($SrO\bullet Al_2O_3$), but strontium hexaluminate may also be present. Strontium aluminate cement hydrates relatively slowly, yielding to strontium aluminate hexahydrate phase of the approximate composition $SrO\bullet Al_2O_3\bullet 7\text{-}10H_2O$ as the main hydration product. In later stages of hydration this phase gradually converts to $6SrO\bullet Al_2O_3\bullet 6H_2O$ and hydrous alumina. This reaction is associated with the loss of strength. The cement has the potential to be used in the production of refractory concretes [12].

Calcium and strontium aluminate cement contains Ca and Sr as the chemical elements from the second group of the periodic table (alkaline earth metals): Be (beryllium), Mg (magnesium), Ca (calcium), Sr (strontium), Ba (barium) and Ra (radium). Therefore, well-founded assumption exists to prepare similar binding system with barium, i.e. barium aluminate cement based on hydraulic properties of barium aluminate or monobarium aluminate ($BaO\bullet Al_2O_3$, BA).

Barium aluminate cement can be produced by solid-state sintering of equimolar blend of $BaCO_3$ (witherite) and Al_2O_3 at the temperature of 1400 – 1500°C. The sintering temperature can be reduced by the addition of suitable mineralizer such as MgO. The main constituent of such cement is barium aluminate, but small amount of tri-barium aluminate ($3BaO\bullet Al_2O_3$, B_3A) and barium hexaaluminate ($BaO\bullet 6Al_2O_3$, BA_6) is formed. At low water to cement ratios ($w/c <$ 0.35), the monobarium aluminate heptahydrate ($BaO\bullet Al_2O_3\bullet 7H_2O$, BAH_7) is formed as the main hydration product. In the case that w/c ratio is higher; the tribarium aluminate hexahydrate ($3BaO\bullet Al_2O_3\bullet 6H_2O$, B_3AH_6) is formed together with hydrous alumina. The reaction of barium aluminate cement with water is fast. Whole BA hydrates completely within three days at sufficiently high water to cement ratio. The hardened cement pastes show only a small strength loss when heated to high temperatures. Barium aluminate cements can be used for the production of refractory materials and they absorb radiation effectively [12,15,19].

The initiation of hydration reactions occurs with leaching of Ba^{2+} ions and with formation of hydroxides following the reaction [14]:

$$BA + 2\,H_2O \rightarrow Ba^{2+} + 2\,AlO^{2-} + H_3O^{+} + OH^{-} \tag{2}$$

Under low w/c ratio the following reactions take place:

$$BA+6\,H \rightarrow BAH_6 \tag{3}$$

$$2\,BAH_6 \rightarrow B_2AH_9+2\,AH_3 \tag{4}$$

$$B_2AH_9+BA \rightarrow B_3AH_6+2\,AH_3 \tag{5}$$

Chaterje [14] described the pilot production process for the preparation of both BAC and SrAC cements, where raw materials in required stoichiometric proportions were cogrinded and next milled in a ball mill to the fineness of 5% residue on 212 μm sieve. The ground raw mix was treated with water and the pellets sized about 15 mm were prepared. Dried pellets were fed into a pilot rotary kiln. The sintering temperature was maintained at about 1620 °C and 1640 °C for SrAC and BAC respectively. The clinker was air quenched and ground to 300 ± 10 m^2•kg^{-1} and 280 ± 10 m^2•kg^{-1} for SrAC and BAC, respectively.

Several types of stoichiometric as well as non-stoichiometric aluminates are known. The most studied and also used aluminates of barium are the stoichiometric aluminates, such as: barium monoaluminate (BaO•Al_2O_3, BA), tribarium monoaluminate (3BaO•Al_2O_3, B_3A), barium tetraaluminate (BaO•4Al_2O_3, BA_4) and barium hexa-aluminate (BaO•6Al_2O_3, BA_6) [16,17,24].

All three alkali-earth monoaluminates, namely CA, SrA and BA being the primary cement-forming phases, properties are listed in Table 1. As to melting temperatures of these phases, their high-temperature stabilities sequence is as follows [14]:

CA > SrA > BA.

Phase	Composition[%]			Molarmass	Density	Crystalsystem	Meltingtemperature/
	Me	Al	O	[g•mol^{-1}]	[g•cm^{-1}]		behaviour
CA	25.4	34.1	40.5	158.04	2.98	Monoclinic	1605°C/ incongruent
SrA	42.6	26.3	31.1	205.58	3.82	Pseudo-hexagonal	1790°C/ congruent
BA	53.8	21.1	25.1	255.29	3.99	Cubic	1815°C/ congruent

Table 1. Properties of $MeAl_2O_4$ phases in calcium, strontium and barium aluminate cements.

Further, the traces of the liquidus curves around SrA and BA are much flatter than those around the CA phase. Thus the compositional deviations in the CA line cements are more precarious for the CA cements than the compositional shifts in BA and SrA cements [14].

2. Phase relationship in SrO – Al_2O_3 system

In this part, the phase relationships in SrO – Al_2O_3 system (Fig.1) will be discussed. The equilibrium phase diagram is useful tool for the determination of phase composition of the system at given temperature and composition. The phase diagrams also enable to read the temperature of formation of eutectic or peritectic melts and equilibrium phases. Using the "**lever rule**" we can estimate the ratio of formed phases or the amount of melt. Moreover there is the possibility to calculate some thermodynamic parameters as will be described in Chapter 1.2.3.

Complete phase diagram of strontium–alumina binary system was constructed as early as 1979 [18,19]. The field of stability of five binary phases, i.e. tetra-strontium aluminate ($Sr_4Al_2O_7$, Sr_4A), cubic tristrontium aluminate ($Sr_3Al_2O_6$, Sr_3A), monoclinic strontium aluminate ($SrAl_2O_4$, SrA), strontium dialuminate ($SrAl_4O_7$, SrA_2) and strontium hexaaluminate phase ($SrAl_{12}O_{19}$, SrA_6) is given below.

High temperature modification of α-Sr_4A is stable in the temperature range from 1320 to 1690 °C. The transformation of tetrastrontium aluminate into low temperature β-Sr_4A phase takes place below 1320 °C. Under the temperature of 1125 °C α-Sr_4A phase is decomposed to Sr_3A and SrO. The phase diagram shows that the low temperature Sr_4A phase forms a solid solution with Sr_3A [18,19].

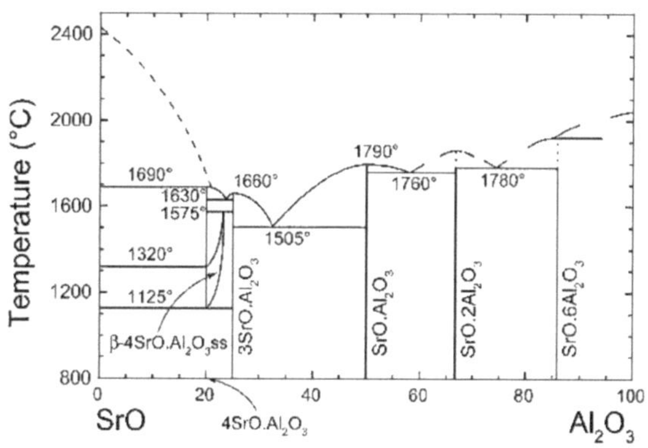

Figure 1. Isobaric equilibrium phase diagram of the system of SrO – Al_2O_3 showing the composition and the temperature region of stability of binary compounds and the position of invariant points.

Liquidus temperature drop rapidly upon the addition of Al_2O_3 to SrO. Thus $SrAl_{12}O_{19}$ melts incongruently to Al_2O_3 at 1690°C. The compositions with the minimum melting temperatures are the eutectics between [18]:

- SrO•2Al_2O_3 and SrO•6Al_2O_3 located at 1780°C;
- SrO•Al_2O_3 and SrO•2Al_2O_3 located at 1760°C;
- 3 SrO•Al_2O_3 and SrO•Al_2O_3 located at 1505°C;
- 4 SrO•Al_2O_3 and 3SrO•Al_2O_3 located at 1630°C.

The following phases occur in the SrO – Al_2O_3 system:

Strontium oxide (strontia, strontium monoxide, SrO, S): is grayish-white colored, has elemental composition of 84.56 % Sr and 15.44 % O and cubic crystalline structure. The molar weight and the density of SrO are 103.63 g•mol^{-1} and 4.70 g•cm^{-3}, respectively. It melts at the temperature of 2430 °C and vaporizes above 3000 °C. It reacts with water to strontium hydroxide ($Sr(OH)_2$) evolving heat. Strontium oxide can be prepared by thermal decomposition of strontium carbonate, hydroxide, or nitrate [91]:

$$SrCO_3(s) \rightarrow SrO(s) + CO_2(g) \tag{6}$$

$$Sr(OH)_2(s) \rightarrow SrO(s) + H_2O(g) \tag{7}$$

$$Sr(NO)_3(s) \rightarrow SrO(s) + N_2O_5(g) \tag{8}$$

SrO is miscible with fused caustic potash, is slightly soluble in alcohol and insoluble in acetone and ether [91]. The methods used for industrial production of strontium carbonate as the intermediate for the preparation of other compounds of strontium is described in Chapter 2.1.1.

Tetrastrontium aluminate ($Sr_4Al_2O_7$, 4SrO•Al_2O_3, Sr_4A): this phase is stable only within the temperature range from 1125 to 1630 °C. It decomposes into mixture of tri-strontium aluminate and free strontium oxide if the temperature is lower than 1125 °C, peritectic melting takes place at the temperature of 1630 °C. In the temperature of 1320 °C the high temperature modification of α-phase is transformed into low temperature β-phase. As can be read from the phase diagram in Fig.1, the low temperature modification of tetra-strontium aluminate forms solid solution with tri-strontium aluminate [14].

Tristrontium aluminate ($Sr_3Al_2O_6$, 3SrO•Al_2O_3, Sr_3A): has cubic "*stuffed-tridimite*" structure with the space group *Pa3* based on $(AlO_3)_6$ rings. The noteworthy feature of this structure is the puckered six-membered tetrahedral AlO_4 rings. There is a front layer and a back layer in this structure, each running in parallel to the other. This structure is typical for strontium aluminates where either tetrahedral AlO_4 or octahedral AlO_6 groups are present (or sometimes both). The lattice parameters are ***a***=15.8556 Å, *V*=3986.3 Å^3, *Z*=24 [19]. The structure of strontium aluminate is shown in Fig.2(a).

Strontium aluminate ($SrAl_2O_4$, SrO•Al_2O_3, SrA): white color, melts congruently at the temperature of 1790°C. The spinel type (AB_2O_4) compound has monoclinic structure that belengs to the $P2_1$ or $P2_1/m$ space group with unit cell parameters *a*=5.1497 Å, *b*=8.836 Å, *c*=8.442 Å, β=93.43° and contains four formula per cell (Z=4). $SrAl_2O_4$ structure is derived from "*stuffed tridimite structure*", where all Si^{4+}cations are replaced by $Al^{3+}$and the charge compensating cations Sr^{2+}occupy the large open channels in the framework[2]. Strontium aluminate is an

2 Analogically to the structure of $CaAl_2O_4$ and $BaAl_2O_4$ that belong to the family of tridymite stuffed derivates of MAl_2O_4, where M = Ca, Sr, Ba. This structure is also typical for compounds such as $BaNSiO_4$ (N = Co, Mg and Zn) and (Na,K)$ZXGeO_4$ (Z = Aland Ga and X = Ge and Si [22]. The structure is similar to zeolites, but differs in the orientation of corner-connected tetrahedral units in the way that some of 6-membered rings convert to 4- and 8-mebered rings [23].

indirect-band-gap oxide and its band gap was estimated to ~6.9 eV [19-23]. The structure of strontium aluminate is shown in Fig.2(b). The high temperature hexagonal polymorph is of $P6_3$ space group with the cell parameters: ***a***=8.926 Å, *c*=8.4985 Å, *γ*=120° and ***Z***=6 [24].

Strontium dialuminate ($SrAl_4O_7$, $SrO•2Al_2O_3$, SrA_2): has two polymorphs, α-SrA_2 is stable at normal pressures, while high pressure form of β-SrA_2 can be prepared only at high pressure and temperature. The crystals of β-SrA_2 are orthorhombic with the space group *Cmma*, ***a***=8.085 Å, ***b***=11.845 Å, *c*=4.407, ***Z***=4 and calculated density of 4.84 g cm^{-3}. The lattice consists of three-dimensional (Al_4O_7) network of AlO_6 octahedra (length bond from 1.795 to 1.968 Å) and AlO_4 tetrahedra (length bond from 1.449 to 1.537 Å). The structure has certain similarities to the structure of tetra-strontium aluminate. Each atom of Sr is surrounded by 10 O atoms with the distances ranging from 2.503 to 2.664 Å [19,25]. Monoclinic α-SrA_2 has the space group C12/c1 with the lattice parameters ***a***=13.0389 Å, ***b***=9.0113 Å, ***c***=5.5358 Å, *β*=106.12° and ***Z***=4 [26].

Strontium hexaaluminate ($SrAl_{12}O_{19}$, $SrO•6Al_2O_3$, SrA_6): this phase has hexagonal magnetoplumbite structure[3] with space group *P6/mmc*. These compounds have a layer structure composed of spinel blocks and conduction layers which are stacked alternately (Fig.2(c)). There are five distinct Al^{3+}sites in the lattice: one tetrahedral (Al(IV), AlO_4 with C_{3v} site symmetry) one trigonal bipiramidal (Al(V), AlO_5 with C_{3v} site symmetry) and three octahedral (Al(VI), AlO_6). The octahedral sites are of different symmetry: regular octahedron (D_{3d}, Al(VI)-1), antiprism (C_{3v}, Al(VI)-2) and distorted octahedron (C_s, Al(VI)-3). The Al(V) sites are present only in so called *"central atom model"* (residual factor 0.0340) while half occupied site of distorted tetrahedron (Al(IV)-d of C_{3v} site symmetry) results from later *"split atom model"* (residual factor 0.0331). The unit cell is composed of spinel-structured slabs containing Al^{3+}cations separated by mirror planes (*2b*[4] site in the central atom model and *4e* with half occupancy in the split atom model) which contain $Sr^{2+}$and three oxygen ions per unit cell. The band gap of strontium aluminate is about 7.6 eV [19,27,28,29,30].

The unit cell parameters of $SrAl_{12}O_{19}$ hexagonal cell are: *a*=5.562, *c*=21.972 Å and *Z*=2 [31,32]. Small divalent cations Mg^{2+}can be incorporated in the spinel block with replacing trivalent Al^{3+}ions and the difference of their ionic valences causes the defects in the structure or sometimes the modification of the structure [33]:

- Mg replaces Al in the Al-spinel block, inducing a valence deficiency. A charge compensation with positively charged conduction layer is required.
- Spinel unit ($Mg_2Al_4O_8$) will be inserted in the Al-spinel block.
- Al^{3+}ions in octahedral sites should be replaced by Fe^{3+}[34,35], Cr^{3+}[36], Co^{3+}[36,37], Mn^{2+}[38] and rare earth elements ions such as Y^{3+}[39] or Ga^{3+}[40].

---

3 MP-type alkaline earth (AE) aluminum oxide ($AEAl_{12}O_{19}$). These compounds belong to the same space group as β-alumina (Chapter 4.2).

4 Wyckoff position denotes the point belonging to a set of points for which the site symmetry groups are conjugate subgroups of the space group.

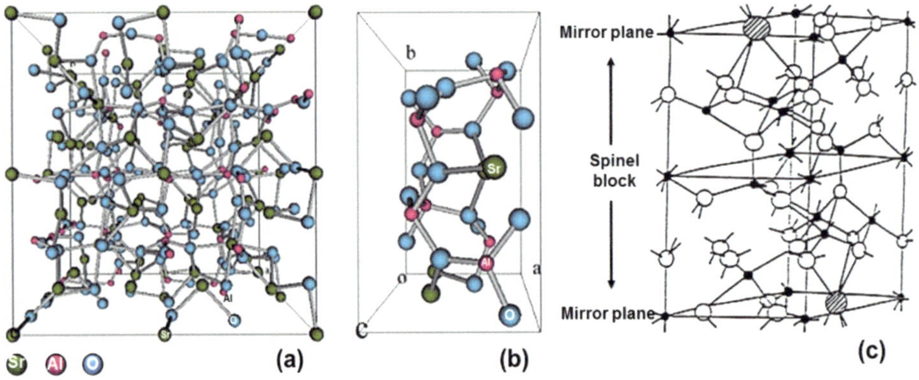

Figure 2. Structure of tristrontium aluminate (a) and strontium aluminate (b), schematic drawing of the structure of magnetoplumbite compounds (c) according to [24], where large shaded circles represent large cations (Na^+, K^+, Sr^{2+} ...), white and black circles represent $O^{2-}$and Al^{3+}, respectively.

Aluminium oxide (Al_2O_3, A): belongs to the R $\bar{3}$ c space group and forms hexagonal crystal (in the only thermodynamically stable crystallographic modification of corundum, α-Al_2O_3) with the lattice cell parmeters (a=4.754 Å and c=12.99 Å). O^{2-}ions are arranged in close hexagonal arrangement, with cations Al^{3+}occupying two-thirds of the octahedral interstitial positions. Corundum has the density of 3.97 g•cm^{-3} and the elemental composition of 52.91% Al and 47.08% O.

The molar weigh of aluminium oxide is 101.96 g•mol^{-1}. Al_2O_3 is available or prepared in several forms for various commercial applications. Some of these are α-alumina (corundum), activated aluminas (such as, γ-alumina, η-alumina and ρ-alumina), hydrated aluminas including aluminum oxide monohydrate, Al_2O_3•H_2O and aluminum oxide trihydrate, Al_2O_3•$3H_2O$ (natural gibbsite) and acidic, neutral and basic aluminas (no definite chemical compositions, they are prepared by the addition of various amounts of water to activated aluminas). Alumina exhibits amphoteric behaviour [91,424]. The properties and the methods of industrial production of Al_2O_3 are described in Chapter 2.1.2.

Other phases found and described within the SrO – Al_2O_3 oxide system, which aren't plotted in the equilibrium phase diagram (Fig.1) are:

- **$Sr_{12}Al_{14}O_{33}$ (12SrO·7Al_2O_3)**: has cubic structure of space group I43d and following cell parameters: a=12.325 Å and Z=2 [41].

- **$Sr_{10}Al_6O_{19}$ (10SrO·3Al_2O_3)**: has monoclinic structure (space group C12/c1, a=35.5823 Å, b=7.8460 Å, c=15.7485 Å, β=103.68° and Z=8, Fig.3), which belongs to the group of oligoaluminates, where the prominent structural features are [Al_6O_{19}] – groups. The arrangement of tetrahedra within hexamers is not linear but highly puckered. In detail, the geometry can be characterized as open five-membered tetrahedral rings with one additionally attached sixth tetrahedron. The Al-O bond distances in $Sr_6Al_{10}O_{19}$ range between 1.73 and 1.81 Å [42]. $Sr_6Al_{10}O_{19}$ is isotypic with strontium gallate α-$Sr_{10}Ga_6O_{19}$ [43].

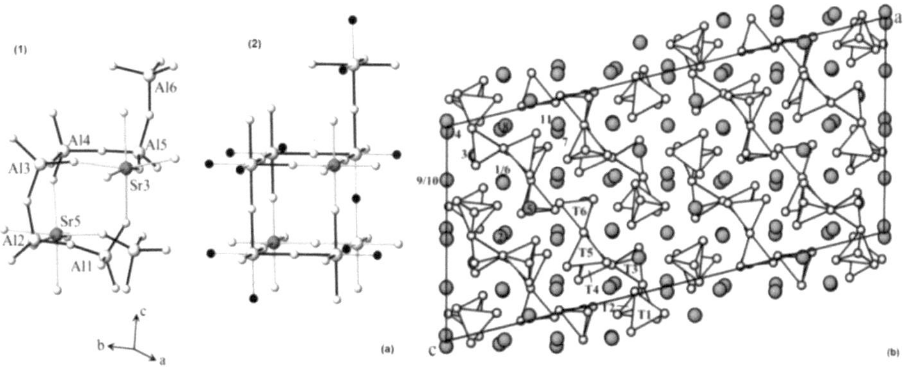

Figure 3. Single $[Al_6O_{19}]$ hexamer in the structure of $Sr_{10}Al_6O_{19}$ with two linking tetrahedra (1) and the arrangement of cations, anions and vacancies in the corresponding idealized unit of perovskyte (2) containing atoms of Sr (big dark spheres) and Al (light gray spheres), oxygen ligands (small gray spheres) and vacancies (small dark spheres). Al-O bonds within $[Al_6O_{19}]$ hexamer are represented by dark rod (a). The structure of $Sr_{10}Al_6O_{19}$, projection is parallel with [010] (b). Big gray spheres corresponding to Sr cations and tetrahedra (T) belonging to one of single $[Al_6O_{19}]$ hexamers are marked [42].

- **$Sr_7Al_{12}O_{25}$ ($7SrO\cdot6Al_2O_3$)**: is trigonal phase with the space group P3 and following cell parameters *a*=11.91 Å, *γ*=120° and *Z*=3 [44].
- **$Sr_4Al_{14}O_{25}$ ($4SrO\cdot7Al_2O_3$)**: is orthorhombic phase of *Pnna* space group with following cell parameters: *a*=24.7451 Å, *b*=8.4735 Å, *c*=4.8808 Å and *Z*=2 [45].
- **$Sr_2Al_6O_{11}$ ($2SrO\cdot3Al_2O_3$)**: is orthorhombic phase of *Pnnm* space group with following cell parameters: *a*=21.914 Å, ***b***=4.884 Å, *c*=8.4039 Å and *Z*=4 [46].

2.1. Application of Gibbs phase law on the phase diagram

The Gibbs phase rule [47] enables to calculate the number of degrees of freedom of the system. The most general and well known formulation is:

$$v = s - f + 2 - C \tag{9}$$

where *v*, *s* and *f* denote the number of degrees of freedom, species (constituents) and phases, respectively. Number "two" represents the temperature and pressure. The value of parameter *C* is derived from other restriction conditions in given thermodynamic system, for example from the number of independent chemical reactions (Gibbs stoichiometric law, Chapter 1.4), from the Dalton law, from the principle of electroneutrality, etc.).

The Gibbs phase law can be derived from the following consideration which includes the system formed from *s* species and *f* phases. There are certain minimal numbers of parameters which are necessary to describe the state of this system ($f(s-1)+2$):

$$\begin{aligned} x_s^1 &= 1 - \sum_{j=1}^{s-1} x_j^1 \\ &\vdots \\ x_s^f &= 1 - \sum_{j=1}^{s-1} x_j^f \end{aligned} \tag{10}$$

There is also certain minimal number of parameters which are necessary to describe the equilibrium in the system. The equilibrium is defined as the equivalence of chemical potential of s components across $f - 1$ phases, i.e. $s\ (f - 1)$ variables.

$$\begin{aligned} \mu_1^1 &= \mu_1^2 = \ldots = \mu_1^{f-1} \\ \mu_2^1 &= \mu_2^2 = \cdots = \mu_2^{f-1} \\ &\vdots \\ \mu_s^1 &= \mu_s^2 = \cdots = \mu_s^{f-1} \end{aligned} \tag{11}$$

For example, if we consider a system that consists of two components and three phases (e.g. some of eutectic points or the formation of peritectic melt in Fig.1), the condition can be explained as follows: if μ_{11}=μ_{12} and μ_{12}=μ_{13} then μ_{11}=μ_{13}. Therefore, only 2×(3-1)=4 terms are necessary for the description of equilibrium.

The Gibbs phase law is a simple difference between the number of parameters defining the state and the equilibrium in the system:

$$v = f\ (s-1) + 2 - s\left(f-1\right) = fs - f + 2 - fs + s \tag{12}$$

$$v = s - f + 2 \tag{13}$$

In the case of isobaric type [p] of phase diagram in Fig.1 without other restricting condition (C=0) it is possible to write:

$$v = s - f + 1 \tag{14}$$

2.2. Phase diagram and lever law

The equilibrium phase diagram or isobaric equilibrium phase diagram is a useful toll, which enables to predict the system behaviour during the thermal treatment. It is possible to estimate:

- Final phase composition of material;
- Field of stability (temperature and composition) of formed phases;

- Formation and decomposition of solid solutions;
- Melting temperature;
- Composition and temperature of invariant points (eutectic and peritectic points);
- Behaviour during melting (congruent or incongruent);
- Immiscibility gap in melt;
- Behaviour during crystallization;
- Ratio of formed phases.

Moreover is possible to predict the structure of cooled product (for example grain of precipitate dispersed in matrix of solidified eutectic melt) or to calculate melting enthalpy from the liquids line of diagram (described in Chapter 2.3).

It should be pointed out, that the equilibrium phase composition is not often established during industrial production of materials, especially during fast firing process of ceramics. Moreover metastable phases exhibit higher reactivity, therefore are often prepared purposely. For example, fast cooling of clinker of ordinary Portland cement (OPC) is applied in order to prevent it from the decomposition of tricalcium silicate (C_3S, $3CaO{\cdot}SiO_2$) into dicalcium silicate and free lime (C, CaO), as well as to keep the formed dicalcium silicate (C_2S, $2CaO{\bullet}SiO_2$) in the hydraulic reactive polymorph of β-C_2S. Stable modification of α-C_2S is not capable of hydration and contribution to hardening of concretes and mortars.

Five binary compounds are present in the phase diagram of SrO – Al_2O_3 system (Fig.1). Since 4SrO Al_2O_3 and SrO 6Al_2O_3 melt incongruently, there are two peritectic points. Other known strontium aluminate phases melt congruently. Five eutectic points are present in the phase diagram. Both, the eutectic and the peritectic points are invariant points without any degree of freedom (Gibbs Phase Rule, Chapter 2.1), but the system may contain thermodynamically stable melt at the temperature lower than the temperature of peritectic point. There is no stable melt present at the temperature lower than the temperature of eutectic point in the binary system.

Melt or liquid phase can be formed via congruent or incongruent melting process. Incongruent melting means the formation of peritectic melt and other solid phase. Example for strontium hexaaluminate it can be written as:

$$SrO{\cdot}6\,Al_2O_3(s) \xrightarrow{T_P} Al_2O_3(s) + \text{peritectic melt} \tag{15}$$

It means that the peritectic point is shifted to the left from the stoichiometric composition of strontium hexaaluminate. The peritectic liquid then contains 1 % less alumina than strontium hexaaluminate. The entire surplus of alumina stays in solid phase, therefore the temperature cannot rise if $SrO{\cdot}6Al_2O_3$ solid is present (invariant point). Once $SrO{\cdot}6Al_2O_3$ melts the system gets one degree of freedom and the composition of melt can change according to the line of liquidus until the rest of alumina is dissolved. The temperature as well as final composition of melt phase can be determined by the extrapolation of $SrO \cdot 6Al_2O_3$ composition to the line of liquidus of Al_2O_3 (1920 °C, x_{SrO}=14 %). The liquidus line can be defined as the borderline separating heterogeneous system, which consists of precipitate and melt from the field of stability of homogeneous melt.

The ratio of formed phases can be derived by the "**lever rule**" as is described below (Al_2O_3 : peritectic melt=1 : 14). On the contrary, congruently melting strontium dialuminate forms the liquid phase of the same composition as original solid:

$$SrO \cdot 2Al_2O_3(s) \xrightarrow{T_T} \text{melt} \tag{16}$$

The melting process turns the complex structures into the melt, negative charge of $[AlO_4]^{5-}$ anions is compensated by cations of Sr^{2+}. Basic oxides also provide O^{2-}anion, that is needed for the modification of melted compounds into the simpler structures [48,49] and affects the equilibria of redox reactions (e.g. Fe^{3+}/Fe^{2+}equilibrium) [50,51,52].

The system composition can be expressed by the molar or weight ratio of constituents. The rule must be fulfilled that the sum of molar (x_j) as well as weight (w_j) ratio is equal to one (or 100 %) (Eq.17), therefore x_2=1 – x_1 or w_2=1 – w_1 in the binary system.

$$\sum_{i=1}^{j} x_j = 1 \tag{17}$$

$$\sum_{i=1}^{j} w_j = 1 \tag{18}$$

These ratios can be recalculated one to another as follow:

$$w_i = \frac{M_i x_i}{\sum_i M_i x_i} \tag{19}$$

$$x_i = \frac{\frac{w_i}{M_i}}{\sum_i \frac{w_i}{M_i}} \tag{20}$$

where M is the molecular mass of i-th species.

The behaviour of the system during cooling of melt phase will be explained on the examples in Fig.4 with the composition given by line (1) and (2). Starting with composition (1) when melted phase contains 90 % SrO, the first SrO solid appears at the temperature corresponding to the point A (2200 °C). The infinitesimal amount of SrO(*s*) (*A´*) is in the equilibrium with the melt of composition (*A*). This equilibrium is described by Eq.39[5]. Using general formula of Gibbs phase law, the system has one degree of freedom. Therefore, the composition of the system may change with decreasing temperature according to the line of liquidus. Strontium oxide precipitating from the melt leads to the liquid phase enriched with Al_2O_3 (dark red arrows).

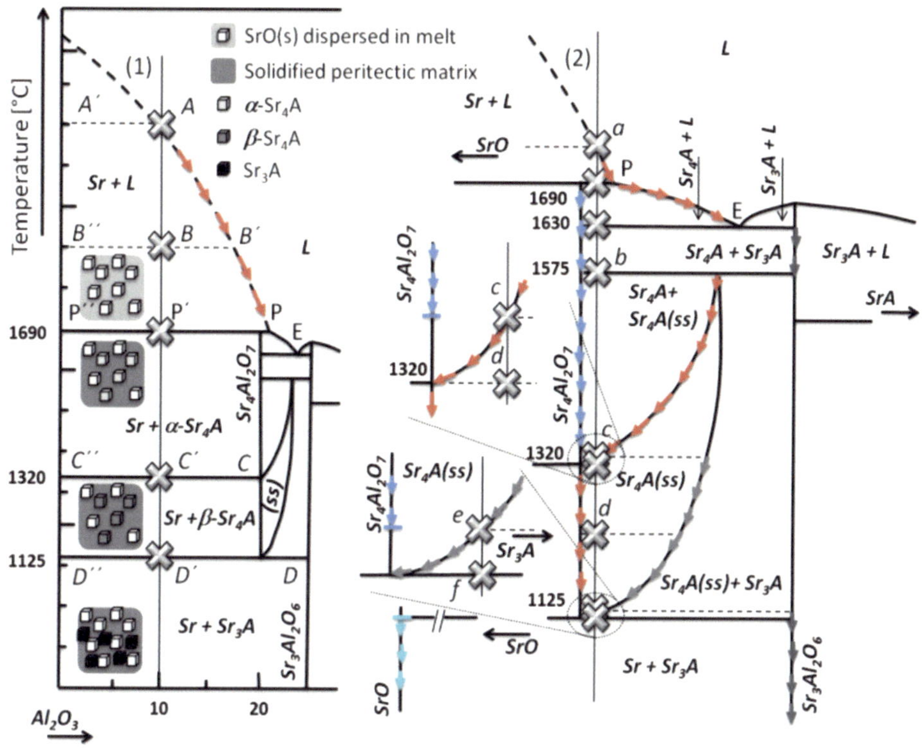

Figure 4. Crystallization pathway for the system A and B (a) with graphical illustration of "lever rule" and cooling curve.

At the temperature of 1900 °C that corresponds to the point *B*, the system consists of SrO(s) and melt in the ratio given by the lever rule:

$$\frac{\mathrm{SrO(s)}}{\mathrm{melt}} = \frac{|BB'|}{|B''B'|} \Rightarrow \mathrm{SrO(s) : melt} = 1:1.\overline{3} \tag{21}$$

5 the line of liquidus in the phase diagram given in Fig.4 is a function of this equation, but increasing non-ideal behaviour of melt leads to the differences between experimental and calculated data as the content of the second constituent increases.

Therefore pure SrO is formed (B′′), the system consists of SrO precipitate dispersed in the equilibrium melt, the composition of which is given by the line of liquidus. Three phases are present if the system reaches the composition and the temperature corresponding to the peritectic point: peritectic melt (P), tetra-strontium aluminate (Sr_4A) and strontium oxide (Sr). There is zero degree of freedom and the temperature cannot be changed until the solidification of whole melt proceeds.

Within the temperature interval from 1690 to 1320 °C, the solid system consists of strontium oxide and tetra-strontium aluminate in the ratio $|CC| : |C''C'|=1:1$. The ratio stays unchanged to the temperature 1125°C. Below the temperature of 1125 °C, Sr_4A is decomposed to SrA and Sr_3A in the ratio 1 : 1.5.

The crystallization of the system with the composition given by line (2) begins at the temperature of point (*a*) where the first infinitesimal amount of strontium oxide appears. The composition of melt changes according to the line of liquidus as the temperature decreases. On the contrary in the system (1) the free strontium oxide disappears at the temperature of peritectic point (P). The Sr_4A solid and melted phase, the composition of which changes according to the line of liquidus, is the equilibrium as the temperature decreases from P to E. At the temperature of eutectic point, the system becomes invariant again and the temperature stays constant until whole eutectic melt is solidified. Sr_4A crystals appear in the fine crystallized matrix of eutectic melt, but both phases are pure crystalline phases.

Below the temperature of 1575 °C (b) the Sr_3A is dissolved in S_4A phase. Because there are three equilibrium phases (Sr_4A, Sr_4A(ss) and Sr_3A) in the two component system at constant pressure, the system becomes invariant until entire Sr_3A is dissolved. Formed solid solution is being enriched by dissolved Sr_4A as the temperature decreases from point (*b*) to (*c*). That can be easily proved by the lever rule. At the temperature of point (c) Sr_4A dissolves in the solid solution and line (2) heads to the homogeneous field of solid solution, but at the temperature of 1320 °C (d) the Sr_4A phase is formed again.

Tri-strontium aluminate starts precipitating from the solid solution at the temperature of point (*e*) and this process continues to the temperature of point (*f*) where the solid solution does not exist any longer. Below this temperature, the decomposition of Sr_4A to free SrO and Sr_3A takes place.

Tri-strontium aluminate and strontium aluminate are neighboring congruently melting compounds. Therefore, they form relatively simple binary subsystem of the SrO – Al_2O_3 system that can be made as separate phase diagram (Fig.5). From this point of view, the phase diagram in Fig.1 consists of four simple systems, which are formed between oxide species and congruently melting phase or two congruently melting phases: SrO – Sr_3A, Sr_3A – SrA, SrA – SrA_2 and SrA_2 – SrA_6.

Fig.5 shows the lever rule applied to the system containing slightly higher amount of SrO with regard to strontium aluminate stoichiometry heated to the temperature of 1600 °C. The solid to melt ratio has the value:

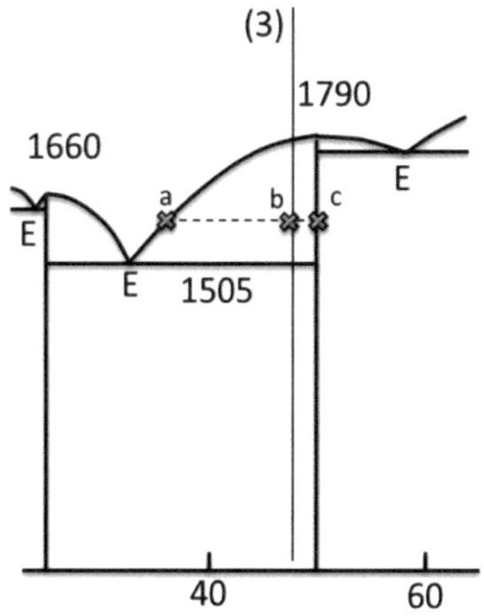

Figure 5. Binary system formed between tri-strontium aluminate and strontium aluminate.

$$\frac{\text{solid}}{\text{melt}} = \frac{|ab|}{|bc|} = \frac{47.5 - 35}{50 - 47.5} = \frac{5}{1} \tag{22}$$

The original system containing 45.7 % Al_2O_3 and 100 – 45.7=54.3 % SrO consists of 1 piece of melt and 5 pieces of solid. That can be verified as follows:

$$\begin{array}{llll} \text{Melt} & Al_2O_3\,[\%] & 35 & & 2.5 \rightarrow & 1 \\ & & & 45.7 & & \\ \text{Solid} & Al_2O_3\,[\%] & 50 & & 12.5 \rightarrow & 5 \end{array} \tag{23}$$

The same rule can be used to calculate the ratio for two systems mixed in order to reach required composition. In the case, that x_{Al2O3} is recalculated to w_{Al2O3} (Eq.19), the following relation can also be used for these purposes:

$$m_1 w_1 + m_2 w_2 + \ldots m_i w_i = W(m_1 + m_2 + \cdots m_1) \tag{24}$$

For example, if the system containing 50 % Al_2O_3 should be prepared by mixing 100 g of system with 30 % Al_2O_3 with pure Al_2O_3 (100 %), the solution is:

$$\begin{aligned} &100 \cdot 30 + m_{Al_2O_3} \cdot 100 = 50(100 + m_{Al_2O_3}) \\ &m_{Al_2O_3} = 40\ g \end{aligned} \tag{25}$$

2.3. Calculation of melting enthalpy from equilibrium phase diagram

The equilibrium phase diagram constructed from the experimental data on the behaviour of system during heating can be used for the calculation of melting enthalpy (Δh_m) of the constituents. The solution is named as "Schröder and Le Chatelier's law" and it will be demonstrated on the SrO – Al_2O_3 diagram (Fig.1). It must be pointed out, that the following solution is derived for the equilibrium of pure crystalline solids with melt where no solid-solution is formed. That means that presented solution is applicable for all congruently melting solids in the binary system of SrO – Al_2O_3.

The liquidus line in the phase diagram represents the equilibrium state between solids and melt phase at given temperature. In the equilibrium state the chemical potential of both phases must fulfill the term:

$$\mu^*_{A,s} = \mu_{A,\ell} \tag{26}$$

where $\mu^*_{A,s}$ and $\mu_{A\ell}$ denote the chemical potential of species A as the pure (*) solid phase (s) and in the melt (ℓ), respectively. The chemical potential of pure solids is given as follows:

$$\mu^*_{A,s} = f(T, p) \tag{27}$$

$$d\mu^*_{A,s} = \left(\frac{\partial \mu^*_{A,s}}{\partial T}\right)_p dT + \left(\frac{\partial \mu^*_{A,s}}{\partial p}\right)_T dp \tag{28}$$

$$d\mu^*_{A,s} = -s^*_A dT + v^*_A dp \tag{29}$$

where s^* and v^* are the molar entropy and volume.

The chemical potential of species A in the melt with ideal behaviour can be expressed by the equation:

$$\mu_{A,\ell} = f(T, p, x_A) \tag{30}$$

$$d\mu_{A,\ell} = -\bar{s}_{A,\ell} dT + \bar{v}_{A,\ell} dp + RT\, d\ln x_A \tag{31}$$

where $\bar{s}$ and $\bar{v}$ are partial molar entropy and volume.

With respect to Eq.26 the equilibrium state requires:

$$d\mu^*_{A,s} = d\mu_{A,\ell} \tag{32}$$

$$-s^*_A\, dT + v^*_A\, dp = -\bar{s}_{Al}\, dT + \bar{v}_{Al}\, dp + RT\, d\ln x_A \tag{33}$$

$$\left(\bar{s}_{A|}\,dT - s_A^*\right)dT = \left(\bar{v}_{A|} + v_A^*\right)dp + RT\,d\ln x_A \tag{34}$$

The term in the brackets corresponds to the change of entropy and volume during melting. Therefore, the applied equilibrium phase diagram is constructed for the isobaric conditions [p] which can be written:

$$\Delta s_{A,m}\,dT = RT\,d\ln x_A \tag{35}$$

where Δs_m is the melting entropy of the process, which can be expressed using the definition law of Gibbs energy for the equilibrium state (Eq.36).

$$\Delta g_m = \Delta h_m - T\,\Delta s_m = 0 \;\Rightarrow\; \Delta h_m = T\,\Delta s_m \;\Rightarrow\; \Delta s_m = \frac{\Delta h_m}{T} \tag{36}$$

From the combination of Eqs.35 and 36 the following can be derived:

$$\frac{d\ln x_A}{dT} = \frac{\Delta h_{A,m}}{RT^2} \tag{37}$$

$$\int_1^{x_A} d\ln x_A = \frac{\Delta h_{A,m}}{R}\int_{T_A}^{T_{x_A}} \frac{dT}{T^2} \tag{38}$$

$$\ln x_A = -\frac{\Delta h_{A,m}}{R}\left(\frac{1}{T_{x_A}} - \frac{1}{T_A}\right) \approx -\frac{\Delta h_{A,m}}{R}\frac{T_A - T_{x_A}}{T_A^2} \qquad \left[x_A \to 1 \Rightarrow T_A \approx T_{X_A}\right] \tag{39}$$

where T_A and T_{xA} are the melting temperature of pure species A and of the system of composition x_A, respectively. The melting (fusion) enthalpy can be assessed then from the slope (term- $\Delta h_m/R$) of plot $ln\ x_A$ to be the difference (T_{xA}^{-1}-T_A^{-1}).

The value T_{xA} should be the closest possible value to T_A (x_A is close to one) due to the presumption of ideal behaviour of melted phase. In the case that the value of melting enthalpy is known, the liquidus line can be calculated from Eq.39. The example of the estimation of melting enthalpy of SrO from the equilibrium phase diagram is shown in Fig.6.

2.4. Ternary system SrO-CaO-Al_2O_3

The phase relationships of the SrO – CaO – Al_2O_3 system (Fig.7) were described by Massazza at al. [53,54]. The ternary diagram shows the formation of solid solution between corresponding EA_xA_y phases, where EA denotes alkali earth element (Ca or Sr): $CaAl_{12}O_{19}$ (CA_6)-

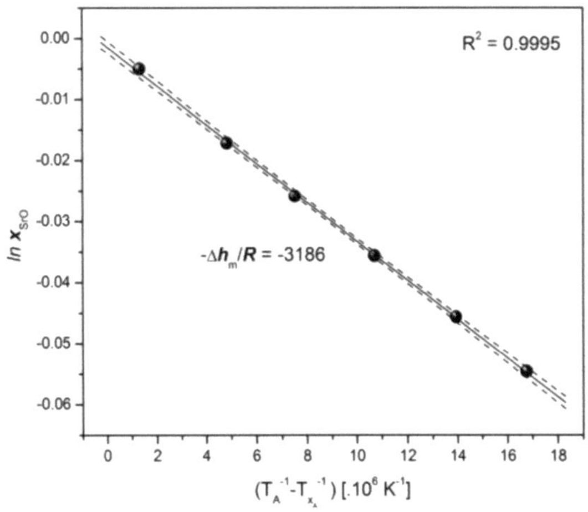

Figure 6. Estimation of melting enthalpy from phase diagram.

$SrAl_{12}O_{19}$ (SrA_6), $CaAl_4O_7$ (CA_2)-$SrAl_4O_7$ (SrA_2), $CaAl_2O_4$ (CA) – $SrAl_2O_4$ (SrA) and $Ca_3Al_2O_6$ (C_3A)-$Sr_3Al_2O_6$ (Sr_3A) [55].

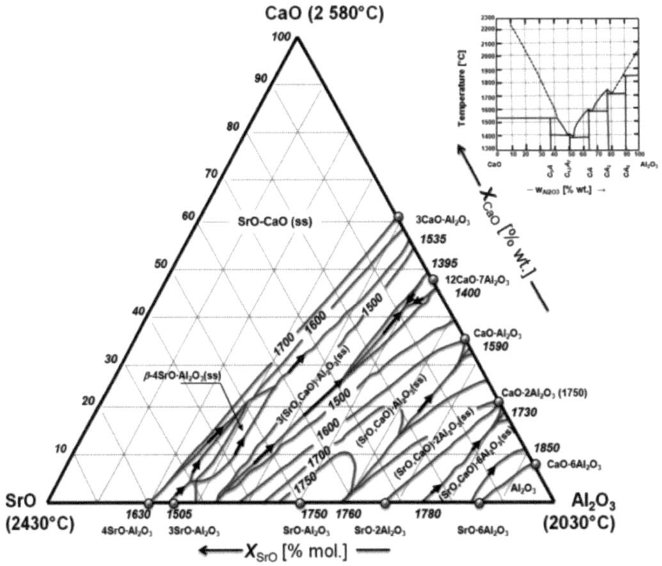

Figure 7. Phase relationship in the ternary system of CaO – SrO – Al_2O_3 [53].

The interesting fact is the stability of β-Sr_4A solid solution up to the melting temperature in the ternary system, while in the binary system SrO – Al_2O_3 the β-Sr_4A is stable only up to the temperature of 1575 °C (Fig.1).

2.5. Quaternary system $CaO - SrO - Al_2O_3 - ZrO_2$

The compounds formed in the system $CaO - SrO - Al_2O_3 - ZrO_2$ (Fig.8) are of great interest for the production of refractories due to high melting temperatures. This system contains 14 binary and ternary compounds, but the $SrO - ZrO_2$ system is very complicated and the literature contains contradictionary information on it. The connection lines in Fig.8 divide the quaternary system into the 21 elementary tetrahedra. There are two series of continuous solid solutions $CaAl_2O_4 - SrAl_2O_4$ and $CaZrO_3 - ZrSrO_3$ [55].

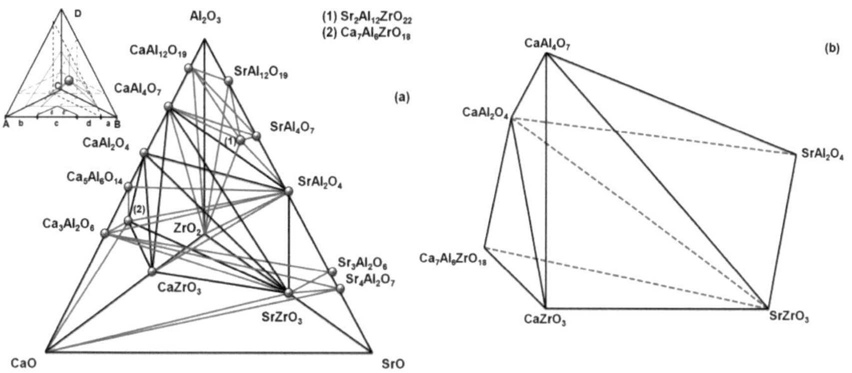

Figure 8. Elementary tetrahedra of the quaternary system $CaO-SrO-Al_2O_3-ZrO_2$ (a) [55] with highlighted elementary tetrahedron including CA-SrA-SrZ region (b) favourable for the production of refractory binders [56].

The region CA – SrA – SrZ is considered as favourable for the production of special cements for refectory purposes. While strontium and calcium aluminates possess the bonding properties, the calcium and strontium zirconates have high melting temperatures. These cements are characteristic of water to cement ratio from 0.20 to 0.31 and rapid hardening [56].

3. Thermodynamics of $SrO-Al_2O_3$ system

3.1. Estimation of molar thermal capacity

The thermal capacity of solid inorganic compounds, which are formed from elements or simple solid compounds (Eq.40), can be estimated by the Neumann-Koop rule according to the relation 41. The method is applicable for intermetallic compounds and mixed oxides.

$$m\ A(s) + n\ B(s) \rightarrow A_m B_n(s) \tag{40}$$

$$c_{pm}^{o}(A_m B_n) = m\ c_{pm}^{o}(A) + n\ c_{pm}^{o}(B) \tag{41}$$

For example, the thermodynamic data of binary compounds from SrO – Al_2O_3 system (Fig.1) can be calculated as follows:

$$\begin{aligned} c_{pm}^{o}(\mathrm{Sr_3Al_2O_6}, 298\ K) &= 3\ c_{pm}^{o}(\mathrm{SrO}, 298\ K) + c_{pm}^{o}(\mathrm{Al_2O_3}, 298\ K) \\ c_{pm}^{o}(\mathrm{SrAl_4O_7}, 298\ K) &= 3 \cdot 45.241 + 79.029 = 214.752\ J \cdot mol^{-1} K^{-1} \end{aligned} \quad (42)$$

The overview of results of other binary compounds and the comparison with experimental value is shown in Table 2. More methods can be found in literature. Kellogg [57] originally suggested a method for the estimation of heat capacity of predominantly ionic, solid compounds at 298 K. It is analogous to the Latimer's method for the estimation of standard entropies [58,59]. The method is based on the summation of contributions from the cationic and anionic groups in the compound[6].

Binary compounds	c_{pm}° (25 °C) [J•mol^{-1}K^{-1}]		*A*	•10^{-3} *B*	•10^{5} *C*	•10^{-6} *D*	Difference [%]
	Estimated [(a)]	Determined	$c_{pm}^{\circ} = A + BT + C/T^2 + DT^2$ [J.mol^{-1}K^{-1}]				
$4SrO \bullet Al_2O_3$	259.686	253.659	308.958	36.507	-58.833	0.008	2.38
$3SrO \bullet Al_2O_3$	214.445	209.152	257.294	31.873	-51.243	-0.012	2.56
$SrO \bullet Al_2O_3$	123.963	119.030	177.192[(1)]	4.937	-53.11	---	4.14
			146.105[(2)]	29.288	---	---	
$SrO \bullet 2Al_2O_3$	202.685	?	269.527[(e)]	48.85	-71.815	-9.84	---
$SrO \bullet 6Al_2O_3$	517.572	?	707.381[(e)]	138.6	-203.60	-30	---

(1) Temperature range 298.15 – 932 K.

(2) Temperature range 932 – 2063 K.

(a) HSC software.

(e) Results based on experimental data are not available, constants A, B, C and D were determined from estimated data.

Table 2. Estimation of molar thermal capacity of binary compounds from SrO-Al_2O_3 system from their oxides.

The temperature dependence of c_{pm}° of pure oxides in the SrO – Al_2O_3 system:

6 For example, the cationic contribution of Sr to heat capacity at 298 K is 25.52 J·K^{-1}. The anionic contribution for Al_2O_4 and Al_2O_6 is 98.52 and 135.46 J·K^{-1}, respectively. For strontium aluminate the heat capacities can be then calculated to be 25.52 + 98.52 = 124.04 J·mol^{-1}·K^{-1}. For tri-strontium aluminate the value 3 25.52 + 135.46 = 212.02 J·mol^{-1}·K^{-1} at 298 K. can be found. The results show good agreement with data in Table 2.

$$c^{o}_{pm}(SrO,T) = 46.047 + 14.163 \cdot 10^{-3}T - \frac{3.885 \cdot 10^{5}}{T^{2}} - 7.4 \cdot 10^{-6}T^{2}$$
$$[298.15 - 900\ K] \qquad (43)$$

$$c^{o}_{pm}(SrO,T) = 83.372 - 36.439 \cdot 10^{-3}T - \frac{65.775 \cdot 10^{5}}{T^{2}} + 12.175 \cdot 10^{-6}T^{2}$$
$$[900 - 1700\ K] \qquad (44)$$

$$c^{o}_{pm}(Al_2O_3,T) = 109.474 + 22.414 \cdot 10^{-3}T - \frac{32.949 \cdot 10^{5}}{T^{2}} - 4.155 \cdot 10^{-6}T^{2} \qquad (45)$$

the $c_{pm}{}^{\circ}$ of binary compounds at required temperature can be calculated (Fig.9).

3.2. Estimation of enthalpy and entropy

Using the Hess′ Law, the concept of formation reaction is employed for the numerical determination of enthalpy changes. In general, the reaction of the formation of pure substance B from the elements (E_i) is:

$$B = \sum v_i E_i \qquad (46)$$

where v_i is the stoichiometric coefficient. The standard enthalpy of formation if defined by the formula:

$$\Delta H^{o}_{f,B}(T) = H^{o}_{f,B}(T) - \sum v_i H^{o}_{E,i} \qquad (47)$$

Standard enthalpies of elements in their reference phases at ambient standard conditions[7] and their reference phases are set to zero ($\Delta H_{f,i}{}^{\circ}$=0) [303].

Various empirical methods were developed to estimate the enthalpy of formation of mixed oxides from either the elements or the constituent binary oxides [60-63]. The Aronson's method [60], which is based on the Pauling's relation between the enthalpy of formation and the differences between the electronegativities of elements forming the compound, is the most general one:

$$\Delta H^{o}_{f} = -96.5\, n_O (X' - X_O)^2 \quad \left[kJ \cdot mol^{-1}\right] \qquad (48)$$

7 Standard ambient conditions are defined in Chapter 3.3.

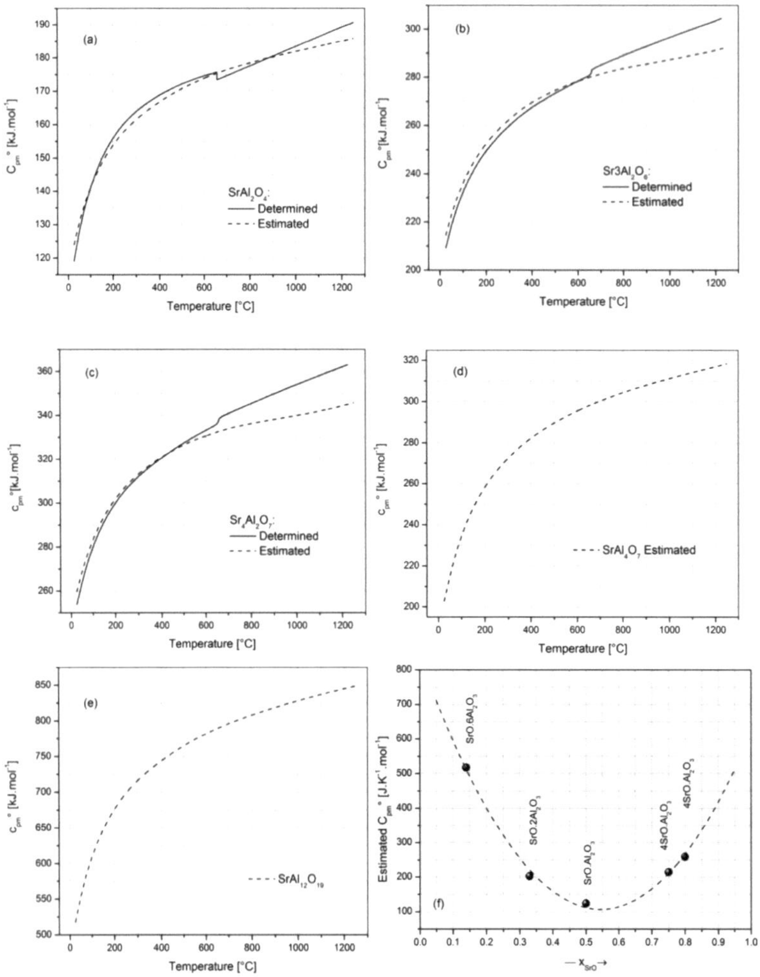

Figure 9. Temperature dependence of thermal capacity of binary compounds from SrO – Al_2O_3 system (a-e) and dependence of c_{pm}° on the system phase composition at temperature of 298.15 K (f).

where n_O is the number of oxygen atoms in the formula unit, X_O is the Pauling´s electronegativity of oxygen and $X'$is the weighted geometrical mean of so-called "pseudoelectro-negativities" of the oxide forming elements. The pseudoelectronegativity values are derived from known values of enthalpy of formation of relevant binary oxides. The enthalpy of formation of oxide can be then calculated as [61]:

$$X' = X_O + \sqrt{\frac{-\Delta_f H}{96.5\, n_O}} \tag{49}$$

Another method, also based on the concept of electronegativity, was proposed by Zhuang et al. [62]. The method calculates the enthalpy of formation of binary oxides according to the relation:

$$\Delta_f H^o = (n_1 + n_2)\, x_1 x_2 \lambda \tag{50}$$

Where n_i and x_i are the number of moles and the mole fraction of *i*-th constituent of mixed oxide and λ is the constant similar to the interaction parameter used in regular solution model. Assuming the formation of double oxide from the two simple oxide species:

$$n_1\, A_aO_x + n_2\, B_bO_y \rightarrow n_1 A_a X \cdot n_2 B_b Y \tag{51}$$

Once the formation enthalpy is known for any double oxide $n_1(A_aO_x)\, n_2(B_bO_y)$, the value of λ can be calculated from the formation enthalpy of double oxide with given stoichiometry and used for oxides of another stoichiometry. Otherwise, λ can be estimated using the following formula [61,62]:

$$\lambda = -96.5\, z\, (X_A - X_B)^2 \quad [kJ \cdot mol^{-1}] \tag{52}$$

where X_A and X_B are the Pauling's electronegativities for *A* and *B* elements and *z* is the stoichiometric factor (double number of oxygen atoms or ions given by oxide formula (Eq. 51)) is given by the relation:

$$z = 2\left(\frac{x}{a} + \frac{y}{b}\right) \tag{53}$$

A new method of estimation of the enthalpy of Al_2O_3 – Ln_2O_3 mixed oxides formation was derived by Voňka and Leitner [61]. The method is based on the Pauling's concept of electronegativity and, in particular, on the relation between the enthalpy of formation of binary oxide and the difference between the electronegativities of the oxide-forming element and oxygen. This relation can be extended also for the calculation of enthalpy of other types of mixed oxides.

When the Pauling´s electronegativity of the central cations is ≥ 1.9, the method suggested by Šesták at al [64] enables to calculate the formation enthalpy of mixed oxide as follows:

$$\Delta_f H^o = -29.288\, M \tag{54}$$

where M is the number of oxygen atoms or ions in the double oxide. The Pauling´s electronegativities of Al and Sr are 1.61 and 0.95, respectively. Therefore the assumption for Šesták method is not fulfilled in SrO – Al_2O_3 system.

Le Van [65] described the method based on the assumption of additivity of bond energies for the estimation of $\Delta_f H°$:

$$\Delta_f H^o(298\ K) = n_-\ P + n_+\ Q + 4.184\,(4n_+^2) + 4.184\,n_-^2 \quad [kJ \cdot mol^{-1}] \tag{55}$$

where $n_+$and n_-denote the number of cations and anions, respectively. The P and Q are characteristic parameters[8] for cations and anions.

As was mentioned in Chapter 1.3.1, the Latimer's method of estimation of standard entropies (predominantly ionic compounds) uses the empirically found values of anion and cation contributions[9] [58,59].

The comparison of experimental and calculated standard enthalpy of formation and the standard entropy for known binary compounds (mixed oxides) in the SrO-Al_2O_3 system is given in Table 3.

Binarycomponds	$\Delta H_{f,B}$°(298K)[kJ•mol^{-1}]		S_m°(298K)[kJ•K^{-1}mol^{-1}]	
	Determined	Estimated[(a)]	Determined	Estimated[(a)]
SrO	-591.999	---	55.580	---
4SrO•Al_2O_3	-4118.725	-4407.794	322.570	283.947
3SrO•Al_2O_3	-3544.681	-3717.308	255.375	229.160
SrO•Al_2O_3	-2338.897	-2336.337	108.800	120.219
SrO•2Al_2O_3	?	-3994.452	?	180.945
SrO•6Al_2O_3	?	-5424.412	?	423.852
Al_2O_3	-1675.690	---	50.949	---

[(a)] HSC software

Table 3. Standard enthalpy of formation and standard entropy for binary compounds in the SrO-Al_2O_3 system.

Fig.10 (a) shows that estimated value of $\Delta_f H°$(298 K) is quite different from the extrapolation of fitted experimental data.

8 The characteristic values of Q(Sr^{2+}) = -862 and P($Al_2O_4^{2-}$) = -1494 enable to estimate the formation enthalpy of strontium aluminate to be -862 -1494 + [4.184 42] + 4.184 = -2285 kJ·mol^{-1}.

9 The anionic and cationic contributions of Sr^{2+} and $Al_2O_4^{2-}$ ions are 48.7 and 57 J·K^{-1} mol^{-1}, respectively. The entropy of strontium aluminate can be then estimated to 52.9 + 57 = 109.9 J·K^{-1} mol^{-1}. Using the contribution of O^{2-} for divalent cation (2.5 J·K^{-1}·mol^{-1}), the value of (3·52.9) + (2·2.5) + 57 = 220.7.

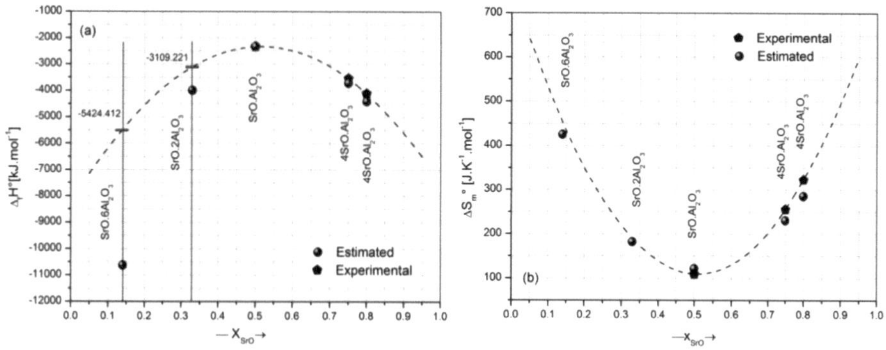

Figure 10. Estimated enthalpy (a) and entropy (b) of binary compounds in the $SrO\text{-}Al_2O_3$ system.

Therefore, the values expected from fitted dependence were applied for the calculation of temperature phase stability (Chapter 1.3.3). The estimated entropy (Fig.10 (b)) shows good agreement with predicted results (dashed line). The enthalpy and the entropy reach the maximum at the composition that corresponds to strontium aluminate.

The enthalpy and the entropy of species can be recalculated to required temperature using the following relations:

$$H(T) = H(298.15) + \int_{298.15}^{T_{\alpha\to\beta}} c_{pm}^{o,\alpha}\, dT + \Delta H_{\alpha\to\beta} + \int_{T_{\alpha\to\beta}}^{T} c_{pm}^{o,b}\, dT \tag{56}$$

$$S(T) = S(298.15) + \int_{298.15}^{T_{\alpha\to\beta}} \frac{c_{pm}^{o,\alpha}}{T}\, dT + \frac{\Delta H_{\alpha\to\beta}}{T} + \int_{T_{\alpha\to\beta}}^{T} \frac{c_{pm}^{o,b}}{T}\, dT \tag{57}$$

The definition law can be then used for the calculation of Gibbs energy as follows:

$$G(T) = H(T) - TS(T) \tag{58}$$

3.3. Temperature phase stability

There are five binary compounds in the phase diagram in Fig.1: $4SrO\bullet Al_2O_3$, $3SrO\bullet Al_2O_3$, $SrO\bullet Al_2O_3$, $SrO\bullet 2Al_2O_3$ and $SrO\bullet 6Al_2O_3$. The thermodynamic considerations enable to evaluate the stability of these phases at given temperature. For example, all compounds mentioned above are stable at the temperature of 600 °C. This question can be solved by the calculation of reaction Gibbs energy ($\Delta_r G°$) related to the formation of compounds from simple oxides (SrO and Al_2O_3):

$$Al_2O_3 + 4\,SrO \rightarrow Sr_4Al_2O_7 \tag{59}$$

$$Al_2O_3 + 3\,SrO \rightarrow Sr_3Al_2O_6 \tag{60}$$

$$Al_2O_3 + SrO \rightarrow SrAl_2O_4 \tag{61}$$

$$2\,Al_2O_3 + SrO \rightarrow SrAl_4O_7 \tag{62}$$

$$6\,Al_2O_3 + SrO \rightarrow SrAl_{12}O_{19} \tag{63}$$

While the reactions 59-61 are exothermic ($\Delta_r H° < 0$), the formation of $SrAl_{12}O_{19}$ is endothermic process at the temperature of 600 °C. Since $\Delta_r G° < 0$ (Table 4), all these reactions lead to thermodynamically stable products. The question is, if the products are stable than basic oxides SrO and Al_2O_3.

Eq.	$\Delta_r H°$ [kJ·mol⁻¹]		ΔS_m [J·K⁻¹mol⁻¹]	$\Delta_r G°$ [kJ·mol⁻¹]		$\Delta_r G^{o(bo)}$ [kJ·mol⁻¹]	x_{SrO} [%]
59	-75.664	EXO	47.509	-117.147	$\Delta_r G° < 0$	-23.429	80.0
60	-94.403		34.681	-124.685		-31.129	75.0
61	-70.513		3.230	-73.333		-36.667	50.0
62	-54.211		17.517	-69.506		-46.337[e]	33.3
63	10.342	ENDO	46.971	-30.670		-26.289[e]	14.3

[bo] The reaction Gibbs energy recalculated to one mol of basic oxides ($SrO+Al_2O_3$)

[e] Calculated from estimated thermodynamic data (Chapter 1.3.1 and 1.3.2.).

Table 4. Reaction thermodynamics of the formation of binary compounds in the $SrO-Al_2O_3$ system at 600 °C.

The value of $\Delta_r G°$ was recalculated to one mol of basic oxides $\Delta_r G^{o(bo)}$ and plotted as the function of x_{SrO} in the compound (Fig.11(a)). The value calculated for $SrO\cdot 6Al_2O_3$ phase is placed over dashed line. That means that the compound is unstable at given conditions and should be transformed into neighboring stable phase:

$$SrAl_{12}O_{19} \rightarrow SrAl_4O_7 + 4\,Al_2O_3 \tag{64}$$

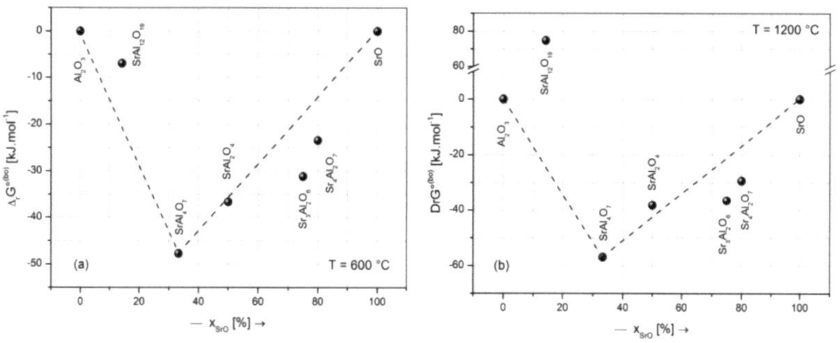

Figure 11. Thermal stability of binary compounds in the SrO – Al_2O_3 system at the temperature of 600 °C (a) and 1200 °C (b).

The negative value of $\Delta_r G°$ (-63.321 kJ) of the process provides further evidence about the thermodynamical instability of $SrO\cdot 6Al_2O_3$ phase[10].

3.4. Equilibrium composition in ternary system

The calculation of equilibrium composition of ternary system will be demonstrated with the example for the temperature of 1000°C. The first step of this solution includes the construction of ternary plot with the points corresponding to the composition of all known phases (Fig. 12). Now it is possible to draw all possible connecting lines between all points. The intersectional points present the chemical reaction of corresponding phases.

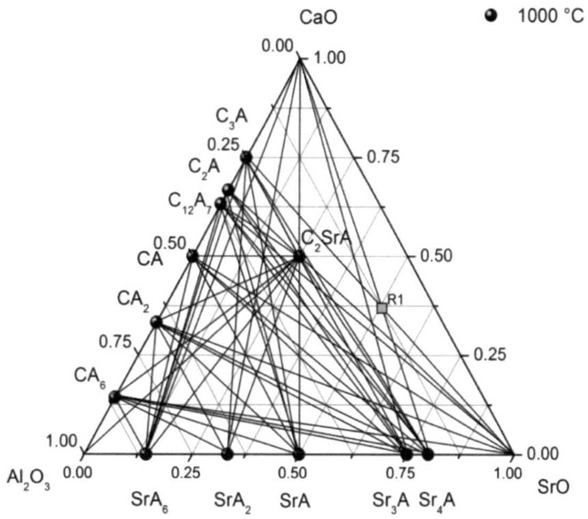

Figure 12. Possible chemical reactions in the ternary system CaO-SrO-Al_2O_3.

10 Thermal treatment of raw meal (Chapter 4) shows that formation SrA_6 is preferred from the formation of SrA_2 due to reasons discussed in Chapter 1.5.

For example, the intersectional point R1 in Fig.12, corresponds to the chemical reaction, which can be expressed by usual notation in cement chemistry as follows:

$$Sr_4A + 3\,C \xleftrightarrow{?} C_3A + 4\,Sr \tag{65}$$

The decision, which pair of compounds is stable under given conditions (are the species on the left side of Eq.65 or the products on the right side stable?). This decision is based on the standard Gibbs energy of reactions ($\Delta_r G°$) at given temperature:

$\Delta_r H°$(Eq.65, 1000°C)=67 580 J ($\Delta_r H° > 0 \rightarrow$ endothermic reaction)

$\Delta_r S°$(Eq.65, 1000°C)=-7.531 J K^{-1}

$\Delta_r G°$(Eq.65, 1000°C)=67 580+(1273,13 7.531)=77 168 J

$$Sr_4A + 3\,C \xleftarrow{1000°C} C_3A + 4\,Sr \tag{66}$$

Therefore if $\Delta_r G° > 0$, the products of reaction according to Eq.65 are thermodynamically unstable under given temperature (Eq.66) and the connection line between C_3A and Sr must be withdrawn from the diagram. These calculations should be repeated until all interconnection points are solved. The ternary diagram in Fig.13 shows the equilibrium phase composition of Al_2O_3 – SrO – CaO system calculated for the temperature of 1000°C.

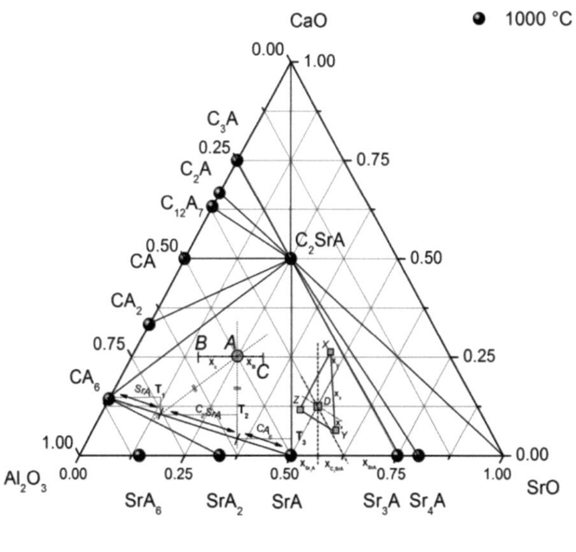

Figure 13. Phase stability in the ternary system Al_2O_3 – SrO – CaO at the temperature of 1000°C.

The equilibrium ternary diagram in Fig.13 can be a useful tool to predict the phase composition of the mixture of raw materials under the thermal treatment, e.g. by thermal treatment of mixture, the composition of which corresponds to the **point A** (50 % Al_2O_3, 25 % SrO and 25 % CaO). This composition provides the system consisting of SrA, C_2SrA and CA_6 in the ratio 1:1.5:1. It is also possible to assess the ratio in which the constituents should be mixed in order to obtain the mixture of required composition. For example, in order to get the system A from the systems B and C, they must be mixed in the ratio |AC| : |BA|=x_B : $x_{C=}$1:1.55.

Another example is the system D prepared in the corresponding ratio of systems X, Y and Z, which, after the thermal treatment to the temperature of 1000 °C provides compounds SrA. Sr_3A and C_2SrA in the ratio 2.3:1:1.

4. System with chemical reaction

The number of independent chemical reactions and their stoichiometry can be derived from the Gibbs Stoichiometric Law. The system is assumed, which contains *N* chemical compounds consisting of *M* elements (C_1, C_2... C_N) where the following reaction takes place:

$$\sum_{i=1}^{N} \nu_i C_i = 0 \tag{67}$$

where ν_i are the stoichiometric coefficients. The reaction can be expressed in the form of matrix equation:

$$\nu^T [1, N]\, C[N,1] = 0 \tag{68}$$

where ν^T is the stoichiometric coefficient matrix (1 row, *N* column) and C is the species matrix (*N* row, 1 column). The composition of chemical compounds can be expressed as:

$$\sum_{j=1}^{M} A_{ij}\ B_j = C_i \tag{69}$$

Where A_{ij} is the constitution coefficient related to the amount of atoms of *j*-element in the molecules of *i*-compounds. Therefore, it can be written:

$$A^T [N, M]\, B = C \tag{70}$$

Where A^T is the transposed matrix of constitution coefficient (N row, M column). The combination of Eq.70 and Eq.68 gives the following formula:

$$A^T[N,M]\,B = C \tag{71}$$

$$B^T A\nu = 0 \tag{72}$$

Since $B \neq 0$ and thus $B^T \neq 0$, the following term must be fulfilled:

$$A[M,N]\,\nu_N = 0 \tag{73}$$

For example we may consider the following system consisting of seven compounds and five elements:

N=7: $SrCO_3$, SrO, Al_2O_3, $Al(OH)_3$, $SrAl_2O_4$, CO_2 and H_2O;

M=5: C, Sr, Al, O and H.

	ν_1	ν_2	ν_3	ν_4	ν_5	ν_6	ν_7
	$SrCO_3$	SrO	Al_2O_3	$Al(OH)_3$	$SrAl_2O_4$	CO_2	H_2O
C	1	0	0	0	0	1	0
Sr	1	1	0	0	1	0	0
Al	0	0	2	1	2	0	0
O	3	1	3	3	4	2	1
H	0	0	0	3	0	0	2

Table 5. Constitution coefficients in reaction system.

Via the substitution from the table of constitution coefficients (Table 5) to Eq.73, the following expression should be derived:

$$\begin{bmatrix} 1 & 0 & 0 & 0 & 0 & 1 & 0 \\ 1 & 1 & 0 & 0 & 1 & 0 & 0 \\ 0 & 0 & 2 & 1 & 2 & 0 & 0 \\ 3 & 1 & 3 & 3 & 4 & 2 & 1 \\ 0 & 0 & 0 & 3 & 0 & 0 & 2 \end{bmatrix} \begin{bmatrix} \nu_1 \\ \nu_2 \\ \nu_3 \\ \nu_4 \\ \nu_5 \\ \nu_6 \\ \nu_7 \end{bmatrix} = \begin{bmatrix} 0 \\ 0 \\ 0 \\ 0 \\ 0 \end{bmatrix} \tag{74}$$

The order of matrix A $[M,N]$ can be now determined[11]:

$$\begin{bmatrix} 1 & 0 & 0 & 0 & 0 & 1 & 0 \\ 1 & 1 & 0 & 0 & 1 & 0 & 0 \\ 0 & 0 & 2 & 1 & 2 & 0 & 0 \\ 3 & 1 & 3 & 3 & 4 & 2 & 1 \\ 0 & 0 & 0 & 3 & 0 & 0 & 2 \end{bmatrix} \Rightarrow \begin{bmatrix} 1 & 0 & 0 & 0 & 0 & 1 & 0 \\ 0 & 1 & 0 & 0 & 1 & -1 & 0 \\ 0 & 0 & 1 & \frac{1}{2} & 1 & 0 & 0 \\ 0 & 0 & 0 & 1 & 0 & 0 & \frac{2}{3} \\ 0 & 0 & 0 & 0 & 0 & 0 & 0 \end{bmatrix} \tag{75}$$

The last row of matrix A is a liner combination of other rows and the order of matrix of constitution coefficients (h) is equal to 4. From the **Gibbs Stoichiometric Law** the following can be now calculated:

$$R = N - h = 7 - 4 = 3 \tag{76}$$

Therefore, there are three independent chemical reactions in the system. Eq.74 and the solution 75 lead to the following results:

$$\nu_1 + \nu_6 = 0 \tag{77}$$

$$\nu_1 + \nu_2 - \nu_6 = 0 \tag{78}$$

$$\nu_3 + \frac{1}{2}\nu_4 + \nu_5 = 0 \tag{79}$$

$$\nu_4 + \frac{2}{3}\nu_7 = 0 \tag{80}$$

The stoichiometry of these processes can be calculated choosing three parameters:

a. The choice ν_7=1, ν_6=0 and ν_5=0 leads to the solution:

$$\nu_4 = -\frac{2}{3} \tag{81}$$

---

11 Except for the switch of the row with column (order of column must remain unchanged), all usual mathematical operation can be used. These calculation includes: The change of the order of row (1), The multiplication of row by a constant (2), The summation of row with other row or their linear combination (3), The exclusion of row which is a linear combination of other rows (e.g. last row in Eq.75).

$$\nu_3 = \tfrac{1}{3} \tag{82}$$

$$\nu_2 = 0 \tag{83}$$

$$\nu_1 = 0 \tag{84}$$

$$\begin{aligned} &H_2O + \frac{1}{3}\,Al_2O_3 = \frac{2}{3}\,Al(OH)_3 \\ &3\,H_2O + Al_2O_3 = 2\,Al(OH)_3 \end{aligned} \tag{85}$$

From the calculation of the reaction Gibbs energy it can be estimated, that the reaction 85 proceeds to right side (product) if the temperature is lower than 113 °C ($\Delta_r G < 0$) and further, the dehydration of gibbsite takes place if the temperature is higher ($\Delta_r G > 0$).

b. The choice ν_7=0, ν_6=1 and ν_5=0 leads to the solution:

$$\nu_4 = 0 \tag{86}$$

$$\nu_3 = 0 \tag{87}$$

$$\nu_2 = 1 \tag{88}$$

$$\nu_1 = -1 \tag{89}$$

$$SrCO_3 = SrO + CO_2 \tag{90}$$

This reaction reaches the equilibrium state at the temperature of 1071°C ($\Delta_r G$=0).

c. The choice ν_7=0, ν_6=0 and ν_5=1 enables to find the last independent reaction in the system where ν_4=0, ν_3=0, ν_2=1 and ν_1=-1:

$$SrO + Al_2O_3 = SrAl_2O_4 \tag{91}$$

The formation of strontium aluminate from the oxide species (Eq.91), the thermal decomposition of gibbsite (Eq.85) and strontium carbonate (Eq.90) are three independent reactions in the above mentioned system. The solution represents the base vectors (ν_1, ν_2, ν_3) of 3D reaction space (N-h=3) which come through the beginning of 7D (N) vector space:

Eq.85: (001);

Eq.90: (-110);

Eq.91: (0-1-1).

All other solutions are their linear combinations. All these reactions also increase the value of parameter C in Eq.1.9.

The system base consists of first (N-h) species in Table 5. Therefore, we can choose different order of species and solve different three independent reactions. The number of possible system bases in this system can be calculated from the equation:

$$\binom{N}{h} = \frac{N!}{h!(N-h)!} = \frac{7!}{4!3!} = 35 \tag{92}$$

For example, if we rearrange the system as follows (system base is marked bold):

$SrAl_2O_4$, SrO, H_2O, CO_2, $Al(OH)_3$, $SrCO_3$, Al_2O_3,

the following independent reaction is determined:

d. The choice ν_7=1, ν_6=0 and ν_5=0 leads to the solution:

$$SrAl_2O_4 = SrO + Al_2O_3 \tag{93}$$

e. The choice ν_7=0, ν_6=1 and ν_5=0 leads to the solution:

$$SrO + CO_2 = SrCO_3 \tag{94}$$

f. The choice ν_7=0, ν_6=0 and ν_5=1 leads to the solution:

$$\begin{gathered} \frac{1}{2}SrAl_2O_4 + \frac{3}{2}H_2O = \frac{1}{2}SrO + Al(OH)_3 \\ SrAl_2O_4 + 3H_2O = SrO + 2\,Al(OH)_3 \end{gathered} \tag{95}$$

But not all possible system bases are valid; the stoichiometry of independent reaction can be calculated only from the system base that contains all elements occurring in the system. For example the choice:

$SrAl_2O_4$, SrO, Al_2O_3, CO_2, $Al(OH)_3$, $SrCO_3$, H_2O;

is not acceptable, because the compounds in the system base do not contain hydrogen.

In order to better explain the geometric meaning of the Gibbs Stoichiometric Law, the following simple reaction system is applied:

- N=3: SrO, Al_2O_3 and $SrAl_2O_4$.
- M=3: Sr, Al and O.

$$\begin{matrix} & \nu_1 & \nu_2 & \nu_3 \\ Sr \\ Al \\ O \end{matrix} \begin{bmatrix} 1 & 0 & 1 \\ 0 & 2 & 2 \\ 1 & 3 & 4 \end{bmatrix} \begin{bmatrix} \nu_1 \\ \nu_2 \\ \nu_3 \end{bmatrix} = \begin{bmatrix} 0 \\ 0 \\ 0 \end{bmatrix} \tag{96}$$

$$\begin{bmatrix} 1 & 0 & 1 \\ 0 & 2 & 2 \\ 1 & 3 & 4 \end{bmatrix} \Rightarrow \mathsf{K} \Rightarrow \begin{bmatrix} 1 & 0 & 1 \\ 0 & 1 & 1 \\ 0 & 0 & 0 \end{bmatrix} \quad \Rightarrow \quad h = 2 \tag{97}$$

$$R = N - h = 3 - 2 = 1 \tag{98}$$

Therefore we search for only one independent reaction, where ν_3=1:

$$\nu_2 = -\nu_3 = -1 \tag{99}$$

$$\nu_1 = -\nu_3 = -1 \tag{100}$$

$$\mathrm{SrO} + \mathrm{Al_2O_3} = \mathrm{SrAl_2O_4} \tag{101}$$

The results represent the coordinates vector (-1-11) related to the line passing through the beginning P[0,0,0] of 3D reaction space (Fig.14). As results from Eq.92, there are three options for the system base.

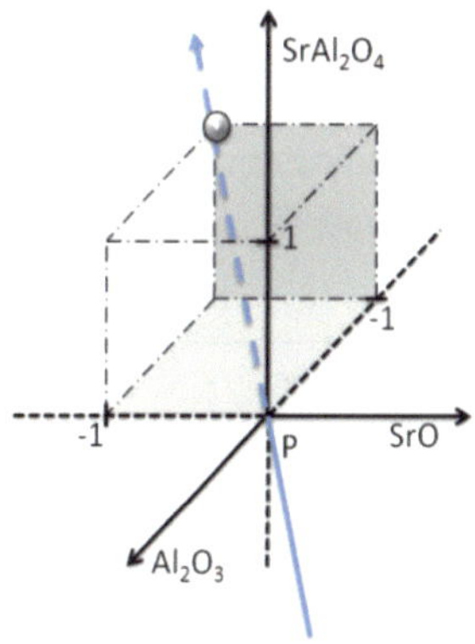

Figure 14. Geometric drawing of the solution resulting from the Gibbs Stoichiometric Law.

5. Density of cement phases

The densities of cement phases are fundamental parameter with many uses, such as calculating the space filling in pastes by solid cement substances as well as the assessment of the potential for changes of size and porosity within hardened pastes undergoing the phase changes. The substances containing cement can be divided into two groups: crystalline and amorphous. The principal amorphous substance of Portland cement is C–S–H. It poses certain problems in determining its density as it is insufficiently crystalline for the application of conventional crystallographic methods: only experimentally-determined measurements are useful. Crystalline solids, on the other hand, are amenable for the calculation of density from crystallographic constants using the relationship [66]:

$$\rho = \frac{ZM}{N_A V} \quad [kg \cdot m^{-3}] \tag{102}$$

where ρ is the calculated density, Z is the number of formula units per unite cell, M is the formula weight, V is the unite cell volume ($V=M/\rho$) and N_A=6.02214 10^{23} mol^{-1} is the Avogadro ´s number.

The volumes of the unit cell can be calculated as follows:

- Isometric /cubic crystal system: $V=a^3$;
- Tetragonal crystal system: $V=a^2c$;
- Orthorhombic crystal system: $V=abc$;
- Hexagonal /trigonal crystal system: $V=a^2c\ sin\ (60°)$;

- Monoclinic crystal system: $V=abc \sin(\beta)$;
- Triclinic crystal system: $V=abc \sin((1-\cos^2\alpha - \cos^2\beta - \cos^2\gamma)+2(\cos\alpha \cos\beta \cos\gamma))^{1/2}$

where *a, b, c* are the unit cell axial dimensions and α, β, γ are the relevant angles. The parameters of basic seven lattice systems are shown in Fig.15.

PhaseinSrO-Al_2O_3 system	**Sr_3A** $3SrO{\cdot}Al_2O_3$ 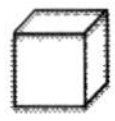	**SrA** $SrO{\cdot}Al_2O_3$ 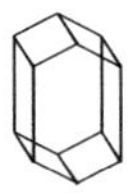	**α-SrA_2** $SrO{\cdot}2Al_2O_3$ 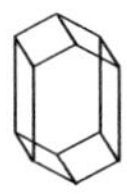
Crystal system	Cubic	Monoclinic	Monoclinic
M [kg·mol^{-1}]	0.41282	0.20558	0.30754
Z	24	4	4
a [m]	$1.586{\cdot}10^{-9}$	$5.150{\cdot}10^{-10}$	$1.304{\cdot}10^{-9}$
b [m]	---	$8.836{\cdot}10^{-10}$	$9.011{\cdot}10^{-10}$
c [m]	---	$8.442{\cdot}10^{-10}$	$5.536{\cdot}10^{-10}$
β [°]	---	93.43	106.12
***V*[m^3]**	$3.986{\cdot}10^{-27}$	$3.384{\cdot}10^{-28}$	$6.249{\cdot}10^{-28}$
***ρ*[kg.m^{-3}]**	4127	3561	3269
Phase in SrO-Al_2O_3 system	**β-SrA_2** $SrO{\cdot}2Al_2O_3$ 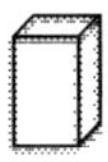	**SrA_6** $SrO{\cdot}6Al_2O_3$ 	**$Sr_{10}A_3$** $10SrO{\cdot}3Al_2O_3$ 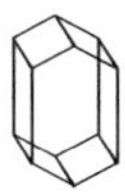
Crystal system	Orthorhombic	Hexagonal	Monoclinic
M [kg·mol^{-1}]	0.30754	0.71539	1.34208
Z	4	2	8
a [m]	$8.085{\cdot}10^{-10}$	$5.562{\cdot}10^{-10}$	$3.558{\cdot}10^{-9}$
b [m]	$1.185{\cdot}10^{-9}$	---	$7.846{\cdot}10^{-10}$
c [m]	$4.407{\cdot}10^{-10}$	$2.197{\cdot}10^{-9}$	$1.575{\cdot}10^{-9}$
β [°]	---	60	103.68
***V*[m^3]**	$4.220{\cdot}10^{-28}$	$5.886{\cdot}10^{-28}$	$4.272{\cdot}10^{-27}$
***ρ*[kg.m^{-3}]**	4840	4036	4174

Table 6. Calculation of the density of strontium aluminate cement phases.

Some examples including calculated volume and the density of compounds from SrO – Al_2O_3 system are given in Table 6.

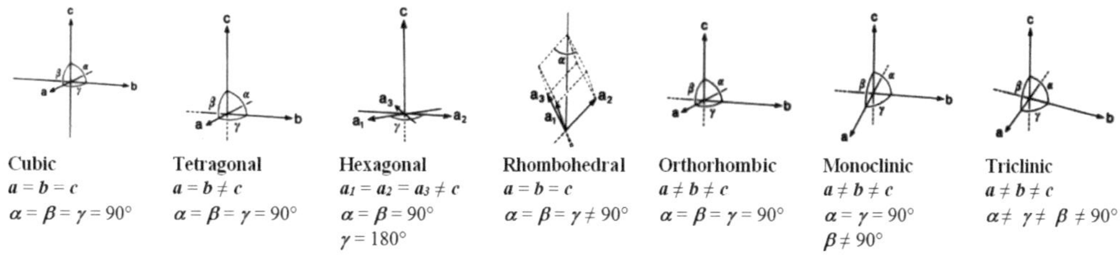

Figure 15. Parameters of basic lattice system.

Calculated data from Table 6 enable to explain how the density of compounds formed in the binary system of SrO-Al_2O_3 depends on the SrO/ Al_2O_3 ratio. Fig.16(a) shows that there are two exponential functions, which are cross each other in the composition corresponding to the SrA_2 phase (please consult with Fig.11).

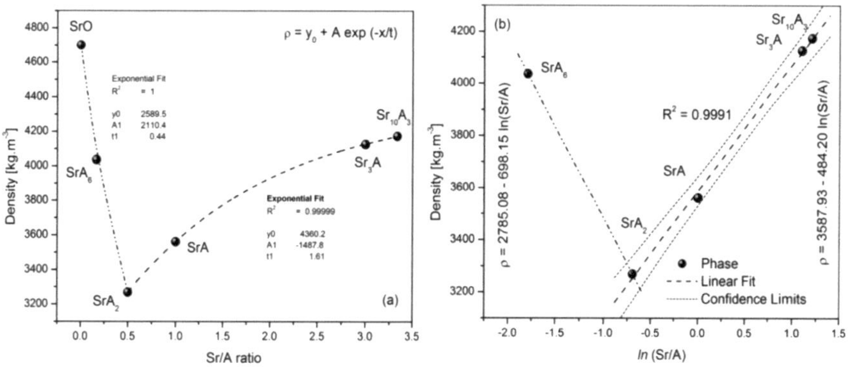

Figure 16. The influence of Sr/A ratio on the density (a) and its linearized dependence (b).

These exponential functions should be linearized as is shown in Fig.16(b). It is also a possible reason, why the formation of SrA_6 phase is preferred from the formation of SrA_2; lesser energy is required for the deformation of surroundings by newly formed crystal. Therefore the phase should be prepared via precipitation form melt, but the solid state synthesis is difficult [67,68].

6. Kinetics of heterogeneous processes

Since the thermal decomposition of solids is the most significant field of heterogeneous kinetics, the common analytical techniques include the methods such TG, DTA, DSC, TDA, etc., but other techniques could be also used by kinetic methods. These measurements are usually applied for the investigation of kinetics and mechanism of heterogeneous processes using the data measured as the function of time or temperature [69,70].

6.1. Mechanism dependent methods

Mechanism dependent or reaction model fitting method should be divided into two large groups [63,67]:

- Isothermal method;
- Non-isothermal method.

Both use the reaction models, which describe the dependence of degree of conversion (fractional conversion or the extent of conversion, α) on time (isothermal method) or temperature (non-isothermal method). The reaction model usually refers to the rate limiting step of process and the reaction system geometry (Table 7)

The fractional conversion is usually defined as the ratio of some system additional properties (mass, length, area...), the change of volume or mass of the system are mostly used:

$$\alpha = \frac{N_0 - N_t}{N_0 - N_\infty} \tag{103}$$

where N_0, N_t and N_∞ are the initial value, the value reached in time t and the final value, respectively. The value α is normalized in the range from 0 to 1 (or 0 to 100 %).

The following differential kinetic equation was used for the description of kinetics of investigated process via the **isothermal method** [71-73,456].

$$\frac{d\alpha}{dt} = k(T) \cdot f(\alpha) \tag{104}$$

The integration of Eq.104 leads to the formula:

$$\int_0^\alpha \frac{d\alpha}{f(\alpha)} = k \cdot \int_{t=0}^{t} dt \Rightarrow g(\alpha) = kt \tag{105}$$

where k is the constant of reaction rate. In the case that correct kinetic function was applied, the plot of $g(\alpha)$ versus t provides straight line. The rate constant k can be then calculated as the slope of this plot.

The value of k for several temperatures must be determined for the calculation of activation energy (E_a) and frequency factor (A) via the logarithmic form (Eq.107) of Arrhenius law (Eq. 106):

$$k = A \exp\left(-\frac{E_a}{RT}\right) \tag{106}$$

$$\ln k = \ln A - \frac{E_a}{R}\frac{1}{T} \tag{107}$$

where R denotes the universal gas constant (8.314 J•K^{-1}•mol^{-1}) and other symbols have their usual meaning. The plot (so-called Arrhenius plot) of $ln\ k$ versus T^{-1} provides straight line with the slope $-E_a/R$ and the intercept with y-axis equal to $ln\ A$. Several experiments should be performed in order to obtain the line with insufficient number of points, which enables to determine reliable slope of the Arrhenius plot (please see Fig.18 in Chapter 4 for example of Arrhenius plot).

Sign.	f(α)	g(α) = kt	Name of kinetic function	Rate limiting step	Description
$F_{1/3}$	$(3/2)(1-\alpha)^{1/3}$	$1-(1-\alpha)^{2/3}$	One-third order	Chemical reaction	Chemical process or mechanism non-invoking equations
$F_{3/4}$	$4(1-\alpha)^{3/4}$	$1-(1-\alpha)^{1/4}$	Three-quarters order		
$F_{3/2}$	$2(1-\alpha)^{3/2}$	$(1-\alpha)^{-1/2}-1$	One and half order		
F_2	$(1-\alpha)^2$	$(1-\alpha)^{-1}-1$	Second order		
F_3	$(1/2)(1-\alpha)^3$	$(1-\alpha)^{-2}-1$	Third order		
$P_{3/2}$	$(2/3)\alpha^{-1/2}$	$\alpha^{3/2}$	Mampel power law	Nucleation	Acceleratory rate equations
$P_{1/2}$	$2\alpha^{1/2}$	$\alpha^{1/2}$	Mampel power law		
$P_{1/3}$	$3\alpha^{2/3}$	$\alpha^{1/3}$	Mampel power law		
$P_{1/4}$	$4\alpha^{3/4}$	$\alpha^{1/4}$	Mampel power law		
E_1	α	$ln\ \alpha$	Exponential law		
A_1,F_1	$1-\alpha$	$-ln(1-\alpha)$	KJMA or JMAYK (Johanson-Mehl-Avrami-Yerofeyev-Kolgomorov equation)	Random nucleation, $n=1$	Sigmoidal rate equations or random nucleation and subsequent growth
$A_{3/2}$	$3/2(1-\alpha)[-ln(1-\alpha)]^{1/3}$	$[-ln(1-\alpha)]^{2/3}$		$n=3/2$	
A_2	$2(1-\alpha)[-ln(1-\alpha)]^{1/2}$	$[-ln(1-\alpha)]^{1/2}$		$n=2$	
A_3	$3(1-\alpha)[-ln(1-\alpha)]^{2/3}$	$[-ln(1-\alpha)]^{1/3}$		$n=3$	
A_4	$4(1-\alpha)[-ln(1-\alpha)]^{3/4}$	$[-ln(1-\alpha)]^{1/4}$		$n=4$	
A_u	$\alpha(1-\alpha)$	$ln[\alpha/(1-\alpha)]$	Prout-Tomkins equation	Branching nuclei	
R_1	$(1-\alpha)^0$	α	Power law	Contracting disc	Phase boundary deceleratory rate equations
R_2	$2(1-\alpha)^{1/2}$	$1-(1-\alpha)^{1/2}$	Power law	Contracting cylinder	
R_3	$3(1-\alpha)^{2/3}$	$1-(1-\alpha)^{1/3}$	Power law	Contracting sphere	
D_1	$1/2\ \alpha$	α^2	Parabola law	1D diffusion	Diffusion limited deceleratory reaction
D_2	$[-ln(1-\alpha)]^{-1}$	$\alpha+(1-\alpha)ln(1-\alpha)$	Valensi equation	2D diffusion	
D_3	$3/2(1-\alpha)^{2/3}[1-(1-\alpha)^{1/3}]^{-1}$	$[1-(1-\alpha)^{1/3}]^2$	Jander equation	3D diffusion	

Sign.	f(a)	g(a) = kt	Name of kinetic function	Rate limiting step	Description
D_4	$3/2[(1-\alpha)^{-1/3}-1]^{-1}$		Ginstling-Brounstein equation		
D_5	$3/2(1-\alpha)^{4/3}[(1-\alpha)^{-1/3}-1]^{-1}$		Zhuravlev, Lesokin, Tempelman eq.		
D_6	$3/2(1+\alpha)^{2/3}[(1+\alpha)^{1/3}-1]^{-1}$		anti-Jander eq.		
---	$3/2\ [(1+\alpha)^{1/3}-1]^{-1}$		anti- Ginstling-Brounstein eq.		
---	$3/2(1+\alpha)^{4/3}[(1+\alpha)^{-1/3}-1]^{-1}$		anti- Zhuravlev, Lesokin, Tempelman eq.		

Table 7. Overview of known kinetic functions representing applied kinetic model for the process.

The non-isothermal method enables to determine the kinetic parameters of the process from one experiment[12], therefore it is sometime termed as the single run method. The non-isothermal method is based on the assumption that the kinetics of processes should be described by the constitutive formula Eq.104. The function $f(\alpha)$ depends on applied kinetic model (Table 7) and the temperature dependence of the rate constant $k(T)$ is given by the Arrhenius law (Eq.106).

If the constant heating rate (CHR methods) Θ is applied, i.e. $dT=\Theta\ dt$, the combination of Eqs. 106 and 107 leads to the relationship [73-75,87]:

$$\frac{d\alpha}{f(\alpha)} = \frac{A}{\Theta} \cdot \exp\left(-\frac{E_a}{RT}\right) dT \tag{108}$$

The integration of Eq.108 yields to the formula:

$$g(\alpha) \cong \frac{AE_a}{\Theta R} \cdot p(x) \tag{109}$$

where the function $g(\alpha)$ represents the kinetic function and A, E_a, R and Θ are the temperature independent parameters. The determination and accurate calculation of the $p(x)$ function is a marginal problem in the non-isothermal kinetic [64]. The Coats and Redfern approximation of the $p(x)$ function is usually used [76].

$$\ln\left(\frac{g(\alpha)}{T^2}\right) = \left[\ln\frac{AR}{\Theta E_a}\left(1-\frac{2RT}{E_a}\right)\right] - \frac{E_a}{R} \cdot \frac{1}{T} \cong \ln\left(\frac{AR}{\Theta E}\right) - \frac{E_a}{R} \cdot \frac{1}{T} \tag{110}$$

12 As was demonstrated in Chapter 4.2.1.

A straight line was obtained by plotting $ln[g(\alpha)/T^2]$ versus reciprocal temperature (T^{-1}) for correct mathematical model $g(\alpha)$ of the reaction mechanism. The overall activation energy and pre-exponential factor were calculated from the slope and the intercept with y-axis of the plot, respectively.

The expression of $p(x)$ according to Schlomilch can be written as [77,456]:

$$p(x) = \exp\left(-\frac{E_a}{RT}\right)\left(\frac{1}{x}\right)\left(\frac{1}{x+2}\right) \tag{111}$$

6.2. Model free methods

The Kissinger`s kinetic approach is simple and often applied kinetic equation [78,79]:

$$\ln\left[\frac{\Theta}{T_m^2}\right] = \ln\left[\frac{AR}{E_a} n(1-\alpha_m)^{n-1}\right] - \frac{E_a}{RT_m} = const. - \frac{E_a}{RT_m} \tag{112}$$

where T_m is the peak temperature measured under applied heating rate Θ, n is the empirical reaction order (kinetic exponent), α_m is the fractional conversion reached for the temperature T_m and R is the universal gas constant. The plot of $ln\ (AR/\ E_a)$ versus T_m^{-1} was fitted by the straight line with the slope equal to-E_a/R whereas the intercept yielded to the constant term of Eq.112.

The peak methods approximate the fixed state of reaction to the stage at which the maximum rate of the process is achieved, i.e. to the peak of thermoanalytical curve. It must be pointed out that the Kissinger approach can be used for the rate (differential)-isoconversion (Friedman or FR-[80]) type as well as for the $p(y)$-isoconversion methods in principle, but the method's popularity place the Eq.112 into a group of maximum rate methods [81].

6.3. Estimation of the mechanism of process

The mechanism was estimated from the shape of DTG peak via the value of kinetic exponent (n) which was related to the empirical order of reaction [78,82] The exponent can be calculated from the equation [83,84]:

$$n = \frac{2.5\, R\, T_m^2}{w_{1/2}\, E_a} \tag{113}$$

where $w_{1/2}$ is the half-width (width at a half high) of peak. The value of kinetic exponent is typical for various mechanisms of investigated process [85].

Based on the foundation that the peak asymmetry increases with decreasing value of n Kissinger [78] has proposed the method that uses the shape index (SI) of TA curve peak that can be analytically expressed as:

$$SI = \frac{\left(\partial^2 N/\partial t^2\right)_{T_1}}{\left(\partial^2 N/\partial t^2\right)_{T_2}} \tag{114}$$

where N denotes the value of measured property, t is the time, T_1 and T_2 are the first (frontal) and second (terminal) inflection point, respectively. The shape index is only the function of reaction order.

The value of kinetic exponent can be then calculated according to the first [78] and the second Kissinger approach [76]:

$$n_{\mathbf{K}1} = 1.26\ SI^{1/2} \tag{115}$$

$$n_{\mathbf{K}2} = 1.88\ SI \tag{116}$$

6.4. Thermodynamic consideration

The correlation between kinetic and thermodynamic parameters of the investigated process results from the combination of Arrhenius with Eyring or Wertera and Zenera laws related to the temperature dependence of rate constant ($k(T)$) [86]:

$$\begin{aligned} k(T) = A\exp\left[-\frac{E_a}{RT}\right] = \frac{k_B T}{h}\exp\left[\frac{\Delta S^{\#}}{R}\right]\exp\left[-\frac{\Delta H^{\#}}{RT}\right] \\ = \nu\exp\left[-\frac{\Delta G^{\#}}{RT}\right] = \nu\ K^{\#} \end{aligned} \tag{117}$$

where k_B, h and $\nu = k_B T/h$ are the Boltzmann, the Plank constant and the vibration frequency, respectively. The thermodynamic parameters of activated complex, including free energy ($\Delta G^{\#}$), enthalpy ($\Delta H^{\#}$) and entropy ($\Delta S^{\#}$) of the process were calculated using Eyring equations [86,87,88]:

$$\Delta H^{\#} = E_{a,\alpha} - RT_{\alpha} \tag{118}$$

$$\Delta S^{\#} = R\left[\ln\left(\frac{h\ A_{\alpha}}{k_B T_{\alpha}}\right) - 1\right] \tag{119}$$

$$\Delta G^{\#} = \Delta H^{\#} - T_{\alpha}\ \Delta S^{\#} \tag{120}$$

The thermodynamic parameters of activated complex are often calculated using the peak temperature T_m so that the value of $\Delta G^{\#}$, $\Delta H^{\#}$ and $S^{\#}$ is related to the highest rate of process.

6.5. Calculation of theoretical value of activation energy

The approach is known as the "Congruent dissociative vaporization mechanism" (CDV). In the case of a solid compound *S* decomposed into gaseous products *A* and *B* with simultaneous condensation of low-volatility of species *A*, that is [89,369]:

$$S(s) \rightarrow a\ A(g) \downarrow + b\ B(g) \tag{121}$$

The theoretical value of activation energy E_a^{T} should be different for equimolar (in the absence of gaseous product *B* in the reactor atmosphere) and isobaric (in the presence of the excess of gaseous product B) modes of decomposition. The relations for the equimolar (E_a^{Te}) mode are:

$$E_a^{Te} = \frac{\Delta_{rc}H^{o}(T)}{v} = \frac{\Delta_r H^{o}(T)}{a+b} \tag{122}$$

and for the isobaric mode (E_a^{Ti}) they are:

$$E_a^{Ti} = \frac{\Delta_{rc}H^{o}(T)}{v-b} = \frac{\Delta_r H^{o}(T)}{a} \tag{123}$$

$$\Delta_r H^{o}(T) = \sum v_i \Delta_f H_i^{o} \tag{124}$$

v denotes total number of moles of gaseous product (*a+b*) and $\Delta_r H°$ is the reaction enthalpy for given temperature (Eq.124). The temperature dependence of enthalpy is given by Eq.21 in Chapter 4. In both cases, the E_a^{T} parameter corresponds to the specific enthalpy, i.e. the enthalpy of the decomposition reaction reduced to one mole of primary products without including components present in excess.

In order to take into account the partial transfer of energy released in the condensation of low-volatility product *A* to the reactant, the calculations of enthalpy of decomposition reaction 121 require an additional term (τ a $\Delta_c H°$(A,T)) where the coefficient τ corresponds to the fraction of condensation energy transferred to the reactant at the interface. The reaction enthalpy is then given by the relation:

$$\begin{aligned}\Delta_{rc}H^{o}(T) &= a\,\Delta_{f}H^{o}(A,T) + b\,\Delta_{f}H^{o}(B,T) \\ &-c\,\Delta_{f}H^{o}(C,T) + \tau\, a\,\Delta_{c}H^{o}(A,T)\end{aligned} \tag{125}$$

For the majority of substances, the condition τ=0.5 can be applied [89,90].

An essential difference between the CDV mechanism and the Arrhenius activation mechanism is that during the interface reactions, a proportion of the energy released on condensation of a non-volatile product is transferred to the solid reactant, reducing the energy barrier for further reactant volatilization. Thus, 'recycled' energy is responsible for the autocatalytic behaviour, justifying the following important generalizations [70]:

- Models identifying the preferred occurrence of chemical change at reactant/product contact interfaces with (uncharacterized, qualitative) 'strain', 'catalysis of by product', etc., can now be discarded as providing no insights into the reaction controls and mechanisms.
- Autocatalytic behaviour, resulting from the redistribution of product condensation energy, occurs when the decomposition proceeds at a reactant/product contact interface. Reactions yielding non-condensed (non-volatile) product transfer no energy, so that τ=0.
- Because the energy transfer is responsible for the autocatalysis, the variations in τ, found for a number of diverse substances, are identified with supersaturation of the non-volatile component rather than the chemical properties of the different original reactants.

The simplest presumption for the energy redistribution at an interface is that the condensation energy is shared equally between the reactant and the solid products, which is expressed as τ=0.5. The deviations, where $\Delta_c H°$ is distributed unequally between the solid reactant and the product phases in the ratio $\tau/(1-\tau)$, are ascribed to the degree of supersaturation of the non-volatile vapor.

The Arrhenius model is often represented by the familiar graph of energy variations as the reaction progresses by the "Advance along the Reaction Coordinate". This shows an initial rise to a maximum value, to form the 'transition complex', being followed by a decline thereafter. The activation energy is then the energy required for forming the 'activated' transition complex in an assumed "rate-determining step". However, in CDV theory, the value of parameter E represents the vaporization enthalpy [70].

Chapter 2

Raw Materials for Production of SrAC

1. Raw Materials and Raw Material Treatment

For the synthesis of strontium aluminate cement it is necessary to find the proper source of strontium (SrO) and aluminium oxide (Al_2O_3).

Two major strontium minerals are its carbonate, strontianite ($SrCO_3$) and more abundant sulfate mineral celestite ($SrSO_4$). William Cruickshank in 1787 and Adair Crawford in 1790 independently detected strontium in the strontianite mineral, small quantities of which are associated with calcium and barium minerals. They determined that the strontianite was an entirely new mineral and was different from barite and other barium minerals known in those times. In 1808, Sir Humphry Davy isolated strontium by the electrolysis of a mixture of moist strontium hydroxide or chloride with mercuric oxide, using mercury cathode. The element was named after the town Strontian in Scotland where the mineral strontianite was found [91].

The **strontium oxide** (SrO) is the first substantial component of strontium aluminate clinker. Therefore, the strontium carbonate ($SrCO_3$) is the most appropriate input material for the synthesis of strontium aluminate clinker. In nature $SrCO_3$ occurs as rare orthorhombic mineral **strontianite**[1] (space group *Pcmn*) and together with isostructural minerals aragonite ($CaCO_3$), witherite ($BaCO_3$) and cerussite ($PbCO_3$) it belongs to anhydrous carbonates from the group of aragonite[2] [92,93].

The structure of strontianite (Fig.1(a)) is based on isolated $[CO_3]^{2-}$triangles which are placed in layers perpendicular to *c*-axis. The layer has two structural planes where $[CO_3]^{2-}$ions are oriented in the opposite direction. Cations with the coordination number of 9 are placed between these layers.

Natural and artificially synthesized binary (**aragonites** up to 14 mol. % Sr [94], **strontianites** up to 27 % Ca [94], **witherites** [94], **baritocalcites** [95]) or ternary solid-solutions (**alstonites** [94]) of these carbonates are intensively studied in order to elucidate the mechanism of their formation, their structure, the thermodynamic stability and the luminescence properties.

1 Discovered in 1787 (Strontian, Scotland). Originally was considered the barium bearing mineral; which was disproved by Crawford and later by Klaproth and Kopp. Named in 1791 by Friedrich Gabriel Sulzer after the locality Strontian in Scotland.

2 There are three main groups of anhydrous carbonates without additional anions. The group of calcite (trigonal, space group $R\bar{3}c$, A = Ca, Mg, Mn, Fe, Co, Ni, Zn and Cd) and aragonite (orthorhombic, space group Pmnc, A = Ca, Sr, Ba and Pb) has the composition given by general formula ACO_3. The trigonal group of dolomite (space group $R\bar{3}$ where A = Ca and B = Mg, Fe, Mn and Zn) has general composition given by the general formula $AB(CO_3)_2$.

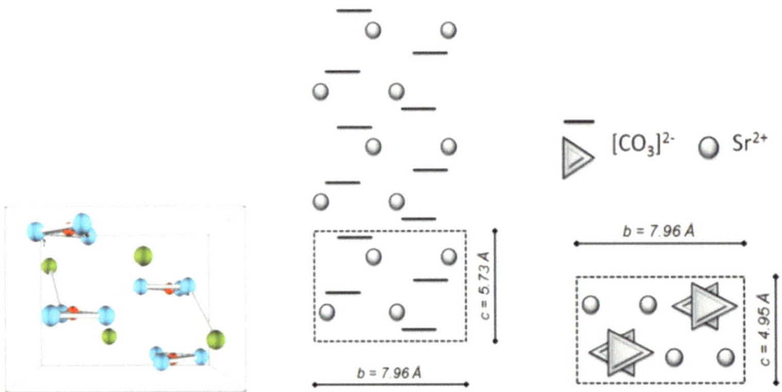

Figure 1. Structure of strontianite consists of isolated $[CO_3]^{2-}$triangles arranged into layers with Sr^{2+}ions in the space between layers.

Calcium carbonate minerals include considerable amount of strontium from seawater as they precipitate. It stands to reason that the solid-solutions of strontianite with calcite and aragonite ($Ca_xSr_{1-x}CO_3$) are the most explored. There is an immiscibility gap in the range 0.12 (aragonites) $< x <$ 0.87 (strontianites) under ambient conditions, which disappears at the temperature of ~107 °C [92,94,96,97-100].

Therefore natural sources of $SrCO_3$ are rare and have no industrial importance, strontium carbonate as well as other compounds such as strontium nitrate, strontium oxide and chloride are prepared from the orthorhombic mineral **celestite**[3] ($SrSO_4$, space group *Pnma* with cell unit parameters ***a***=8.359 Å, ***b***=5.352 Å, ***c***=6.686 Å and ***Z***=4) using the techniques described in Chapter 2.1.1. The structure of celestite consists of isolated $[SO_4]^{2-}$tetrahedrons and Sr^{2+}ions (Fig.2).

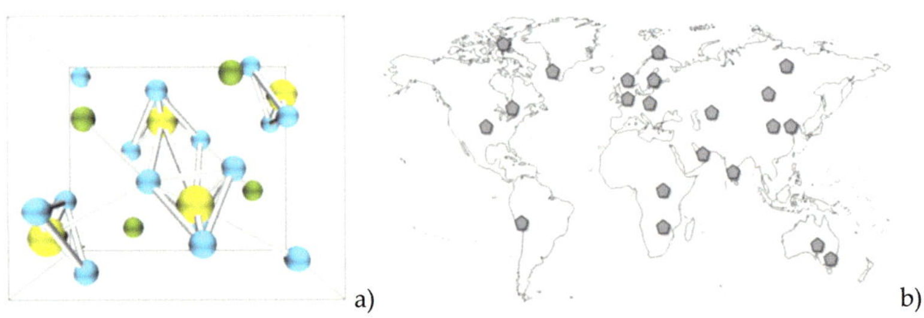

Figure 2. Structure of celestite (a) and distribution of large celestite deposits in the world (b).

Celestite together with isostructural barite ($BaSO_4$) and anglesite ($PbSO_4$) belong to anhydrous sulfates from the group of barite[4]. Similarly to the solid solutions of carbonates men-

3 Discovered in 1791 and named in 1799 by Abraham Gottlieb Werner from the Greek "cœlestis," for celestial, in allusion to the faint blue color of the original specimen.

tioned above, also celestite and barite ($BaSO_4$) coexist in the marine environment with significant fractions of Sr and Ba in solid solutions. Therefore it is better to identify barite suspended in seawater as the strontian barite ($Sr_xBa_{1-x}SO_4$) [101].

The second substantial component of strontium aluminate cement is **aluminium oxide** (Al_2O_3). The most stable crystalline form of Al_2O_3 is the polymorphic modification of hexagonal corundum (α-Al_2O_3) from the R $\bar{3}$ C space group[5]. The crystal structure of corundum type is typical for other oxides, such as **hematite** (Fe_2O_3), **eskolaite** (Cr_2O_3), **karelianite** (V_2O_3) and **tistarite** (Ti_2O_3). Therefore, naturally occurring minerals are often colored by the admixture of these elements, e.q. ruby is red colored by Cr and blue sapphire by Fe and Ti. That means that these elements are also interesting from the point of view of modification the properties (course of sintering, hydration and setting) of strontium aluminate cement. Chapters 4 and 5 are dedicated to this topic [424].

Pure aluminium oxide is relatively rare, but single crystals of gemstones such as sapphire (colorless) or ruby (red due to the content of chromium) can be found in nature [424]. Industrial production of Al_2O_3 is based on the Bayer process of bauxite. The main part of produced alumina is used in metal industry for the production of aluminium by Hall-Heroult Process [102-105].

The application of Al_2O_3 in ceramics includes the production of alumina porcelain and alumina oxide ceramics, ZTA (Zirconia Toughened Alumina) ceramics and the applications such as electroceramics, construction ceramics, shaped and unshaped refractory products, abrasive materials, etc [106-112]. From the point of view of the volume of production, polycrystalline alumina is the most frequently used material as ceramics for the structural applications. However, in comparison with for example, silicon nitride (Chapter 6), where the influence of various additives on the microstructure and properties is well characterized and understood, alumina remains the material with many unknown factors yet to be revealed. Alumina based materials can be roughly divided into three groups [424]:

- **Solid-state sintered aluminas:** enable to prepare nanocrystalline materials with excellent mechanical properties and well-sintered ceramics being transparent to visible light [113,114].
- **Liquid-phase sintered aluminas (LPS):** are substantial part of industrially produced alumina-based materials. Silica, alkali oxides and oxides of alkali earth metals are used as sintering additives [115-117].
- **Alumina-based composites:** ZTA and alumina based nanocomposites with non-oxide phases such as SiC or TiC [118-123].

The preparation of Al_2O_3 mono-crystals is based on **Verneuil process** consisting in the flame fusion in high temperature region from 1500 to 2500 °C [124-127]. Bauxite (Fig.3) is also used

4 Barite – celestite series from the group of barite. Anhydrous sulfates without additional anions with the composition given by general formula ASO_4, where A = Pb, Ba, Sr.

5 Structure and lattice parameters of corundum are described in Chapter 4.1.

for the production of calcium aluminate cements [128] or is calcined and used as opening material for the refractory products [129-131].

Figure 3. SEM image of calcined bauxite grain.

In order to obtain good quality in abrasive, refractory and pottery products, the content of impurities should be reduced. Chemical processes include the pyrochemical techniques, acid leaching methods or reductive dissolution alternatives. The pyrochemical techniques involve the treatment of bauxite at high temperature with gases such as H_2, Cl_2 or anhydrous HCl [132,133]. The acid leaching methods are based on the application of strong inorganic acids such as HCl or H_2SO_4 [134-137].

A serious problem with these techniques is that leaching of iron is often accompanied by substantial co-dissolution of aluminium hydroxides, particularly during the treatment of gibssitic and boehmitic ores. Selective dissolution of iron can be obtained applying mild reducing conditions. In this case the dissolution of Fe(III) oxides takes place via the reduction of ferric iron to the divalent state. It is widely accepted that biological mechanisms are often involved in the mobilization of iron in natural systems. For the particular case of bauxites the biological activity of iron reducing microorganisms is most probably involved in the generation of gray-colored iron depleted bauxites [138,139].

Since the production of alumina from bauxite ores consumes large amount of caustic soda, and generates large amount of "**red mud**" slurry waste, the alternative processes for the production of aluminium and aluminoalloys via carbothermic reduction of bauxite ores was investigated. The reduction sequence of metal oxides in bauxite ores is iron oxides then silica and titania and then alumina (Fig.4). Metallic iron is formed at the temperatures below 1100 °C. At 1200 °C or above the ferroalloy phase with silicon and aluminium is formed. Carbides of titanium, silicon and aluminium were formed by the carbothermal reduction. The metals were formed and dissolved in the ferroalloy phase, which after saturation, was segregated as metal carbides distributed inside the alloy phase as inclusions or around the alloy particles [140,141].

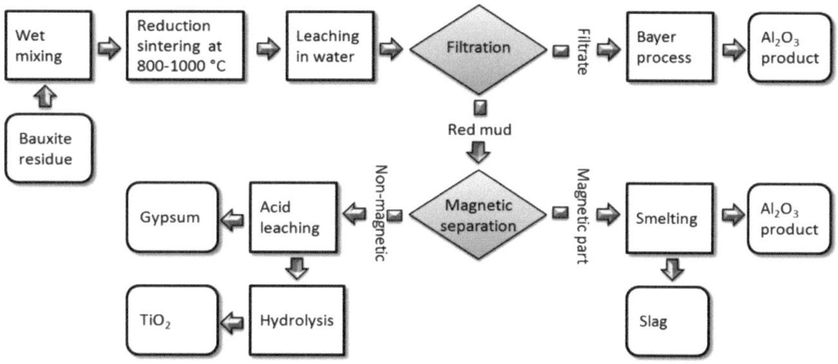

Figure 4. Process suggested for simultaneous recovery of iron, aluminium and titanium from red mud [146].

The utilization of Bayer's process residues in the cement production is also studied. Previous works proposed a method of treating red mud with saturated $Ca(OH)_2$ solution followed by 3% H_2SO_4, in order to remove Na. After heating, the treated material is suggested for the application in cement manufacturing. The major parts of red mud are hematite and alumina-rich phases (Fig.7), participating in the production of hydraulic crystal phases C_3A and C_4AF. Fe-rich waste could be then used for the production of sulphate resistant cements [142]. Other option includes the applications such as catalysts and adsorbents, ceramics, coatings and pigments, waste water and gas treatment, recovery of major and minor metals [143-146].

Bayer suggested that [143]: "Red, iron-containing residue, that occurs after digestion, settles well and, with sufficient practice, can be filtered and washed. Due to its high iron content and low aluminium oxide content, it can be, in an appropriate manner, treated or melted with other iron ores to iron". The concept of bauxite residue as an iron resource was tested by a number of workers over the intervening 120 years, however, the "appropriate manner" of treatment remains elusive [144].

The aluminium gels, salts (sulphates, nitrates or chlorides) or alkoxides and advanced ceramic fabrication techniques can be applied for the preparation of high purity products (please referee to Chapter 9). Bauxite is the mixture of aluminium hydroxides and oxyhydroxides such as boehmite, diaspore and gibbsite, with varying content of admixture minerals. Goethite, lepidocrocite, hematite, magnetite, kaolinite, chlorites, calcite, anatase, phosphastes, etc are the major ones [148].

Bauxite, as the primary source of aluminum, represents a typical accumulation of weathered continental crust [147,148]. Bauxites are usually considered to be of three major genetic types [149-152]:

1. **Lateric bauxites** (sometimes called equatorial) are formed from weathered primary aluminosilicate rocks in equatorial climates comprising ~90% of the world's exploitable bauxite reserves. Lateritic bauxite is generally formed by in-situ lateritization, therefore, the most important factors in determining the extent and grade of it are thought to be

the parent rock composition, climate, topography, drainage, groundwater chemistry and movement, location of water table, microbial activity, and the duration of weathering processes.

2. **Sedimentary bauxites** are primarily formed by the accumulation of lateritic bauxite deposits during the mechanical transportation of surface flows. In addition, the consequent weathering and transfer of Al and Fe play substantial roles in bauxitization, which not only supports the formation of bauxite from kaolin clays but also refines the primary clastic ores.

3. **Karst bauxites** are named for their confinement to karst zones with karstified or karstifying carbonate rocks. Karst-type deposits originate from a variety of different materials, depending on the source area.

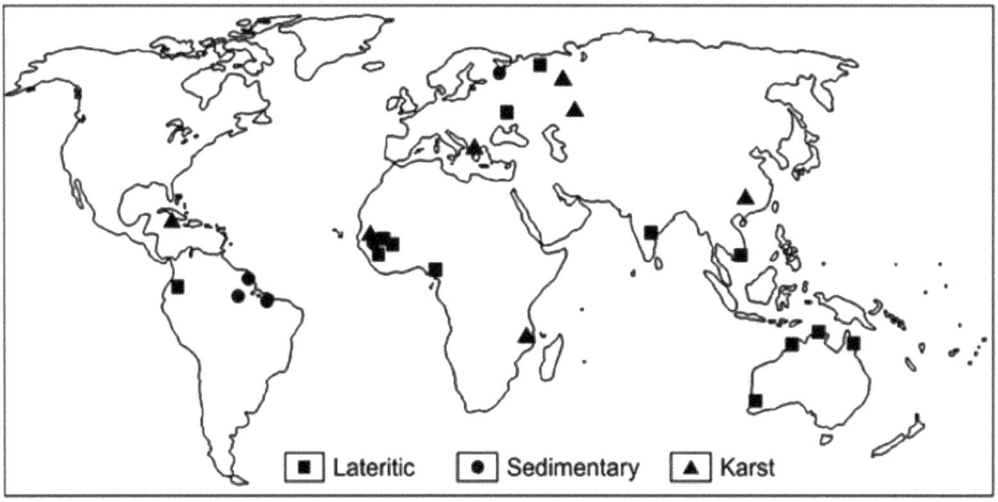

Figure 5. Distribution of superlarge bauxite deposits worldwide [152].

Each genetic group of bauxite experienced the separation of aluminum (Al) and silicon (Si) by the accumulation of Al, and the removal of Si, alkali metals, and rare earth elements from parent rock (sediment) during its weathering [148].

Bauxite deposits (Fig.5) form mainly at ambient pressure and temperature on the (sub)surface of continents. Abundant bioavailable irons, nutrient elements, sulfurs, and organic carbons make bauxite suitable for microorganisms to inhabit so they become rare geological sites that can preserve records of microbiological activity on the surface of continents under strong weathering effects. The microorganism activities can produce a family of minerals with special morphologies and stable isotope compositions. Bauxite deposits were studied in detail because of their economic value. They play an important role in the study of paleoclimate and paleogeography of continents because they contain scarce records of weathering and evolution of continental surfaces [148].

1.1. Industrial and laboratory production of $SrCO_3$

Chemical industry consumes over 95 % mined celestite for the conversion to other strontium compounds. The main admixtures in celestite ores are calcite ($CaCO_3$), gypsum ($CaSO_4$ $2H_2O$), quartz (SiO_2) and clay minerals. The gravity separation techniques and the flotation[6] are mostly used for the separation of those admixtures due to high efficiency and low operating costs. Moreover, the process does not require the usage of other chemicals for the purification and has low environmental impact. On the other hand, the efficiency of these techniques for the preparation of celestite concentrate depends on the texture of ore as well as the type and quantity of associated impurities [153-155]. The particle size is other most important factor. Extremely fine particle sizes must be achieved by grinding in order to release celestite and calcite [156]. The difference in grindabilities makes it possible to separate celestite from gypsum by differential grinding [157].

The shear flocculation[7] of fine celestite suspension with sodium dodecyl sulfate (SDS, $C_{12}H_{25}SO_4Na$) or with anionic alkyl succinate surfactant can be performed in broad pH range (3 – 11) but the highest efficiency is reached at pH 7. Increasing concentration of surfactant has positive effect on the course of process. The most common inorganic dispersants used are sodium silicate, sodium phosphate and sodium polyphosphate. The investigation of mutual influence of additives shows that sodium silicate strongly prevents celestite with sodium dodecyl sulfate from shear flocculation, but the dispersive effect of SDS is low when anionic alkyl succinate surfactant is used. In the presence of sodium polyphosphate, the shear flocculation of celestite suspension increases slowly for both surfactants. The similar increase can also be observed for sodium phosphate in the presence of SDS. However, sodium phosphate dispersed the celestite suspension in the presence of anionic alkyl succinate surfactant [158]. Sodiumoleate (cis-9-Octadecenoic acid sodium salt) and tallow amine acetate (TAA) were more effective for celestite suspensions in the pH ranges 7–11 and 6–10, respectively [159].

The surface of celestite becomes hydrophobic by the adsorption of dodecyl sulfate on the surface. Sodium dodecyl sulfate is also effective for the flotation of celestite in the solution free of carbonate species over the broad pH range of 3-11. The surface transformation of celestite to strontium carbonate which takes place at pH $\geq$ 7.8 causes that the zeta potential of celestite begins to be more negative and subsequently resembles that of strontium carbonate. Sulfate ions are exchanged by carbonate ions in the celestite crystal lattice, so CO_3^{2-} and HCO_3^- species are probably responsible for the negative increase in zeta potential. The surface transformation of celestite to strontium carbonate has no effect on floatability up to the pH of 10. Once the pH is higher than 10, the concentration of CO_3^{2-} and HCO_3^- species in

6 Flotation is an industrial process for the treatment of raw materials. The constituents of fine powdered raw materials are separated from the mixture according to the different wettability of individual solid species. The foam flotation process uses the interaction of gas bubbles with suspended material, which is next concentrated on the liquid level as foam.

7 Flocculation is a special case of coagulation where suspended particles of colloids form flake-like aggregates spontaneously or after the addition of clarifying agents.

aqueous solution is very intrinsic and the decrease of floatability is probably caused by the absorption of these species on carbonated surface of celestite [155].

The coagulation and flocculation characteristics of celestite by inorganic salts, such as $CaCl_2$, $MgCl_2$ and $AlCl_3$, indicate that magnesium ion was more effective on the celestite suspension than calcium and aluminum ions at high pH levels. The effect varied significantly depending on the concentration. While calcium and magnesium ions were not effective for the suspension below neutral pH, aluminum ion caused the stabilization of the celestite suspension at these pH levels [160].

In general, the aggregation of fine particles can be achieved by neutralizing the electrical charge of interacting particles by coagulation, or flocculation can be carried out by crosslinking the particles with polymolecules [161]. The pH of isoelectric point of celestite determined by the hindered settling technique is 2.6 [160].

There are two basic processes to produce $SrCO_3$ from $SrSO_4$ [162]:

a. Pyro-hydrometallurgical process or black ash method;

b. Hydro metallurgical process.

The "**pyro-hydrometallurgical process** or **black ash method**" is first of them. Celestite is carbothermically reduced to water soluble sulphide (SrS), which is next dissolved in hot water[8] The first solid-state reaction during the carbothermic reduction takes place at up to 400 °C [163]:

$$SrSO_4(s)+4\ C(s) \rightarrow SrS(s)+4\ CO(g) \tag{1}$$

After the formation of surface layer of the product the further progress of reaction 1 is inhibited. Formed carbon dioxide diffuses through the layer and reacts with celestite according to the following reaction:

$$SrSO_4(s)+4\ CO(g) \rightarrow SrS(s)+4\ CO_2(g) \tag{2}$$

Carbon dioxide diffuses further out of reaction zone and generates more CO according to the Boudouard reaction if the temperature is ≥ 720 °C:

$$CO_2(g)+C(s) \leftrightarrow 2\ CO(g) \tag{3}$$

That means that direct reaction of celestite with carbon (Eq.1) has little importance and $SrSO_4$ can be transformed to SrS at the temperature higher than equilibrium of **Boudouard reaction**. The important factor of the process 2 is the reduction potential of gas phase given

8 Strontium salt can be then prepared directly by dissolving SrS in acid.

by the partial pressure ratio of p_{CO}/p_{CO2}. It was also observed that the rate of carbothermic reduction significantly increases if celestite concentrate and carbon are milled together. The temperatures in the range from 1100 to 1300 °C with the excess of metallurgical grade coke are necessary to produce water-soluble strontium sulfide.

The dissolution of strontium sulfide in hot water can be expressed by the following heterogeneous reaction [164]:

$$SrS(s)+H_2O(l) \rightarrow Sr^{2+}(aq)+HS^-(aq)+OH^-(aq) \tag{4}$$

Eq.4 shows that the pH of leaching solution increases from almost neutral to the value of 11.5 – 12.5 as the concentration of OH^- ions increases. Extremely high pH values (pH > 14) should be avoided in order to prevent the system from the precipitation of strontium hydroxide[9] :

$$Sr^{2+}(aq)+2\,H_2O(l) \rightarrow Sr(OH)_2(s)+2H^+(aq) \tag{5}$$

The value of equilibrium constant ***K*** at 25 °C is 3.55 10^{-29}, i.e. *log* ***K***=-28.45. Therefore, the concentration of $Sr(OH)_2$ in leaching solution (Fig.6) can be expressed as:

$$\log_{10}\left[Sr^{2+}\right] = 28.45 - 2\,pH \tag{6}$$

That means that leaching of SrS must be carried out in relatively low alkaline medium in order to ensure high concentration of strontium in the solution. The solubility of strontium hydroxide is enhanced by increased temperature. Therefore leaching and precipitation of $SrCO_3$ at higher temperatures mean that the formation of $Sr(OH)_2$ precipitates is reduced.

On the other hand, leaching at pH < 7 generates hydrogen sulphide gas:

$$HS^-(aq)+H^+(aq) \leftrightarrow H_2S(g) \tag{7}$$

The generation of hydrogen sulphide gas takes place in early stages of leaching when the pH of slurry is relatively low.

Introducing the carbon dioxide gas or carbonating agent such as soda ash leads to the precipitation of strontium carbonate from supersaturated solution (Eq.13). The sequence of reaction steps includes the dissolution of carbon dioxide in solution and *in situ* formation of carbonic acid (H_2CO_3, Eq.8), the dissociation of H_2CO_3 (Eq.9 with the equilibrium constant

---

9 $Sr(OH)_2 \cdot 8H_2O$ precipitates during cooling of hot supersaturated solutions. Strontium hydroxide octahydrate transforms to monohydrate by ageing of the precipitate. Anhydrous hydroxide can be prepared via thermal treatment of the precipitate up to 100 °C.

(***K′***) given by Eq.10), the dissociation of bicarbonate species (Eq.11 with the equilibrium constant (***K′′***) given by Eq.12) and the precipitation of strontium carbonate (Eq.13 with the ion product (***P***) given by Eq.14) [163-165].

$$CO_2(g)+H_2O(l) \leftrightarrow H_2CO_3 \tag{8}$$

$$H_2CO_3 \leftrightarrow H^+(aq)+HCO_3^-(aq) \tag{9}$$

$$K'=\frac{\left[H^+\right]\left[HCO_3^-\right]}{\left[H_2CO_3\right]} \tag{10}$$

$$HCO_3^- \leftrightarrow H^+ + CO_3^{2-} \tag{11}$$

$$K''=\frac{\left[H^+\right]\left[CO_3^{2-}\right]}{\left[HCO_3^-\right]} \tag{12}$$

$$Sr^{2+}(aq)+CO_3^{2-}(aq) \rightarrow SrCO_3(s) \tag{13}$$

$$P=\left[Sr^{2+}\right]\left[CO_3^{2-}\right] \tag{14}$$

The solubility of strontium carbonate is 5.6 10^{-10} at the temperature of 25 °C and decreases to 1.32 10^{-10} at the temperature of 100 °C. The hydrolysis reaction leads to alkaline character of aqueous solution of $SrCO_3$.

Eqs.8-14 show that one mole of gaseous CO_2 is required for the precipitation of each mole of $SrCO_3$. The concentration of CO_3^{2-} ions in leaching solution for given pH is expressed by the following law:

$$\log_{10}\left[CO_3^{2-}\right]=\log_{10}K''+pH+\log_{10}\left[HCO_3^-\right] \tag{15}$$

If the pH of leaching solution is higher than 7, H^+ ions formed by the reaction 11 neutralizing OH^- anions are released during leaching of SrS (Eq.4).

In general, the black ash method is concluded to be more the more economical than other alternatives [165].

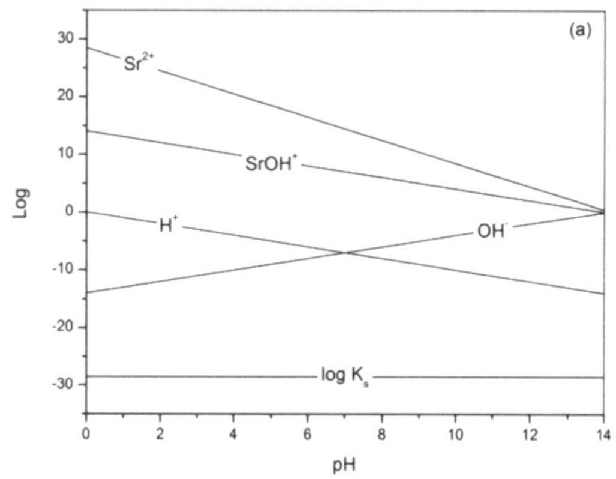

Figure 6. Influence of pH on the equilibrium concentration of Sr^{2+} and $SrOH^{+}$ ions in solution.

The second technique for the preparation of strontium carbonate is the **direct conversion method** or **hydrometallurgical method**. Strontium carbonate is prepared by introducing $SrSO_4$ powder into hot solution of Na_2CO_3, where the following conversion process takes place:

$$SrSO_4 + Na_2CO_3 \rightarrow SrCO_3 + Na_2SO_4 \quad (16)$$

or better:

$$SrSO_4(s) + CO_3^{2-}(aq) \rightarrow SrCO_3(s) + SO_4^{2-}(aq) \quad (17)$$

The effect of experimental conditions on the process includes the influence of temperature, solid to liquid ratio, particle size, stirring rate, Na_2CO_3 : $SrSO_4$ molar ratio, etc. The conversion rate of celestite to strontium carbonate increases with temperature up to 70 °C [165-170]. Prepared carbonate or sulphide is further converted to other strontium salts [91].

It is also possible to use ammonium carbonate ($(NH_4)_2CO_3$)[10] and bicarbonate (NH_4HCO_3) instead of soda ash for the conversion [165,171-174]:

$$SrSO_4 + (NH_4)_2CO_3 \rightarrow SrCO_3 + (NH_4)_2SO_4 \quad (18)$$

(in boiling mixture) $SrSO_4 + (NH_4)_2CO_3 \rightarrow SrCO_3 + 2NH_3 + SO_3 + H_2O$

10 Since NH^{4+} species as well as Na^{+} ion in Eq.16 does not participate in the reaction, the same equation 17 will be obtained for both processes.

$$SrSO_4(s)+2HCO_3^-(aq) \rightarrow SrCO_3(s)+SO_4^{2-}(aq)+H_2O(l)+CO_2(g) \tag{19}$$

There is also an alternative in mechanochemical synthesis, where the mixture of $SrSO_4$ and NH_4HCO_3 is intensively milled. The soluble ammonium sulphate is next removed by leaching of the product in water [165].

Moreover lots of special techniques for the preparation of $SrCO_3$ were described in current literature. These methods include the preparation of strontium carbonate via solid-state decomposition route from inorganic precursor [746]. Simple solution techniques [175], solvothermal synthesis [176-178], refluxing method [188] hydrothermal synthesis [179-182], ultrasonic method or sonochemical-assisted synthesis [183,184], microwave assisted synthesis [185,186] and mechanochemical synthesis [168,187] were described. Depending on applied preparation technique, the strontium carbonate particles of different shape can be prepared, such as spheres, rods, whiskers and ellipsoids, needles, flowers, ribbons, wires, etc. [188].

The solubility of strontium salts is mostly either higher or lower than for corresponding calcium and barium salts (Table 1).

Cations	**Ca^{2+}**	**Sr^{2+}**	**Ba^{2+}**
Anions	Solubility [g•100 cm^{-3}]		
OH-	0.160	0.810	3.890
F-	0.0017	0.0175	0.1600
Cl-	74.5	53.1	36.2
NO_3^{2-}	128.8	70.4	9.10
CO_3^{2-}	0.0014 (25 °C)	0.00155 (25 °C)	0.0022 (18 °C)
SO_4^{2-}	0.20	0.0148	0.00023
$C_2O_4^{2-}$	0.00058	0.0048	0.0125
CrO_4^{2-}	2.25	0.204	0.00037

Table 1. The solubility of salts of alkaline earth metals at the temperature of 20 °C.

Metallic strontium can be prepared by the electrolysis of mixed melt of strontium chloride and potassium chloride in a graphite crucible using iron rod as cathode. The upper cathode space is cooled and metallic strontium collects around cooled cathode and forms a stick. Metallic strontium can also be prepared by thermal reduction of its oxide with aluminum. Strontium oxide-aluminum mixture is heated at high temperature in vacuum. Strontium is collected by the distillation in vacuum. Strontium is also a reducing agent. It reduces oxides and halides of metals at elevated temperatures to the metallic form.

Strontium is also obtained by the reduction of its amalgam, hydride, and other salts. Amalgam is heated and the mercury is separated by the distillation. If hydride is used, it is heated at 1 000°C in vacuum for the decomposition and removal of hydrogen. Such thermal reductions yield high–purity metal which, when exposed to air, oxidizes to SrO. The metal is pyrophoric, both SrO and SrO_2 (strontium peroxide) are formed via ignition in air. When heated with chlorine gas or bromine vapor, strontium burns brightly, forming its halides ($SrCl_2$ or $SrBr_2$). When heated with sulfur, it forms sulfide (SrS) [91].

Strontium reacts vigorously with water and hydrochloric acid forming hydroxide $Sr(OH)_2$ or chloride ($SrCl_2$) with liberation of hydrogen [91]:

$$Sr+2\,H_2O \rightarrow Sr(OH)_2+H_2 \tag{20}$$

$$Sr+2\,HCl \rightarrow SrCl_2+H_2 \tag{21}$$

When heated under hydrogen it forms ionic hydride (SrH_2), a stable crystalline salt. Heating metallic Sr in a stream of nitrogen above 380°C forms nitride (Sr_3N_2).

1.2. Bayer process

Pure alumina, which is required for the production of aluminum by the Hall process, is made by the Bayer process [91]. The Bayer process was developed in 1887 by Carl Josef Bayer (1847-1904). It is the method for industrial production of aluminium oxide from bauxite. This method replaces earlier techniques developed by Henri Étienne Sainte-Claire Deville (1818-1881). Fine milled bauxite powder is leached in the solution of sodium hydroxide in autoclave under the temperature range from 160 to 250 °C and the pressure from 0.4 to 0.8 MPa. The basic components of bauxite are dissolved and soluble salts according to the following reaction scheme are formed [189]:

$$Al(OH)_3+NaOH \xrightarrow{T,p} NaAl(OH)_4 \tag{22}$$

$$SiO_2+2\,NaOH \xrightarrow{T,p} Na_2SiO_3+H_2O \tag{23}$$

$$Fe_2O_3+2\,NaOH \xrightarrow{T,p} Na_2FeO_2+H_2O \tag{24}$$

$$TiO_2+2\,NaOH \xrightarrow{T,p} Na_2TiO_3+H_2O \tag{25}$$

Significant amount of impurities which remains in the dissolved solid rest, so called "**red mud**" (Fig.7), is next separated from the solution by filtration[11]. The IEP values vary with red mud ranging from 6.35 to 8.70 [347]. The liquid filtrate is then diluted so that the concentration of Al_2O_3 in the solution reaches the value of 150 kg $Al_2O_3 \bullet m^{-3}$ and the nuclei of $Al(OH)_3$ are introduced [190].

$$[Al(OH)_4]^- \xrightarrow{\text{seed}} Al(OH)_3 + OH^- \tag{26}$$

The dilution means decreasing the pH of alkaline solution[12] and the precipitation of aluminium hydroxide. Precipitated gibbsite (Eq.27), which is the main product of Bayer process, is washed and calcined to Al_2O_3 in the rotary kiln (Eq.28).

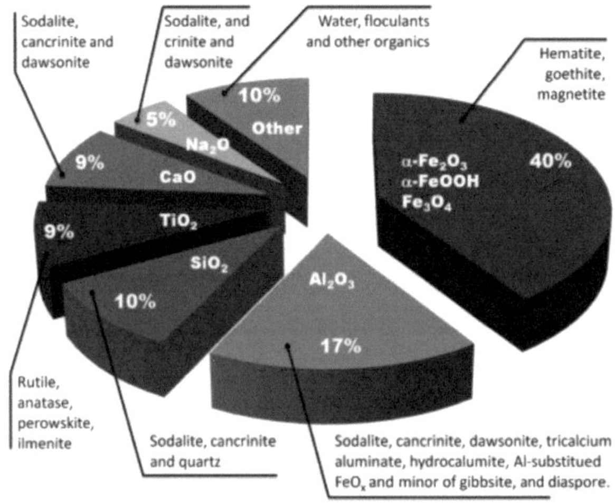

Figure 7. Composition of red mud [145].

The purity of prepared aluminium oxide is about 99.5 % and Na_2O is the main admixture in the product.

$$NaAl(OH)_4 \rightarrow \gamma\text{-}Al(OH)_3 + NaOH \tag{27}$$

$$2\ Al(OH)_3 \xrightarrow{950\text{-}1250°C} Al_2O_3 + 3\ H_2O \tag{28}$$

11 The deposition lagoon of red mud may be significant ecological load as was demonstrated by industrial accident in Hungary (Aika, 2010).

12 According to the definition law pH = -log $[H_3O^+]$ = 14 – pOH = 14 – log$[OH^-]$, the teen times dilution causes the decrease of pH by 1.

The flowing diagram of the process is shown in Fig.9. For the applications where high content of α-Al_2O_3 is necessary, the mineralization is accelerated by AlF_3 (Eq.29). Several micrometer sized plate-like corundum crystals are formed.

$$2\,Al(OH)_3 \xrightarrow{1250°C, AlF_3} \alpha\text{-}Al_2O_3 + 3\,H_2O \tag{29}$$

Tricalcium aluminate hexahydrate (hydrogarnet) is used as a filter aid during the purification of sodium aluminate liquors. Furthermore, C_3AH_6 reduces the TiO_2 content of precipitated gibbsite and the formation of hydrogarnet at high-temperature (250 °C) leaching minimizes the soda content of red-mud waste [191,192].

Fig.8 reveals that aluminium hydroxide can precipitate from the solution by introducing the carbon dioxide gas. The process can be expressed by the following reaction scheme:

$$2\,Al[OH]_4^- + CO_2 \rightarrow 2\,Al(OH)_3 + CO_3^{2-} + H_2O \tag{30}$$

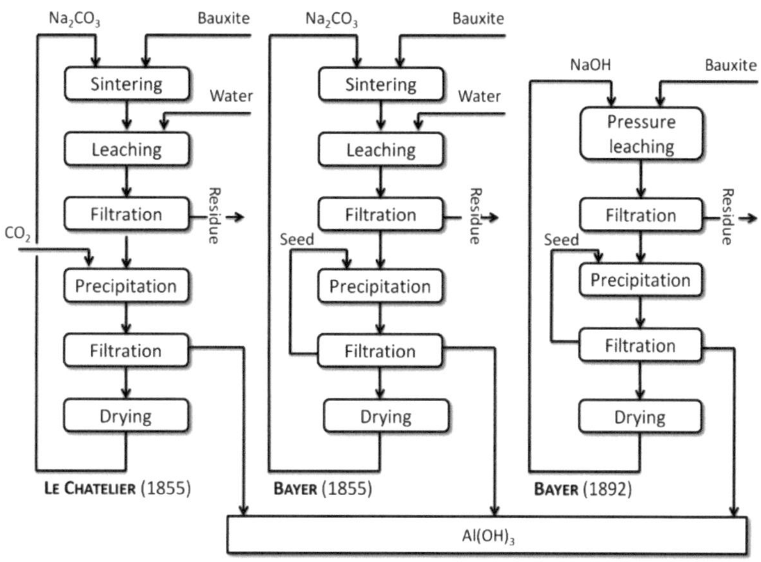

Figure 8. Bayer process of the production of alumina.

The ultrasound [193] and the addition of organics such as methanol [194] and crown ether [195] intensify the nucleation and crystallization of sodium aluminate solution, which has the potential to enhance the throughput of a Bayer process. On the contrary, polyols [196], oleic acid [197] and alditols of hydroxycarboxili acids [198] inhibit the gibbsite precipitation from seeded sodium aluminate liquors.

1.3. Utilization of red mud

The treatment and utilization of red mud waste are major challenge for the alumina industry. The main environmental risks associated with bauxite residue are related to high pH and alkalinity and minor and trace amounts of heavy metals and radionuclides. Many efforts are being globally made to find suitable applications for red mud so that the alumina industry may end up with no residue [199].

The possible applications of red mud include [199-204]:

- Building/construction materials such as bricks, stabilized blocks, light weight aggregates and low density foamed products.
- In cement industry as cements, special cements, additives to cements, mortars, construction concretes, repairs of roads, pavements, dykes.
- Colouring agents for paint works for ground floors of industrial and other buildings.
- Foamed paper in wood pulp and paper industry.
- Reinforced red mud polymer products, ceramic/refractory products.
- In metallurgical industries, as raw material in iron and steel industry as a sinter aid (binder) for iron ores, flux in steel making, etc.
- Micro-fertilizer and a neutralizer of pesticides in agriculture.
- Extracting rare-earth metals and alumo-ferric coagulants as technical raw materials.
- Special use as inorganic chemicals, adsorbents, etc.

1.4. Methods of production of high purity Al_2O_3

Aluminium oxide of high purity and high specific surface area can be prepared by thermal decomposition of alum ($NH_4Al(SO_4)_2 \cdot 12\ H_2O$)[13]. Pure ammonium alum crystal is colorless and transparent and belongs to cubic crystal system. Melting temperature of ammonium alum crystal is 94.5 °C with the phase transition enthalpy of 122.2 kJ mol^{-1}. [205-209]. $NH_4Al(SO_4)_2 \bullet 12\ H_2O$ is widely applied in industries and in water treatment [210,211]. Recently, ammonium alum is used as a promising material for Raman laser converters with a large frequency shift [212], for ferroelectricity [213] and phase transitions for storing energy

13 Alums can be described by general formula $M^{+}Me^{3+}(RO_4)_2 \cdot 12H_2O$, where M^+ is monovalent cation such as Na^+, K^+, Rb^+, Cs^+ or NH_4^+, Me^{3+} is trivalent cation such as Al^{3+}, Fe^{3+}, Cr^{3+} and R is S or Se. Each M^+ and Me^{3+} ion is surrounded by an octahedron of six water molecules. A complex network of H-bonds is one of the main features of alum structures. Alums are classified into α ($RbAl(SO_4)_2 \cdot 12H_2O$ and $NH_4Al(SO_4)_2 \cdot 12H_2O$), β ($CsAl(SO_4)_2 \cdot 12H_2O$), and γ ($NaAl(SO_4)_2 \cdot 12H_2O$) modifications depending upon three slightly different arrangements of ions and molecules within the cubic lattice. The different structures are characterized by different orientations of sulfate ions with respect to the trigonal axes of unit cell [208,209].

absorbed by solar collectors [214,215], as the catalyst [216,217] and for rubidium recovery from the processing of zinnwaldite [218].

Ammonium aluminum sulfate (AAlSD[14]) dodecahydrate undergoes the phase transitions at 58 K and 71 K on cooling and heating, respectively. At room temperature $NH_4Al(SO_4)_2 \bullet 12 H_2O$ crystals have a cubic structure and belong to the space group with four molecules per unit cell with the lattice parameters a=12.242 Å. The structural phase transition mechanism is related to the hydrogen-bond transfer involving the breakage of weak part of the hydrogen bond [208].

The process of alum derived synthesis of alumina often produces nanosized powders consisting of amorphous or transition aluminas (Chapter 4.1). The thermal decomposition of aluminium alum can be described according to the following reaction scheme [219]:

$$NH_4Al(SO_4)_2 \times 12\,H_2O \xrightarrow{\sim 225\,°C} NH_4Al(SO_4)_2 + 12\,H_2O \quad (31)$$

$$2\,NH_4Al(SO_4)_2 \xrightarrow{\sim 600\,°C} Al_2(SO_4)_3 + 2\,NH_3 + SO_3 + H_2O \quad (32)$$

$$Al_2(SO_4)_2 \xrightarrow{\sim 900\,°C} Al_2O_3 + 3\,SO_3 \quad (33)$$

The preparation of submicrometer-grained aluminas requires well-defined pure nanopowders which have many exploitable characteristics, such as low-temperature sinterability, greater chemical reactivity and enhanced plasticity. Therefore, whole range of methods was developed for the preparation of nanopowders with desired properties. These can be roughly devided to [424]:

- **High temperature/ flame/ laser synthesis:** the method usually comprises the injection of suitable gaseous or liquid aluminium-containing precursors (e.g. aluminum tri-sec-butoxide) into the source of intensive heat (flame, laser or plasma), where the precursor decomposes and converts into oxide. In most cases, the transient aluminas are formed. Therefore further high-temperature treatment is necessary in order to obtain α-Al_2O_3.
- **Chemical method including the sol-gel process:** obviously utilizes the low-and medium-temperature decomposition of inorganic aluminium salts and hydroxides or metal-organic compounds of aluminium. Typical precursors include aluminium nitrate and hydroxides.
- **Mechanically assisted synthesis:** the method is based on high-energy milling of coarser-grained powder. In this case, the minimum particle size is limited to approximately 40 nm.

14 Abbreviation, another example is CsAlSeD ($CsAl(SeO_4)\cdot 12H_2O$).

- Other options include the combustion synthesis [220-224], the spray pyrolysis of aerosol of nitrate or other aluminum salts [225-227], the sol-gel process [228,229], the emulsion synthesis [230,231], etc.

2. Adjustment of clinker composition

Strontium aluminate is formed via solid-state reaction of equimolar amount of aluminium oxide with strontium oxide:

$$Al_2O_3 + SrO \rightarrow SrAl_2O_4 \tag{34}$$

With regard to the molar weight ratios of SrO/Al_2O_3=1.016 and SrO/Fe_2O_3=0.65, the proper amount of SrO should be calculated as follows:

$$SrO = 1.016\ Al_2O_3 + 0.65\ Fe_2O_3 \tag{35}$$

The mass ratio of used SrO and the theoretical amount calculated according to Eq.34 should be termed as the "**Saturation Degree**" or "**Strontium saturation factor**" of clinker by strontium oxide ($\boldsymbol{SD}_{SrO}$):

$$SD_{SrO} = \frac{100\ SrO}{1.016\ Al_2O_3 + 0.65\ Fe_2O_3} \quad [\%] \tag{36}$$

Analogically to ordinary Portland cement, the hydraulic module ($\boldsymbol{M}_H$) and the alumina module ($\boldsymbol{M}_A$) of strontium aluminate clinker should be defined:

$$M_H = \frac{SrO}{Al_2O_3 + Fe_2O_3} \tag{37}$$

$$M_A = \frac{Al_2O_3}{Fe_2O_3} \tag{38}$$

To use the strontium aluminate cements for the production of refractory materials the value of $\boldsymbol{SD}_{SrO}$ < 100 and low content of Fe_2O_3 are required. These compositions ensure that the first eutectic melt is formed at the temperature of 1760 °C instead of 1505 °C for the cement with $SD_{SrO} > 100$.

The composition of mixture of raw materials is calculated from required composition of clinker according to the following relations:

$$SrCO_3=1.425\ SrO \tag{39}$$

$$Al(OH)_3=0.765\times Al_2O_3 \tag{40}$$

$$others=100-SrCO_3-Al(OH)_3 \tag{41}$$

3. Calculation of the raw meal composition

The preparation of raw meal using only two components is a simple process. The example could be strontium carbonate with 98.4 % $SrCO_3$ and alumina which does not contain any strontium carbonate. The analysing techniques are described in Chapter 2.4. From Eq.34 it can be read that equimolar mixture of $SrCO_3$ and Al_2O_3 (x_{Al2O3}=0.5) should be prepared. That means that x_{SrCO3}=0.5 and $x_{Al2O3}=1 - x_{SrCO3}$=0.5. The molar ratio can be recalculated to the weight ratio as follows[15]:

$$w = \frac{x_i M_i}{\sum_{j=1}^{k} x_j M_j} \tag{42}$$

$$\sum_{i=1}^{k} w_i = 1 \ \Rightarrow \ w_k = 1 - \sum_{i=1}^{k-1} w_i \tag{43}$$

From the relationships introduced above we can calculate:

$$w_{SrCO_3} = \frac{0.5\cdot 147.6}{0.5\cdot 147.6 + 0.5\cdot 102,0} = 0,591 \Rightarrow 59.1\% \tag{44}$$

15 Please see Eqs.18 and 19 in Chapter 1.

$$w_{Al_2O_3} = 1 - 0{,}591 \Rightarrow 0.409 \Rightarrow 40.9\% \tag{45}$$

The amount of raw mixture constituent can be then calculated:

$$\begin{array}{lllll} SrCO_3 & 98.4 & & 59.1 & Þ \; 100\times59.1/(59.1+39.3)=60.1\% \\ & & 59.1 & & \\ Al_2O_3 & 0 & & 39.3 & Þ \; 100\times39.3/(59.1+39.3)=39.9\% \end{array} \tag{46}$$

The raw meal contains 60.1 % of strontium carbonate and 39.9 % of alumina, i.e. both components are mixed in the weight ratio of 1.5 : 1.

3.1. Calculation of required hydraulic module

The preparation of raw meal of given value of hydraulic module (Eq.37) is demonstrated in this chapter. As an example the raw meal for the strontium aluminate clinker with M_H=0.98 will be prepared. There are **two raw materials** with the composition given in Table 2.

Raw material		SrO	Al_2O_3	Fe_2O_3	Z =Al_2O_3+Fe_2O_3
$SrCO_3$	1	98.4	0	0.2	0.2
Calcined bauxite	2	0	98.2	1.6	99.8

Table 2. Composition of two raw materials for the preparation of raw meal.

Hence we have two equations:

$$w_1 + w_2 = 100 \tag{47}$$

$$a_{21}w_1 + a_{22}w_2 = 0 \tag{48}$$

where:

$$a_{2i} = SrO - M_H Z_i \qquad (i = 1, 2) \tag{49}$$

From the relationships introduced above we can calculate:

$$a_{21} = 98.4 - 0.98 \cdot 0.2 = 98.2 \tag{50}$$

$$a_{22} = 0 - 0.98 \cdot 99.8 = -97.8 \tag{51}$$

From the substitution of 48 by 50 and 51 the following relationship results:

$$\begin{aligned} &w_1 + w_2 = 100 \\ &\underline{98.2\, w_1 - 97.8\, w_2 = 0} \\ &w_1 = \frac{100\, a_{22}}{a_{22} - a_{11}} = \frac{100 \cdot (-97.8)}{\text{-97.8-98.2}} = 49.9\% \\ &w_2 = \frac{100\,(-a_{21})}{\mathrm{a}_{22} - a_{21}} = 100 - w_1 = 50.1\% \end{aligned} \tag{52}$$

Now it is possible to check, if the calculated results are correct:

Raw material	Composition			w_i	Raw meal		
	SrO	Al_2O_3	Fe_2O_3	[%]	SrO	Al_2O_3	Fe_2O_3
1	98.4	0	0.2	49.9	48,3	0	0,1
2	0	98.2	1.6	50.1	0	49,2	0,8
Total				100.0	49,1	49,2	0,9
M_H				49,1/ (49.2 + 0,9) = 0.98			

Another case is that the mixture requires the preparation via **mixing of three raw materials**, e.g. we have strontium carbonate, calcined bauxite and corundum. Corundum is necessary to keep the value of hydraulic module (M_H)=0.98 and alumina module (M_A)=57. The composition of raw materials is listed in Table 3.

Raw material		SrO	Al_2O_3	Fe_2O_3	$Z = Al_2O_3 + Fe_2O_3$
$SrCO_3$	1	98.4	0	0.2	0.2
Calcined bauxite	2	0	98.2	1.6	99.8
Corundum	3	0	99.5	0.1	99.6

Table 3. Composition of three raw materials for the preparation of raw meal.

$$w_1 + w_2 + w_3 = 100 \tag{53}$$

$$a_{21}w_1 + a_{22}w_2 + a_{23}w_3 = 0 \tag{54}$$

$$a_{31}w_1 + a_{32}w_2 + a_{33}w_3 = 0 \tag{55}$$

Where:

$$a_{2i} = SrO - M_H Z_i \tag{56}$$

$$a_{3i} = Al_2O_3 - M_A Fe_2O_3 \qquad (i = 1, 2, 3) \tag{57}$$

From the relationships introduced above we can calculate:

$$a_{21} = 98.4 - 0.98 \cdot 0.2 = 98.2 \tag{58}$$

$$a_{22} = 0 - 0.98 \cdot 99.8 = -97.8 \tag{59}$$

$$a_{23} = 0 - 0.98 \cdot 99.6 = -97.6 \tag{60}$$

$$a_{31} = 0 - 50 \cdot 0.2 = -11.4 \tag{61}$$

$$a_{32} = 98.2 - 50 \cdot 1.6 = 7.0 \tag{62}$$

$$a_{33} = 99.5 - 50 \cdot 0.1 = 93.8 \tag{63}$$

The substitution of equations 53-55 by Eq.58-63 yields to:

$$\begin{aligned} &w_1 + w_2 + w_3 = 100 \\ &98.2\, w_1 - 97.8\, w_2 - 97.6\, w_3 = 0 \\ &\underline{-11.4\, w_1 + 7.0\, w_2 + 93.8\, w_3 = 0} \end{aligned} \tag{64}$$

The value of determinant **D** can be calculated as follows:

$$\begin{aligned} &D = a_{22}\, a_{33} - a_{32}\, a_{23} - \left(a_{21}\, a_{33} - a_{31}\, a_{23}\right) + a_{21}\, a_{32} - a_{31}\, a_{22} \\ &D = -9174.0 - 683.3 - \left(9211.5 - 1112.7\right) + 687.4 - 1115.0 \\ &D = -17017.1 \end{aligned} \tag{65}$$

The composition of raw meal is then:

$$w_1 = \frac{-100\left(a_{22}\, a_{33} - a_{32}\, a_{23}\right)}{D} = 49.9\% \tag{66}$$

$$w_2 = \frac{-100\left(a_{21}\, a_{33} - a_{31}\, a_{23}\right)}{D} = 47.6\% \tag{67}$$

$$w_3 = \frac{100\left(a_{21}\, a_{32} - a_{31}\, a_{22}\right)}{D} = 2.5\% \quad \text{or}$$
$$w_3 = 100 - \left(w_1 + w_2\right) = 2.5\ \% \tag{68}$$

Now it is possible, as previously, to check, if the calculated results are correct:

Raw material	Composition			w_i	Raw meal		
	SrO	Al_2O_3	Fe_2O_3	[%]	SrO	Al_2O_3	Fe_2O_3
1	98.4	0	0.2	49.9	49.10	0	0.10
2	0	98.2	1.6	47.6	0	46.74	0.76
3	0	99.5	0.1	2.5	0	2.50	0
Total				100	49.10	49.24	0.86
M_H				49.10/ (49.24 + 0.86) = 0.98			
M_A				49.24/ 0.86 = 57			

3.2. Calculation from the saturation degree

The method of calculation of raw meal for the preparation of strontium aluminate clinker with given Saturation Degree (SD_{SrO}) and alumina module (M_A) is described. As an example the mixture with SD_{SrO}=0.95 and M_A=60 using three raw materials with the composition listed in Table 4 will be prepared.

Raw material		SrO	Al_2O_3	Fe_2O_3
$SrCO_3$	1	97.3	0	0.3
Calcined bauxite	2	0	96.5	2.6
Corundum	3	0	99.7	0.1

Table 4. Composition of three raw materials for the preparation of raw meal.

The composition of raw meal results from the solution of the following set of three equations:

$$w_1 + w_2 + w_3 = 100 \tag{69}$$

$$a_{21}w_1 + a_{22}w_2 + a_{23}w_3 = 0 \tag{70}$$

$$a_{31}w_1 + a_{32}w_2 + a_{33}w_3 = 0 \tag{71}$$

where

$$a_{2i} = \text{SrO-SD}_{\text{SrO}}\left(1.016\ \text{Al}_2\text{O}_3 + 0.65\ \text{Fe}_2\text{O}_3\right) \tag{72}$$

$$a_{3i} = \text{Al}_2\text{O}_3\text{-M}_\text{A}\ \text{Fe}_2\text{O}_3 \qquad \left(i = 1, 2, 3\right) \tag{73}$$

From the relationships introduced above we can calculate:

$$a_{21} = 90.3 - 0.96\left(1.016 \cdot 0 + 0.65 \cdot 0.3\right) = 90.2 \tag{74}$$

$$a_{22} = 0 - 0.96\left(1.016 \cdot 96.5 + 0.65 \cdot 2.6\right) = -95.6 \tag{75}$$

$$a_{23} = 0 - 0.96\left(1.016 \cdot 99.7 + 0.65 \cdot 0.1\right) = -97.2 \tag{76}$$

$$a_{31} = 0 - 60 \cdot 0.3 = -18.0 \tag{77}$$

$$a_{32} = 96.5 - 60 \cdot 2.6 = -59.5 \tag{78}$$

$$a_{33} = 99.7 - 60 \cdot 0.1 = 93.7 \tag{79}$$

The value of determinant D can be calculated as follows:

$$\begin{aligned} D &= a_{22}\,a_{33} - a_{32}\,a_{23} - \left(a_{21}\,a_{33} - a_{31}\,a_{23}\right) + a_{21}\,a_{32} - a_{31}\,a_{22} \\ D &= -8960.3 - 5783.9 - \left(9099.7 - 1749.8\right) - 5778.4 - 1721.3 \\ D &= -29593.8 \end{aligned} \tag{80}$$

The composition of raw meal is then:

$$w_1 = \frac{-100\left(a_{22}\,a_{33} - a_{32}\,a_{23}\right)}{D} = 49.8\% \tag{81}$$

$$w_2 = \frac{-100\left(a_{21}\,a_{33} - a_{31}\,a_{23}\right)}{D} = 24.8\% \tag{82}$$

$$w_3 = \frac{100\left(a_{21}\,a_{32} - a_{31}\,a_{22}\right)}{D} = 25.3\% \quad \text{or} \quad w_3 = 100 - \left(w_1 + w_2\right) = 25.3\,\% \tag{83}$$

Now it is possible, as previously, to check, if the calculated results are correct:

Raw material	**Composition**			**w_i [%]**	**Raw meal**		
	SrO	**Al_2O_3**	**Fe_2O_3**		**SrO**	**Al_2O_3**	**Fe_2O_3**
1	97.3	0	0.3	49.8	48.48	0	0,15
2	0	96.5	2.6	24.8	0	23.97	0,65
3	0	99.7	0.1	25.3	0	25,27	0,03
Total				100	48,48	49,23	0,82
M_H				48.48/ (49.23 + 0.82) = 0.97			
M_A				49.23/ 0.82 = 60			

3.3. Calculation with SD_{SrO}, M_A and other parameter

Using four raw materials for the preparation of raw meal where other parameter can be used for calculation. That can be useful for the preparation of cement with exceeding substitution of alumina. For example, the Saturation Degree of clinker by strontium oxide and alumina module can be redefined as follows:

$$SD_{SrO}^{*} = \frac{100\,\text{SrO}}{1.016\,\text{Al}_2\text{O}_3 + 0.65\,\text{Fe}_2\text{O}_3 + 0.68\,\text{Cr}_2\text{O}_3} \quad [\%] \tag{84}$$

$$M_A^{*} = \frac{\text{Al}_2\text{O}_3}{\text{Fe}_2\text{O}_3 + \text{Cr}_2\text{O}_3} \tag{85}$$

Therefore a new type of module can be applied:

$$M_{Fe} = \frac{Fe_2O_3}{Cr_2O_3} \tag{86}$$

This is the case of raw meal the preparation of strontium aluminate clinker with following parameters SD_{SrO}^{*}=0.90, M_A^{*}=58 and M_F=10. The composition of raw materials is listed in Table 5.

Raw material		SrO	Al_2O_3	Fe_2O_3	Cr_2O_3	$Y=Fe_2O_3+Cr_2O_3$
$SrCO_3$	1	97.3	0	0.3	0	0.3
Calcined bauxite	2	0	96.5	2.6	0	2.6
Corundum	3	0	99.7	0.1	0	0.11
Cr_2O_3	4	0	2.8	1.5	32.6	34.1

Table 5. Composition of four raw materials for the preparation of raw meal.

The composition of raw meal results from the solution of the following set of four equations:

$$w_1 + w_2 + w_3 + w_4 = 100 \tag{87}$$

$$a_{21}w_1 + a_{22}w_2 + a_{23}w_3 + a_{24}w_4 = 0 \tag{88}$$

$$a_{31}w_1 + a_{32}w_2 + a_{33}w_3 + a_{34}w_4 = 0 \tag{89}$$

$$a_{41}w_1 + a_{42}w_2 + a_{43}w_3 + a_{44}w_4 = 0 \tag{90}$$

Where:

$$a_{2i} = \text{SrO-SD}_{\text{SrO}}^{*} \left(1.016\ \text{Al}_2\text{O}_3 + 0.65\ \text{Fe}_2\text{O}_3 + 0.68\ \text{Cr}_2\text{O}_3\right) \tag{91}$$

$$a_{3i} = \text{Al}_2\text{O}_3\text{-M}_\text{A}^{*}\ (\text{Fe}_2\text{O}_3 + \text{Cr}_2\text{O}_3) \tag{92}$$

$$a_{4i} = \text{Fe}_2\text{O}_3\text{-M}_\text{F}\ \text{Cr}_2\text{O}_3 \qquad (\text{i}=1, 2, 3, 4) \tag{93}$$

From the relationships introduced above we can calculate:

$$a_{21} = 97.3 - 0.90\left(1.016 \cdot 0 + 0.65 \cdot 0.3 + 0.68 \cdot 0\right) = 97.1 \tag{94}$$

$$a_{22} = 0 - 0.90\left(1.016 \cdot 96.5 + 0.65 \cdot 2.6 + 0.68 \cdot 0\right) = -89.8 \tag{95}$$

$$a_{23} = 0 - 0.90\left(1.016 \cdot 99.7 + 0.65 \cdot 0.1 + 0.68 \cdot 0\right) = -91.2 \tag{96}$$

$$a_{24} = 0 - 0.90\left(1.016 \cdot 2.8 + 0.65 \cdot 1.5 + 0.68 \cdot 32.6\right) = -23.4 \tag{97}$$

$$a_{31} = 0 - 58.0\left(0.3 + 0\right) = -17.4 \tag{98}$$

$$a_{32} = 96.5 - 58.0\left(2.6 + 0\right) = -54.3 \tag{99}$$

$$a_{33} = 99.7 - 58.0\left(0.1 + 0\right) = 93.9 \tag{100}$$

$$a_{34} = 2.8 - 58.0\left(1.5 + 32.6\right) = -1975.0 \tag{101}$$

$$a_{41} = 0.3 - 10.0 \cdot 0 = 0.3 \tag{102}$$

$$a_{42} = 2.6 - 10.0 \cdot 0 = 2.6 \tag{103}$$

$$a_{43} = 0.1 - 10.0 \cdot 0 = 0.1 \tag{104}$$

$$a_{44} = 1.5 - 10.0 \cdot 32.6 = -324.5 \tag{105}$$

From the relationships introduced above we can further calculate:

$$w_1 + w_2 + w_3 + w_4 = 100 \tag{106}$$

$$97.1\, w_1 - 89.8\, w_2 + -91.2\, w_3 - 23.4\, w_4 = 0 \tag{107}$$

$$-17.4\ w_1 - 54.3\ w_2 + 93.9\ w_3 - 1975.0\ w_4 = 0 \quad (108)$$

$$0.3\ w_1 + 2.6\ w_2 + 0.1\ w_3 - 324.5\ w_4 = 0 \quad (109)$$

The solution based on the Sauruss law is time consuming without specialized software. Nevertheless, it is possible to use the calculation according to Table 6. This solution is based on the Gauss inversion method. The symbols *i* and *k* denote the line and column of matrix for mathematical operation according to given rule, e.g. $2a_1$ is the second line and the first column member of matrix. The solution consists of the following steps:

i. Inserting the coefficients from the left side of Eqs.106-109 into proper line (1-4) of Table 6. For example the coefficient a_{11} from Eq.106 should be written in the first line and first column; the coefficient a_{32} belongs to the third line and second column, etc.

ii. The column I contains the sum of members 1-4 for given line.

iii. The column II contains the coefficients from the right side of Eqs.106-109 for the line from 1 to 4. Other lines refer to the results of mathematical operation defined in the column rule.

iv. The operation on line 6 ($-2a_1 \bullet 5+2$) means:-97.13 • 1.00+97.13=0, where $2a_1$ is the coefficient related to the second line and first column. The operation 1 : $1a_1$ means that all numbers in the first line are divided by given term.

v. The last four lines of column II provide the solution for the composition of raw meal.

Rule		Matrix				Test: 4		Test:
	i^k					Σi	i_b	I+II.
		1	2	3	4	I	II	III
	1	1.00	1.00	1.00	1.00	4.00	100.00	104.00
	2	97.13	-89.76	-91.22	-23.39	-107.25	0.00	-107.25
i a_k	3	-17.40	-54.30	93.90	-1975.0	-1952.8	0.00	-1952.8
	4	0.30	2.60	0.10	-324.50	-321.50	0.00	-321.50
1 :1a_1 XX	5	1.00	1.00	1.00	1.00	4.00	100.00	104.00
-2a_1 • 5 + 2	6	0.00	-186.89	-188.35	-120.51	-495.75	-9712.45	-10208.2
-3a_1 • 5 + 3	7	0.00	-36.90	111.30	-1957.6	-1883.2	1740.00	-143.2
-4a_1 • 5 + 4	8	0.00	2.30	-0.20	-324.80	-322.70	-30.00	-352.7
-5a_2 • 10 + 5	9	1.00	0.00	-0.01	0.36	-1.35	48.03	49.38

Rule	i^k	Matrix				Test: 4		Test:
						Σi	i_b	I+II.
		1	2	3	4	I	II	III
6 : $6a_2$ XX	10	0.00	1.00	1.01	0.64	2.65	51.97	54.62
-$7a_2$ • 10 + 7	11	0.00	0.00	148.49	-1933.8	-1785.32	3657.70	1872.38
-$8a_2$ • 10 + 8	12	0.00	0.00	-2.52	-326.28	-328.80	-149.53	-478.33
-$9a_3$ • 15 + 9	13	1.00	0.00	0.00	0.25	1.25	48.22	49.48
-$10a_3$ • 15 + 10	14	0.00	1.00	0.00	17.77	14.77	27.14	41.91
11 : $11a_3$ XX	15	0.00	0.00	1.00	-13.02	-12.02	24.63	12.61
-$12a_3$ • 15 + 12	16	0.00	0.00	0.00	-359.08	-359.08	-87.51	-446.58
-$13a_4$ • 20 + 13	17	1.00	0.00	0.00	0.00	1.00	48.16	49.16
-$14a_4$ • 20 + 14	18	0.00	1.00	0.00	0.00	1.00	23.79	24.79
-$15a_4$ • 20 + 15	19	0.00	0.00	1.00	0.00	1.00	27.81	28.81
16 : $16a_4$ XX	20	0.00	0.00	0.00	0.00	1.00	0.24	1.24

Table 6. Numerical solution for raw meal four raw materials.

Now it is possible, as previously, to check, if the calculated results are correct:

Raw material	Composition				w_i [%]	Raw meal			
	SrO	Al_2O_3	Fe_2O_3	Cr_2O_3		SrO	Al_2O_3	Fe_2O_3	Cr_2O_3
1	97.3	0	0.3	0	48.16	46.86	0	0.14	0
2	0	96.5	2.6	0	23.76	0	22.96	0.62	0
3	0	99.7	0.1	0	27.81	0	27.72	0.03	0
4	0	2.8	1.5	32.6	0.24	0	0.01	0	0.08
Total					100.00	46.86	50.69	0.79	0.08
SD^*_{SrO}			46.86/ (1.016•50.69 + 0.65•0.79 + 0.68•0.08) = 0.90						
$M_A{}^*$			50.69/ (0.79 + 0.08) = 58.0						
M_{Fe}			0.79/ 0.08 = 10.0						

It is also possible to define another kind of module using V_2O_3, Ti_2O_3 or their mixture with Fe_2O_3 and to calculate the raw meal composition by the same way. This calculation certainly requires the redefinition of Saturation Degree of clinker from strontium oxide to applied system.

4. Method of analysis of raw material and clinker

The calculation mentioned above requires the analysis of raw materials to calculate the correction of working mixture composition. Some techniques applicable for the analysis of raw materials or cements are presented in this chapter. Chemical analysis, microscopy, XRD and other methods of examination should be carried out on the same, representative sample of material [7].

4.1. Determination of SrO

The EDTA disodium salt (Ethylenediaminetetraacetic acid disodium salt dehydrate, $Na_2H_2Y \cdot 2H_2O$) titration is the most common technique used for the determination of strontium oxide in the strontium carbonate:

$$Sr^{2+} + H_2Y^{2-} \rightarrow SrY^{2-} + 2\,H^+ \tag{110}$$

with the stability constant of complex:

$$K\left(SrY^{2-}\right) = \frac{[SrY^{2-}][H^+]^2}{[Sr^{2+}][H_2Y^-]} = 398.1 \tag{111}$$

$$\log K\left(SrY^{2-}\right) = 8.6 \tag{112}$$

The formation of stable complex during the assessment requires the pH ≥ 10 (the same pH range as for $Mg^{2+}$and Ba^{2+}).

4.2. Determination of SrO in OPC

In compliance with the current ASTM Standard Test Methods for chemical analysis of hydraulic cement (C 114) [241], strontium (usually present in Portland cement as minor constituent), is led to precipitate (Table 1) with calcium (CaC_2O_4) as oxalate (SrC_2O_4, monoclinic, space group P21/n [242,243]) and next it is subsequently titrated and calculated as CaO, or alternative correction of CaO for SrO is made, if the SrO content is known. Therefore, the development of a new, direct, sensitive and accurate method for the determination of strontium as minor constituent in cement is of upmost importance [244].

Strontium oxalate exists in two different forms [245]:

1. Neutral strontium oxalate hydrate, $SrC_2O_4\,xH_2O$.

2. Acid salt of strontium oxalate, $SrC_2O_4\,yH_2C_2O_4\,xH_2O$.

Depending on the concentration of oxalic acid and ammonium oxalate as precipitating agents, both forms can be obtained. At sufficiently low pH, the stoichiometric compound $SrC_2O_4 \cdot \frac{1}{2}H_2C_2O_4 \cdot H_2O$ is formed. The morphologies of precipitated particles (bi-pyramids, rods, peanuts, spheres, etc.) depend on the experimental conditions such as pH, temperature, ageing time and concentration of additives [246].

The structure of acidic strontium oxalate is shown in Fig.9(a). Oxalate and hydrogen oxalate anions are present in such a way that each asymmetric unit contains exactly one molecule with the structural formula $Sr(HC_2O_4)\cdot 1/2(C_2O_4)H_2O$ instead of $Sr(C_2O_4)\cdot 1/2(H_2C_2O_4)\cdot H_2O$. Similarly to other known strontium oxalates, strontium is eight-fold coordinated by oxygen. In this coordination sphere, both, oxalate and hydrogen oxalate anions act once as bidentate and once as monodentate. Two remaining positions are occupied by H_2O molecules. The SrO_8 polyhedron can be described as distorted bicapped trigonal prism, with O7…O2…O5…Ow3 forming the square face. These polyhedrons are connected to each other only by edge sharing it to form one-dimensional chains along the c-axis [242].

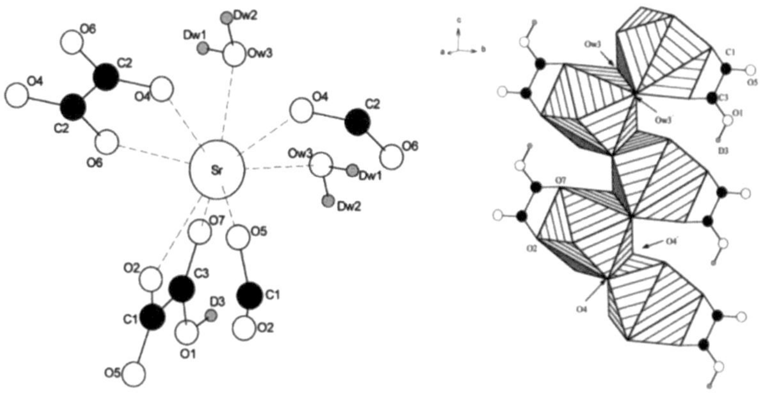

Figure 9. Strontium oxalate: Structure of SrO_8 polyhedron (a) and connection of polyhedrons via shared edge (b) according to [243].

The shared edges are O4…O4′ and Ow3…Ow3′(Fig.9 (b)), which means that H_2O acts as bridging ligand between two strontium atoms. This is in contrast to all other Sr oxalates, where H_2O is also coordinated to Sr, but without any bridging function. In the *ac* plane, the polyhedra chains are connected by the $C_2O_4^{2-}$ groups, while in the *bc* plane the connection is made by the $HC_2O_4^-$ groups. In addition, there is the possibility to form intrachain (Dw2…O5 and D3…O2 along the *bc* plane) as well as interchain (Dw1…O6 along the *ac* plane) hydrogen bridges, which give the whole network an extra stability. Until now the four types of acid strontium oxalates are known, the type 2 and 4 are conformers (Fig.10) with calculated energy difference of ~6.69 kJ mol^{-1} [242,243].

Many analytical techniques were suggested for the determination of strontium in the cement matrix [238,247-250], i.e. under conditions including high concentration of calcium in the sample, based on complicated separation techniques of low selectivity. The atomic absorp-

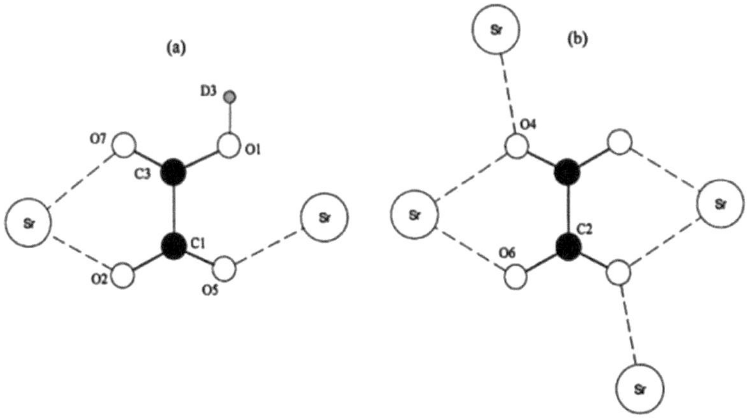

Figure 10. Coordination types of acid strontium oxalate: type 4 (a) and type 2 (b).

tion spectrometric method can be used for the determination of calcium, magnesium and strontium in soils [236] but the assessment requires the removal of the silicon.

Derivative spectrophotometry is analytical technique combining high selectivity [251-255] and sensitivity [244,256-258]. The accuracy of assessment depends on the shape of normal absorption spectra of analyte and interfering substances, as well as on the instrumental parameters and the applied technique of measurement, e.g. peak-to-trough or zero-crossing [259-261]. Salinas et al. [262] developed the derivative spectrophotometric method for resolving binary mixtures when the spectra of components are overlapped. The method uses the first derivation of the spectra. The concentration of other component is then determined from the calibration graph. Later, the method was extended to the resolution of ternary mixtures in combination with zero-crossing method [263].

The determination of strontium and simultaneous determination of strontium oxide, magnesium oxide and calcium oxide content in Portland cement by derivative ratio spectrophotometry uses alizarin Complexone (alizarin-3-methylamine-N, N-diacetic acid, AC) as one of the most common reagents used for the spectrophotometric determination of metal ions. The AC reagent yields five colored acid–base forms in the solutions of pH ~3.2–10.5: H_4L, H_3L^-, $H_2L_2^-$, HL^{3-} and L^{4-}, exhibiting the absorption maxima at 270, 335, 423, 525, and 580 nm, respectively. Distinct isosbestic points are observed for the particular acid–base equilibrium. The formation of SrL^{2-}complex with liberation of one proton occurs in pH range from 7 to 10 [262]:

$$Sr^{2+} + LH^{3-} \rightarrow SrL^{2-} + H^+ \tag{113}$$

The determination of strontium as SrL^{2-} complex was possible in the presence of Li^+, Na^+, K^+, Cs^+, Cd^{2+}, Al^{3+}, Fe^{3+}, Mo^{6+}, SO_4^{2-}, SO_3^{2-}, NO^{3-}, Cl^-, Br^-, I^- and PO_4^{3-} (20.0 mg); Co^{2+},Ni^{2+}, Pb^+, Cr^{3+}, Ti^{4+}, $C_2O_4^{2-}$ and CO_3^{2-} (1.0 mg). Investigated ions Ca^{2+}, Mg^{2+}, $Mn^{2+}$and Zn^{2+}interfered seriously, even when present in amounts higher than 0.1 mg. The interference due to Mn and Zn was eliminated by the addition of ammonium hydroxide, and that of Ca and Mg was overcome by using the derivative ratio zero crossing method. Using the proposed method, it is possible to determine Sr, Mg, and Ca simultaneously in mixtures containing 1.5-18 $\mu g \cdot cm^{-3}$ of strontium, 0.5-5.0 $\mu g \cdot cm^{-3}$ of magnesium and 1.0-8.0 $\mu g \cdot cm^{-3}$ of calcium [262].

4.3. Application of strontium isotopes to determine the origin of cements in concrete

In many disciplines of science it is important to be able to determine the source of material or to characterize its transportation history. The chemical composition has been used extensively to determine the source of materials by fingerprinting the chemical composition of the material to be identified and comparing it to the chemical composition of potential sources. This approach has been used extensively for major elements as well as for trace elements [264,265].

Forensic isotope geochemistry relies on subtle differences in isotopic abundance of element to characterize particular material. These different isotopic abundances give rise to unique isotopic composition that will identify the material come from particular region. Many rocks composed of different minerals have distinctive isotopic compositions and their unique composition can be used to fingerprint them. This distinctive rock/mineral composition usually arises from the decay of radiogenic elements e.g., ^{87}Rb to ^{87}Sr; the transuranic elements to ^{208}Pb, ^{207}Pb, ^{206}Pb; ^{147}Sm to ^{143}Nd [264].

Combined chemical and Sr isotopic analysis may provide the geochemical fingerprints from raw materials, which can be used to identify them in concrete. For successful chemical fingerprinting of cement in concrete, it is important to leach cement without significantly attacking the aggregate, but this can be minimized by using slightly alkaline or neutral EDTA as solvent in preference to weak mineral acids such as HNO_3. Combined chemical and Sr isotopic analysis of commonly used New Zealand cements showed that they contain characteristic fingerprints, which may be used to identify them in concrete of unknown origin. Although cements have typically $^{87}Sr/^{86}Sr$ values similar to their mid-Tertiary limestone source rocks (0.7078 – 0.7085) most are easily distinguishable when their $^{87}Sr/^{86}Sr$ values are plotted against Ca/Sr [266].

4.4. Determination of Al_2O_3 and Fe_2O_3

Different forms of alumina may be identified by x-ray diffraction analysis [91]. Classical, wet analysis gives inaccurate results for Al_2O_3 unless the effects of P_2O_5 and TiO_2 are not taken into account [7].

5. Method for analysis of SrAC cement

The standard test methods for the chemical analysis of hydraulic cement are specified by ASTM C 114-13. The chemical analysis of hydraulic cement is specified by ASTM C 114-88 standard. The mineralogy of cement cannot be determined from the chemical composition because the thermodynamic equilibrium usually is not reached during the production process. The phase (mineralogical) composition of strontium aluminate cement can be in principle determined by the same methods as for aluminous cements:

1. Selective dissolution [269];
2. Electron Microscopy [7,269];
3. Reflected Light Microscopy [7,269];
4. Quantitative X-Ray Diffraction Analysis (QXDA) [7,267-270,599],

Due to recent developments in cement clinker engineering, the optimization of chemical substitutions in the main clinker phases offers a promising approach to improve both reactivity and grindability of clinkers. Thus, the monitoring of chemistry of phases may become a part of the quality control at cement plants, along with usual measurements of the abundance of mineralogical phases [270].

6. Determination of water to cement ratio

The chemical reactions which take place after mixing cement with water are generally more complex than simple conversion of anhydrous compounds into the corresponding hydrates. The mixtures of cement with water, where the hydration reactions, setting and hardening take place are termed as pastes [271], while the hardened material can be termed as cement stone or hardened cement stone. The water to cement ratio (*w/c*) refers to the proportion by mass that is related to water and cement used for the preparation of cement paste [12,271].

The value water-to-cement (*w/c*) ratio is one of the most fundamental parameters in concrete mixture proportioning. The *w/c* ratio has a significant influence on most properties of hardened concrete, in particular on strength and durability due to its relationship with the amount of residual space i.e. capillary porosity, in the cement stone. Since the *w/c* ratio is an indication of quality of concrete mix, the situations often arise in which it is desirable to determine the original *w/c* ratio of particular concrete some time after it has hardened. This often happens when the disputes suspecting the noncompliance with the mix specification arise. The determination of the *w/c* ratio is also important for the quality control during the concrete production and for general quality assurance purposes [273-277].

Unfortunately, once concrete has set, it is very difficult to ascertain the exact amounts of cement and water which were originally added during batching. At any moment after setting, the hardened cement stone can be considered to consist of four main phases [273]:

1. Rest of unreacted cement;
2. Crystalline and semi-crystalline hydration products including their intrinsic gel pores;
3. Capillary pores;

4. Air voids.

The solid hydration products occupy a greater volume than the volume of reacted cement (Fig.11), but slightly smaller volume than the sum of volumes of cement and water due to chemical shrinkage [273,278]. Chemical shrinkage associated with hydration of OPC and AC is about 5 and 10 – 12 ml per 100 g of cement, respectively [12].

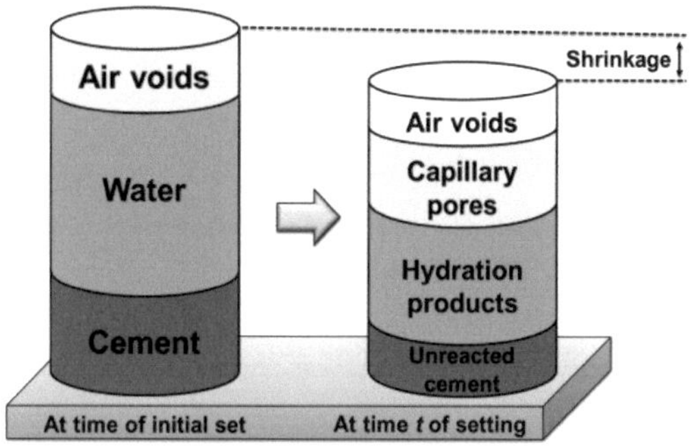

Figure 11. Proportion of main phases in hardened cement stone [273].

The methodology for the estimation of initial cement content, water content and water/ cement ratio of hardened cement-based materials by electron microscopy was developed by Wong and Buenfeld [273] and Sahu at al. [275]. The acoustic-ultrasonic approach for non-destructive determination of *w/c* ratio was described by Philippidis and Aggelis [274]. Betcher at al. [277] published the method using 2.45 GHz microwave radiation which can be conveniently and accurately used for the on-site determination of the water-to-cement (*w/c*) ratio in a batch of fresh rapid-setting concrete.

7. Grinding

Grinding occurs at the beginning and at the end of cement making process [279]. In recent years, the matrix model and the kinetic model, which were suggested by investigators, are used in laboratories and industrial areas. The kinetic model, which is an alternative approach, considers the combination as a continuous process in which the rate of breakage of particle size is proportional to the mass of particles of that size. The analysis of size reduction in tumbling ball mills using the concepts of specific rate of breakage and primary daughter fragment distribution has received considerable attention in the last years [280,320].

To optimize the cement grinding, the standard Bond grinding calculations [281] can be used as well as the modeling and simulation techniques based on the population balance model (PBM) [284,285]. The mill power draw prediction can be carried out using the Morrell power model for tumbling mills [279,282].

The Bonds equation describes the specific power required to reduce the feed from specified feed F_{80} to the product with specified P_{80} [279,281]:

$$W_m = W_i\left(\frac{10}{\sqrt{P_{80}}} - \frac{10}{\sqrt{F_{80}}}\right) \tag{114}$$

where W_m is the mill specific motor output power ($kWh \cdot t^{-1}$), W_i is the Bond ball mill work index ($kWh \cdot t^{-1}$) P_{80} is the sieve size passing 80 % of the mill product (μm), F_{80} is the sieve size passing 80 % of the mill feed (μm). It was found in the crushing area that there are significant differences between the real plant data and the Bond calculations and therefore the empirical corrections were introduced. The following modified Bond equation was proposed for crushing [283]:

$$W_c = \frac{A}{\sqrt{P_c}}W_i\left(\frac{10}{\sqrt{P_c}} - \frac{10}{\sqrt{F_c}}\right) \tag{115}$$

where W_c is the energy consumed for crushing the clinker ($kWh\ t^{-1}$), W_i is the Bond ball mill work index ($kWh\ t^{-1}$), P_c is the sieve size passing 80% of clinker after crushing (μm), F_c is the sieve size passing 80% of clinker before crushing (μm) and A is the empirical coefficient, which depends on clinker and crusher properties.

Based on the above considerations for crushing and grinding, the energy consumption for the clinker pre-crushing and ball milling can be estimated using the following Bond based model:

$$W = W_c + W_m \tag{116}$$

Since the pre-crushing product size P_c is equal to the mill feed size F_{80} then [279]:

$$\begin{aligned} W = {} & \frac{A}{\sqrt{F_{80}}}W_i\left(\frac{10}{\sqrt{P_{80}}} - \frac{10}{\sqrt{F_c}}\right) + 1.3\left(\frac{2.44}{D}\right)^{0.2} \\ & \left\{\left[R_r + (W_i - 7)\left(\frac{F_{80} - F_0}{F_0}\right)\right]\Big/ R_r\right\}\frac{P_{80} + 10.3}{1.145\, P_{80}}W_i\left(\frac{10}{\sqrt{P_{80}}} - \frac{10}{\sqrt{F_{80}}}\right) \end{aligned} \tag{117}$$

where D is the interior mill diameter and $R_r = F_{80}/P_{80}$.

The basis of the population balance model for modeling the two-compartment ball mill is the perfect mixing ball mill model. This model considers a ball mill or a section of it as a perfectly stirring tank. Then the process can be described in the terms of transport through the mill and breakage within the mill. Because the mill or section of it is perfectly mixing a discharge rate, d_i, for each size fraction is an important variable for defining the product [279,284,285]:

$$p_i = d_i s_i \tag{118}$$

Steady state operation conditions can be described by the following relation:

$$f_i - r_i s_i + \sum_{j=1}^{i} a_{ij} r_j s_j - d_i s_i = 0 \tag{119}$$

The substitution of s_i by p_i/d_i leads to the equation:

$$f_i - r_i \frac{p_i}{d_i} + \sum_{j=1}^{i} a_{ij} r_j \frac{p_i}{d_i} - p_i = 0 \tag{120}$$

where f_i is the feed rate of size fraction [t·h^{-1}], p_i is the product flow of size fraction [t·h^{-1}], a_{ij} is the mass fraction of particle of size that appears at size i after breakage, r_i is the breakage rate of particle size i [h^{-1}], s_i is the amount of size i particles inside the mill [t], d_i is the discharge rate of particle size i [h^{-1}]. If the breakage distribution function is known, the calibration of the model to a ball mill involves the calculation of r/d values using the feed and product size distribution obtained under known operating conditions. Where the size distribution of the mill content is available, the breakage and discharge rates can be calculated separately.

The model consists of two important parameters, the breakage function (a_{ij}) that describes the material characteristics and the breakage/discharge rate function (r_i/d_i) which defines the machine characteristics and can be calculated when the feed and product size distributions are known and the breakage function is available. The air classifier controls the final product quality. Therefore, the air classifier has a crucial role in the circuit and a strong attention is paid regarding the design and operation of the air classifier. The classification action is modeled using the efficiency curve approach. The effect of the classifier design and of operational parameters on the efficiency is complicated and the works proceed to improve the current models [279].

7.1. Consumption of grinding media

The consumption of steel grinding media plays an important role in the economics of grinding and as a consequence also in the overall processing of a large variety of ores. The cost associated with grinding media is chiefly determined by two factors; the price and the wear performance of the grinding media. The mass losses of grinding media can be attributed to three basic mechanisms; the abrasion, the impact and the corrosion. These mechanisms can be simultaneously active in given grinding environment, leading to complex interactions [286].

7.2. Grinding aids

The action of grinding media within the rotating mill not only crushes the existing clinker particles, but also sharply compresses them, which in fact leads to the formation of electrostatic surface charges of opposed polarity. The cement particles agglomerate as a result of the forces of attraction acting on them. Consequently, the cement particle agglomeration reduces the efficiency of mill. The extent of agglomeration depends on [287]:

- The specific characteristics of materials to be ground;
- The operating parameters of mill;
- The efficiency and distribution of grinding media;
- The fineness of cement particles;
- The internal operating conditions of mill (humidity, temperature, ventilation, condition of armor plating, etc.).

Additives, such as water, organic liquids and some inorganic electrolytes are used to reduce the surface free energy of the material being ground with a view to improve the grinding efficiency.

In the grinding process, a variety of grinding aids are used. There are aliphatic amines such as triethylenetetramine (TETA), tetraethylenepentamine (TEPA) and aminealcohols such as diethanolamine (DEA), triethanolamine (TEA) and triisopropanolamine (TIPA). Glycol compounds are represented such as ethyleneglycol (EG), diethyleneglycol (DEG). In addition, there are more complex compounds such as aminoethylethanolamine (AEEA) and hydroxyethyl diethylenetriamine (HEDETA). Phenol and phenol-derivates are also used as grinding aids, as well as other compounds, such as amine acetate, higher polyamines and their hydroxyethyl derivates [287,322].

7.3. Grinding kinetics

Assuming that a grinding mill is equivalent to a chemical reactor with a first-order phenomenological rate of reaction kinetics [288], the rate of decrease in particle size during the batch grinding of brittle material in ball mill can be described by the first-order equation. The breakage rate of such material was expressed in literature as [289]:

$$w_{i,t} = w_{i,0} \exp(-S_i t) \tag{121}$$

where S_i is the specific rate of breakage of feed size i and $w_{i,t}$ is the mass fraction of total charge. For single component (i=1) Eq.121 can be rewritten as:

$$\log \frac{w_{1,t}}{w_{1,0}} = -\frac{S_{1,t}}{2.3} \tag{122}$$

The value S_1 for different particle sizes can be estimated by performing the same experiment with uniform-sized material. Different values of S_1 versus the size can then be plotted on *log–log* plot to give a straight line if all the sizes follow the first-order law of grinding kinetics.

The primary breakage distribution ($B_{i,j}$) is also defined in an empirical form in literature as [280,290,319,320]:

$$B_{i,j} = \varphi_i \left(\frac{x_{i-1}}{x_j} \right)^{\gamma} + \left(1 - \varphi_j\right) \left(\frac{x_{i-1}}{x_j} \right)^{b} ; n \geq i \geq j \tag{123}$$

where $B_{i,j}$ is the mass fraction of primary breakage products, x_i is the largest size, and the parameters φ, γ and β define the size distribution of the material being ground. When plotting the size versus $B_{i,j}$ on log paper, the slope of the lower part of the curve gives the value of γ, the slope of the upper part of the curve gives the value of β, and φ is the intercept [319].

8. Granulation

The spheroidal aggregate of particles is called a granule, ball, pellet, or an agglomerate. The nucleation, compaction, size enlargement, and spheroidization of pellets take place in the course of balling and granulation and related agglomeration processes [291]. The granulation converts fine powder and/or sprayable liquids (e.g. suspensions, solutions or melts) into granular solid products with more desirable physical and/or chemical properties than the original feed material. This size enlargement technique constitutes a key process in many industries such as the pharmaceutical, food, ore processing and fertilizers ones. Particularly, the granulation process has clear advantages regarding the storage, handling and transportation of the final product [292,293].

The new king of agglomeration technology is binder-less granulation, where the original cohesiveness of powder material is utilized to arrange them into granules. The strength of product granules can be much weaker. However, if the product granules are just an intermediate product in a larger process, such weakness has significant advantages. In many material-forming processes, the boundary between granules remains even after shaping due to unnecessary strength of granules. With weaker granules, the density of green bodies produced by the application of the same pressure as for conventional granules can be much higher. In many cases, the binder removal cannot be done completely leaving possible defects caused by carbonacious pyrolysis residues. Weaker granules can also be advantageous in pharmaceutical processes depending on the purpose of granulation [294].

Using large pellets for the processing of strontium aluminate clinker (Fig.31 in Chapter 4.5) may change the behaviour during thermal treatment as well as some properties of the product due to increasing influence of partial pressure of carbon dioxide on the thermal decomposition of strontium carbonate. The material forming the diffusion barrier as the reaction zone is shifted from the surface into the deeper zones of pellet. Increasing partial pressure of carbon dioxide slows down the rate of thermal decomposition of strontium carbonate and increases the temperature required for the thermal decomposition (please see the discussion in Chapter 4.2) and temporary lack of SrO in the reaction zone. Therefore, the influence of large pellets on prepared strontium aluminate clinker is similar to the usage of mixture with lower saturation degree (discussed in Chapter 4).

Chapter 3

Technology of Thermal Treatment

Cement is made of clinker and ground gypsum, whereas clinker is produced from fired limestone and clay mixed in particular percentages. Portland cement clinker was first made in 1842 in modified form of traditional static limekiln. Around 1885, the experiments began on the design of continuous kilns. One of the designs was the shaft kiln, similar in design to a blast furnace. The raw meal in the form of lumps and fuel were continuously added at the top, and clinker was continually withdrawn at the bottom. Compressed air was blown through from the base to combust the fuel. The shaft kiln had been used for only short period of time before it was forced out by the rotary kiln, but it has had a limited renaissance from 1970 onward in China and elsewhere, when it has been used for the small-scale, low-tech plants in rural areas away from transport access. A typical shaft kiln can produce 100–200 tons/day. Nowadays, rotary kiln is one of the key equipment in cement industry used to convert calcineous raw meal to cement clinker. Raw meal for the cement production is a mixture of predetermined proportions of limestone, silica, and small quantities of alumina and iron oxide [295,296].

The cement making processes are extremely energy consuming. Typically for the production of one ton of cement, a well-equipped plant consumes nearly 3 GJ. For each ton of produced clinker an equivalent amount of greenhouse gases is emitted. The manufacture of cement is the focus of considerable attention worldwide because of giant amount of used energy and high environmental impact of the process. Considering the recent impetus on the emission of greenhouse gases reductions and on the reduction of energy consumption, a renewed emphasis arises on developing the computational models for cement industry and on applying this understanding for the performance enhancement [297].

The most important plant unit in cement manufacturing is the kiln [298]. The raw material passes sequentially through pre-heater, calciner, kiln and cooler to form the cement clinker. In a pre-heater section the raw meal is pre-heated to the calcination temperature via hot gases coming from calciner. In calciner, the raw meal is partially calcined. The energy required for the endothermic calcination reaction is provided by combusting a suitable fuel. In most cases, coal is used to provide the required energy, especially in India. The calciner is supplied with tertiary air from the cooler and with air coming out of kiln exhaust. The former is to supply coal with O_2 sufficiently during the combustion and the later to utilize the heat of kiln gases to enhance the calcination reaction. Hot gases from calciner are sent to pre-heater assembly to pre-heat the solids. Partially calcined solids from the calciner are fed slowly to a rotary kiln. In rotary kiln, remaining calcination and other reactions of formation of clinker phases proceed [297,299,300].

The energy required for the endothermic clinker reactions is provided by burning coal in the kiln. The pulverized coal along with the pre-heated air (secondary air) is fed to the kiln in a counter current mode with respect to solids. A part of solids melts in the kiln. The melt causes an internal coating on the kiln refractory. The counter current flow of gas entrains solid particles in the free board region. Such entrainment enhances the rates of radiative heat transfer by increasing effective emissivity and conductivity. Hot clinker is discharged from kiln to clinker cooler and hot gases from kiln exhaust are sent to the calciner. In clinker cooler, a part of energy of solids is recovered back by heat exchange with air. The pre-heated air from coolers is passed to kiln and calciner as secondary and tertiary air, respectively. A small part of air may be vented if required [297].

The clinker formation process by the fluidized bed kiln system was developed by Yuko at al. [301].The average size of clinker produced by this system is about 2 mm, that is much smaller than that of a rotary kiln clinker.

1. Thermal treatment of strontium aluminate clinker

The synthesis of strontium aluminate clinker requires the raw material pre-treatment, the calculation of composition, the preparation of mixture of raw materials, the homogenization or granulation, the thermal treatment, the grinding and fine milling of strontium aluminate clinker. The individual steps of this process are discussed in this chapter. The scheme of production technology of clinker is schematically shown in Fig.1.

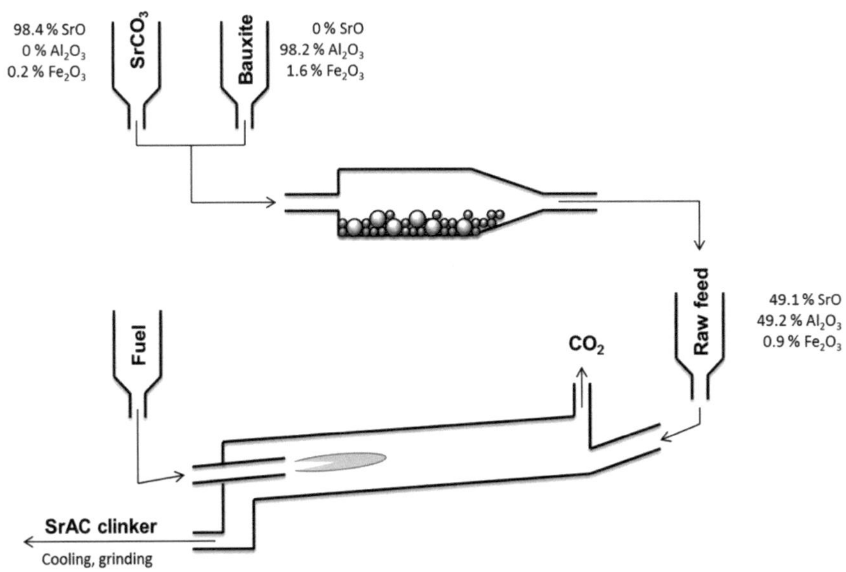

Figure 1. Scheme of clinker production technology.

The heating rate does not play so important role during the processing of strontium aluminate clinker such as for Portland cement, where the decomposition of tricalcium silicate to dicalcium silicate and free water and the formation of γ-dicalcium silicate threaten to devalue the binding properties. On the other hand, increasing cooling rate increases the reactivity of glass phase.

The processes which take place during the thermal treatment of raw meal to strontium aluminate clinker are:

1. Drying: evaporation of free and physically bonded water and adsorbed water without the changes of chemical composition of dry mater. Drying is usually finished up to the temperature of 105 ± 5 °C.

2. Dehydroxylation: thermal treatment of aluminium hydroxide; water is released from hydroxyl groups of mineral.

3. Thermal decomposition of strontium carbonate:

4. Formation of strontium aluminate and the establishment of equilibrium composition.

5. Sintering of solid phase.

6. Formation of melt and sintering of liquid phase.

7. Cooling of clinker.

2. Drying

Drying is the technological process, which removes water from treated material by evaporation. Therefore, the reduction of material moisture does not usually relate to the significant changes in mineralogical composition, the raw material is often considered as the two-component mixture of dry mater and water.

In the dryer, water evaporates from material to flow of air, but the amount of water that can be absorbed by gas phase depends on the conditions of drying process. If the gas phase is supersaturated by water at given temperature, the moisture of material does not decrease any more. Therefore, the drying process must respect certain physical regularities.

There is certain value of maximum amount of water that can be absorbed by air at constant temperature and pressure. Therefore, three parameters are necessary for the characterization of the state of air:

- Temperature (T): the maximum amount of water that can be absorbed by gas phase increases with temperature.
- Pressure (P);
- Moisture: can be characterized by several ways which are described below.
- To specify the content of water in air the following parameters can be used:

- Partial pressure of water vapor in air (P_D [Pa]).
- Absolute air humidity (W_A [kg•m^{-3}]) is defined as the weight of water vapour per cubic meter of air.
- Relative air humidity (W_R) is defined as the dimensionless ratio of parameters mentioned above for actual state of air (P_D or W_A) and air supper saturated by water (P_D^* or W_A^*):

$$W_R = \frac{P_D}{P_D^*} = \frac{W_A}{W_A^*} \tag{1}$$

- Specific air humidity (W_S [kg$_w$•kg$_{air}^{-1}$]) is the weight of water vapours in one kilogram of dry air.
- Temperature of wet point (T_{wp});
- Temperature of wet thermometer (T_{wt}).

The driers use the heat of drying air phase to reduce the moisture of dried material. The temperature of air decreases as water evaporates but the formed water vapor that possesses this heat becomes the part of air phase. In the other words, if the losses aren't taken into account, the enthalpy of air is constant during the drying process. The drier without the heat losses is termed theoretical drier.

Rotary dryers are commonly used to dry and process granular materials. They are applied across a wide range of industries ranging from food production to cement production. Rotary dryers contain lifters to lift the material away from the bottom of the dryer and then to drop it through the gas to promote the mass and heat transfer between the granular material and the gaseous phase [302].

3. Firing process

Firing process includes the technology and equipment which provide heat in order to reach the temperature necessary to achieve required phase composition and properties of the product. Ceramic and concrete technology use furnaces of different type, but it is not necessary to use the pre-calcination unit as for the preparation of strontium aluminate cement. That implies from higher thermal stability of $SrCO_3$ when compared with $CaCO_3$ (discussed in Chapter 4.2). This turns the attention to the technologies of heat regeneration or recuperation from gas processing.

The furnace is heated by the combustion of fuel, the components of which are oxidized according to the following exothermic reactions:

$$H_2 + 1/2\, O_2 \rightarrow H_2O \quad (\Delta_r H° = -241\ 581\ \text{kJ}) \tag{2}$$

$$C{+}O_2 \rightarrow CO_2 \qquad (\Delta_r H^\circ = \text{-}393\ 484\ kJ) \tag{3}$$

$$C{+}1/2\ O_2 \rightarrow CO \qquad (\Delta_r H^\circ = \text{-}110\ 696\ kJ) \tag{4}$$

$$S{+}O_2 \rightarrow CO_2 \qquad (\Delta_r H^\circ = \text{-}296\ 511\ kJ) \tag{5}$$

For these purposes, the dry air is considered as the mixture of ~21 vol. % O_2 and ~79 % N_2. Therefore the $O_2 : N_2$ and O_2 : Air ratios are 21 : 79=1 : 3.76 and 21 : 100=1 : 4.76, respectively. Using the Avogadro law, one kmol of gaseous species under standard conditions corresponds to 22.4 m_N^3 (subscript "*N*" denotes normal condition, i.e. the temperature of 273,15 K and the pressure of 0.1013 MPa).

For example, according to Eq.2, 0.5•22.4•4.76=53.3 m_N^3 of air are necessary for the combustion of 1 kmol of H_2. 1 kmol (22.4 m_N^3) of gaseous water is formed during this process. Air used for the oxidation contains 0.5•22.4•3.76=42.1 m_N^3 of inert N_2 gas. The combustion process is termed as ideal, if the final product contains only CO_2, H_2O and SO_2 species.

The amount of applied oxygen may be stoichiometric, higher or lower than stoichiometric. It can be described by the following dimensionless ratio:

$$n = \frac{\textit{Amount of applied air or oxygen}}{\textit{Stoichiometric amount of applied air or oxygen}} \tag{6}$$

Depending on this, neutral ($\boldsymbol{n} \approx 1$), oxidizing ($\boldsymbol{n} > 1$) or reducing conditions ($\boldsymbol{n} < 1$) can occur in the furnace atmosphere.

The reaction enthalpy for the other species can be calculated from tabulated thermodynamic data, e.g. [303,304], or using the Kirchhoff's laws. For example, the ideal combustion of methane gas is expressed by the equation:

$$CH_4{+}O_2 \rightarrow CO_2{+}2\ H_2O \tag{7}$$

with $\boldsymbol{\Delta_r H^\circ}$=-802 803 kJ. Therefore, the formation of methane (Eq.8) shows the reaction enthalpy of-73 843 kJ:

$$C{+}2\ H_2(g) \rightarrow CH_4(g) \tag{8}$$

That relationship reveals that the value of $\boldsymbol{\Delta_r H^\circ}$(Eq.7)=$\boldsymbol{\Delta_r H^\circ}$(Eq.3)+2 $\boldsymbol{\Delta_r H^\circ}$(Eq.2)-$\boldsymbol{\Delta_r H^\circ}$(Eq. 8)=-802 803 kJ.

Heat released by the combustion of one mol of fuel under standard conditions is termed as the heat of combustion ($\boldsymbol{\Delta_c H^\circ}$)[1]. Depending on the conditions the following expressions of combustion heat apply:

---

1 It stands to reason that $\Delta_r H^\circ = \Delta_c H^\circ$ for Eqs.2 - 5 and Eq. 7.

a. Higher Heating Value (HHV) or High Caloric Value (HCV): heat released during ideal combustion process performed under standard ambient conditions[2]. Formed products were cooled to their original temperature, i.e. the temperature before combustion. All formed water vapor condensed to liquid. For example:

$$CH_4(g)+O_2(g) \rightarrow CO_2(g)+2\ H_2O(l) \qquad HHV=-890\ 564\ kJ \cdot mol^{-1} \tag{9}$$

b. Lower Heating value (LHV), Lower Caloric Value (LCV) or Net Caloric Value (NCV): similar to HHV, but water produced by the combustion process stays in gas phase (Eq. 10). It is obvious that HHV and LHV differ in the enthalpy related to the condensation of water-44 kJ•mol^{-1}. For two mols of water the difference HHV(Eq.9) – LHV(Eq.10)=-88 kJ•mol^{-1}, i.e. HHV=LHV+latent heat of water.

$$CH_4(g)+O_2(g) \rightarrow CO_2(g)+2\ H_2O(g) \qquad LHV=-802\ 556\ kJ \cdot mol^{-1} \tag{10}$$

The bomb calorimeter is used for the determination of heating values of fuels.

The physicochemical environment of cement kilns is ideal for the co-combustion of a number of organic wastes. The economic, as well as environmental benefits are achieved by the substitution of fossil fuels, mainly coal and petcoke. Specific types of alternative fuels (AFs) derived from the mixtures of wastes, such as animal byproduct waste (ABW), agricultural plastic waste (APW), automotive waste plastics (AWP), biological sludge (BIS), cutting oil emulsions (COI), cable plastic waste (CPW), demolition waste (DW), industrial waste fuel (INBRE), refuse derived fuel (RDF or BRAM or SRF), rice husk (RH), refinery sludge (RS), spent solvents (SS), tire derived fuel (TDF), waste oils (WO), etc., are developed [305-308].

Constituents	Raw mixture	Constituents	Raw mixture after recalculation		
	[%] wt.		[g]	[%] wt.	[%] mol.
SrO	49.1	$SrCO_3$	70.0	58.3	49.3
Al_2O_3	49.2	Al_2O_3	49.2	41.0	50.2
Fe_2O_3	0.9	Fe_2O_3	0.9	0.7	0.6
Σ	99.2	Σ	120.1	100	100

Table 1. The composition of raw meal.

Adopting the raw material composition from Fig.1 (calculated for M_H=0.98 as was described in Chapter 2.3.1), it is possible to provide the example for calculation of material and energetic balance for the process of calcination of strontium aluminate clinker.

2 The standard temperature and pressure conditions (STP, 0°C and 100 kPa) are different from standard ambient temperature and pressure (SATP, 25 °C and 100 kPa).

Fuel	Coal	Oil	Gas
Species	Composition [wt.%]		Composition [vol.%]
C	89.0	87,5	---
H_2	5.0	11	0,4
S	1	0,8	---
N_2	3	0,5	2,5
CO_2	---	---	1,6
CH_4	---	---	95.5
SiO_2	2	---	---

Table 2. Composition of coal, oil and gas applied for calculation.

The content of SrO was recalculated to $SrCO_3$ via the multiplication by M_{SrCO3}/M_{SrO} ratio (147.63/103.62=1.42). Other admixtures are neglected and the raw mixture composition is recalculated to 100 % as is shown in Table 1. The composition of coal is given by Table 2.

Assuming the composition mentioned above, the raw material feed of 100 tons per hour, coal used as a fuel to keep the process temperature of 1250 °C and the content of strontium aluminate in clinker at least 90 %, the calculated material and energetic balance of furnace can be seen from Fig.2.

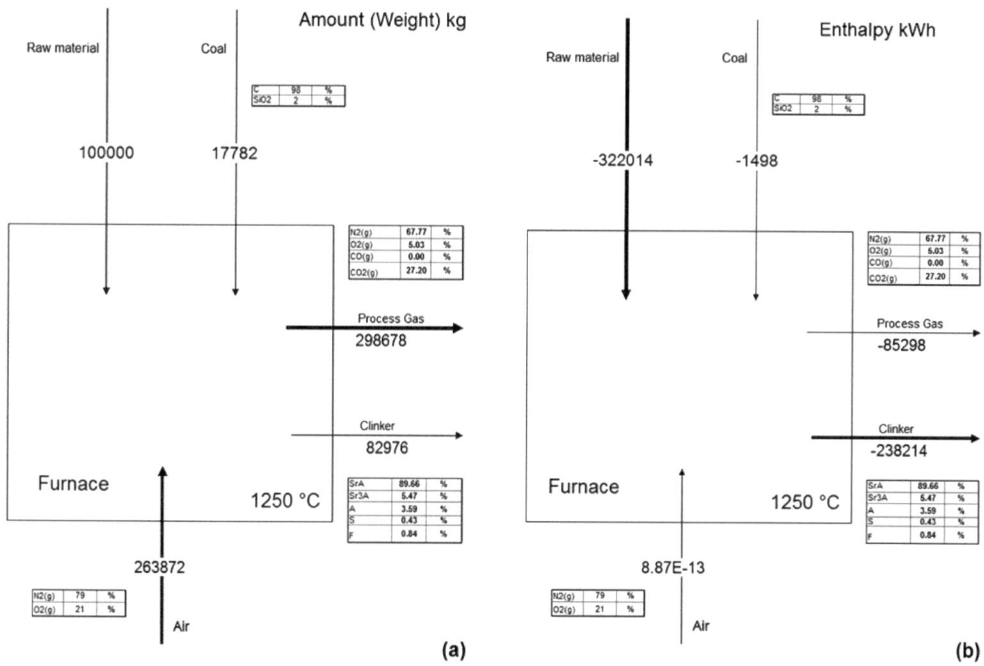

Figure 2. An example materials (a) and enthalpy balance (b) for the strontium aluminate clinker preparation process.

The presence of silica in raw materials or coal used as the fuel may lead to the formation of mullite and strontium bearing feldspar[3]. Both are non-reactive temperature stable compounds without hydraulic behaviour but may serve as microfiller. The reaction of the process which takes place during the calcination of strontium aluminate clinker is discussed in Chapter 4.

Using oil or gas (Table 2) under the same conditions and kiln charge leads to the enthalpy balance shown in Fig.3.

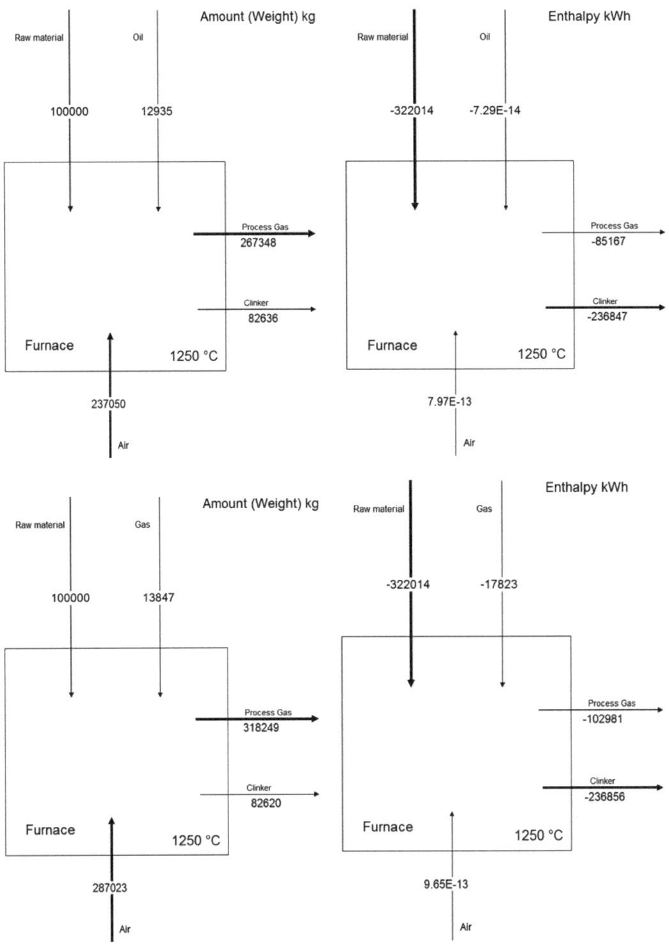

Figure 3. Preparation of strontium aluminate clinker using oil and gas as the fuel.

3 These feldspars have general composition given by formula $Me^{2+}Al_2Si_2O_8$, where $Me^{2+} = Ca^{2+}$, Sr^{2+} and Ba^{2+}. Calcium, strontium and barium bearing feldspars are known as minerals anorthite, slawsonite ($(Sr,Ca)Al_2Si_2O_8$, sometimes termed as Sr-celsian [893,901]) and celsian, respectively. The crystallization of celsian is studied for the preparation of celsian ceramics and glass ceramics [893,895,901]. The formation of celsian was observed by Rezaie at al [15] after the addition of 10 % colloidal silica or kaolin to barium aluminate cement. Regardless the source, the addition of SiO_2 increases the setting time, improves workability and mechanical properties.

Based on the discussion in Chapter 6.1.5, heating to the temperature interval of recrystallization of SrA phase (1300 – 1400 °C) seems to be the optimal conditions for the processing of strontium aluminate clinker. The formation of large crystals due to increasing temperature of thermal treatment over 1400 °C is undesirable since the reactivity of strontium aluminate clinker decreases and the cost of production increases. The sintering and formation of glassy phase at the temperatures over 1450°C (Chapter 4.3.3) increase the demands on grinding of calcinate (Chapter 3.5).

4. Cooling of clinker

Hot solids from the kiln are discharged on the grate of clinker cooler. As the grate moves with uniform speed along the cooler length, the solids lose their heat in cross-flow air. First part of air is generally sent to the kiln as secondary air, second part to calciner as tertiary air and third part is vented to the surroundings (vent air). The clinker comes out of the rotary kiln having high temperature of 1380 °C and is cooled by fresh incoming air to the temperature of 65 °C [297,299].

The clinker cooler serves for two purposes [298]:

i. Supplies the kiln with desired quantity of secondary air at correct pressure and temperature.

ii. Fast cooling the clinker to prevent any further reactions, and far enough to permit further handling.

The effect of OPC cooling rate on the fluidity of mortars was investigated by Nakano at al. [309]. The C_4AF crystallite size calculated by the Rietveld method can be used to measure the cooling rate of clinker and to predict the fluidity of mortar. The fluidity of mortar increased with increasing cooling rate and with decreasing C_3A content. In slowly cooled clinker the calcium aluminoferrite phase is definitely lower in amount and Al_2O_3/Fe_2O_3 ratio than in rapidly cooled clinker [310]. Fast cooling rate results in the higher hydraulic reactivity of cement and reduce the setting time of cement slurry [1010].

5. Grinding of clinker

Rotary kilns produce an intermediate product called "clinker." Clinker is ground to produce cement [311]. More finely ground cement has larger specific surface and the reactive area available for the reaction with water is thus larger. The contact between surfaces of fine additives and sand is improved as well, the initial hydration of clinker minerals is made easier, and the distribution of hydration products is accelerated. It is recommended to keep the particle fineness expressed by the specific surface in the range of 2150–2400 $cm^2 g^{-1}$ if we want to reach the optimal cement properties. Small addition of gypsum or anhydrite retards the start of cement sludge setting [312-315].

The procedure is divided into two sequential phases; crushing of lumps (3 – 25 mm) into loosely connected wafers in a dry roller press and subsequent fine milling of pre-processed clinker wafers in a rotating ball mill [316].

As was mentioned above, the ball mill is commonly used for fine milling of cement clinkers. The type of grinding media exerts significant influence on milling performance in terms of the product size and energy consumption. Various shapes of grinding media including rods, pebbles, and cylinders are used as an alternative to balls. The cylinders received particular attention because they have a greater surface area and higher bulk density than balls of similar mass and size [317-320].

Grinding of clinker represents energetically demanding part of the production process, which consumes about one-third of energy needed to produce cement. The production of fine ground clinker requires grinding stabilizers (grinding activators) which prevent the clinker from re-agglomeration. Such additives, which are well known in both dry grinding and wet grinding processes, help decrease the enormous energy consumption of cement production [285,313,321].

For most of the twentieth century, the dry grinding circuits for the production of finished cement from cement clinker which consist of two-compartment tube mills and air separators, have been used. It is not uncommon to produce cement in an open circuit. The advances in cement grinding technology are slow and limited to more developed countries. Approximately 95% of feed to the cement grinding circuit are clinker and the rest of feed are "additives" which include grinding aids. The quality of cement is measured by the surface area or the Blaine index. The unit of the Blaine index is m^2/kg, and this index is determined by the Blaine air permeability test. In the past 20 years, high pressure grinding roll (HPGR) technology has been used in pre-crushing the clinker. Presently, many American and European cement grinding circuits have HPGR which increases the grinding capacity and energy efficiency [279].

The effectiveness of seven grinding aids, namely triethanolamine, mono-and diethylene glycols, oleic acid, sodium oleate, sulphite waste liquor and dodecylbenzene sulphonic acid on the batch grinding of Portland cement clinker, limestone and quartz was investigated in a laboratory ball mill by Sohoni at al. [322]. While the grinding aids had only a marginal effect on grinding of quartz, they had significantly beneficial effect on the grinding of limestone and cement clinker. Triethanolamine appeared to be the most effective of all aids studied. In grinding of cement clinker, gypsum also acted as very effective grinding aid. The action of these additives was attributed to their ability to prevent the agglomeration and balls and mill from coating by powder.

The presence of easily ground limestone leads to a wider particle size distribution of mixture. Limestone is concentrated in fine fractions (<8 μm) while the clinker is concentrated in coarser ones. However, limestone content over 30% obstructs the grinding of both clinker and limestone [323].

Also calcinate formed during the processing of strontium aluminate clinker must be finely ground in order to prepare the strontium aluminate cement. The increase in specific surface area improves the reactivity of binder with water. Moreover, the admixtures such as calcined bauxite and alumina can be introduced into the clinker. These mineral admixtures serve as micro-fillers, similarly to calcite[4] milled with clinker during the preparation of Portland cement [315,320,323,324]. The grinding kinetics, the effect of grinding media and the additives were already described in Chapter 2.7.

4 The use of Portland limestone cements has many benefits, both technical and economical. The European Prestandard prEN 197-1 identifies two types of Portland limestone cement containing 6–20% limestone and 21–35% limestone, respectively [320].

Chapter 4

Processes during Thermal Treatment

The processes discussed in this chapter include the mechanisms and temperature ranges of chemical reactions and phase transformations, which take place during the treatment of strontium aluminate raw materials as the mixture of strontium carbonate and alumina. Fig.1 shows the equilibrium phase composition during the preparation of calcium (a), strontium (b) and barium aluminate clinker (c). These systems show many similarities, therefore alumina react with carbonate in the molar ratio close to one. Therefore, is interesting to compare the changes of equilibrium composition which take place with increasing temperature for clinkers and pure carbonates (d).

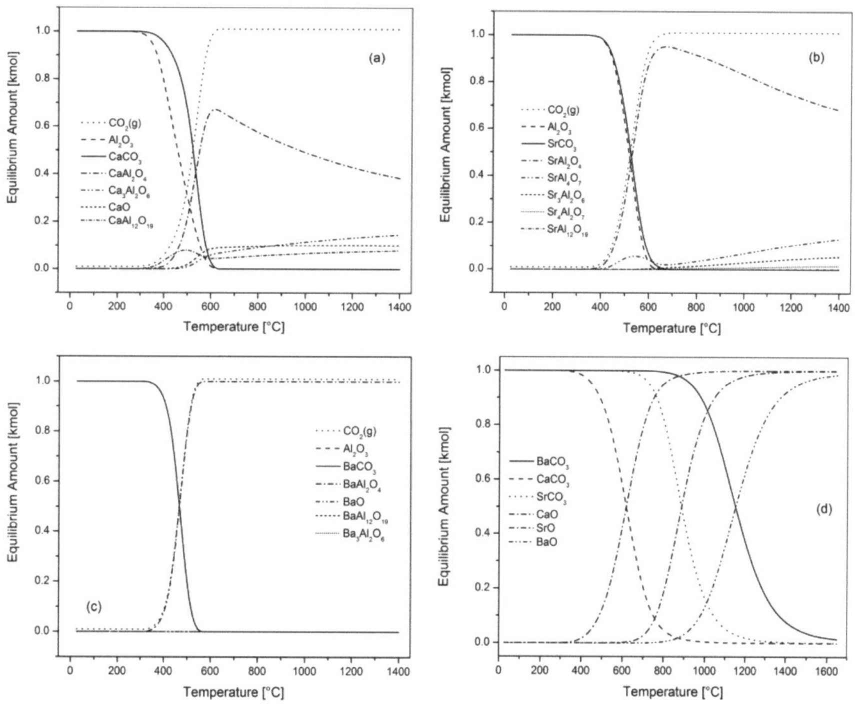

Figure 1. Calculated equilibrium composition during the thermal treatment of calcium (a), strontium (b) and barium aluminate cement (c) and thermal decomposition of pure carbonates (d).

It can be seen that strontium dialuminate phase should be formed first via the reaction of alumina with strontium oxide after the thermal decomposition of strontium carbonate. The maximum amount of strontium dialuminate occurs at the temperature of 540 °C. The amount of $SrO{\bullet}2Al_2O_3$ phase decreases with increasing temperature and formed amount of strontium aluminate. This behaviour is in agreement with experimental results (Chapter 4.3), but the temperature of the process is lower. On the other hand, the differences between the equilibrium temperature and the temperature determined by experiment are common fact for heterogeneous systems, where the kinetic obstacles e.g. related to the nucleation of product occur. The content of strontium aluminate reaches the maximum at the temperature of 675 °C. Except for strontium hexaaluminate, the content of other phases from $SrO - Al_2O_3$ system increases with increasing temperature.

The similar behaviour can be observed for calcium aluminate cement, where higher aluminate phases (SrA_2 and SrA_6) are formed prior to calcium aluminate. On the contrary the formation of higher aluminate phases did not occur during the formation of barium aluminate. The formation of all three aluminates depends on the thermal decomposition of carbonates where the equilibrium constant of the process is not changed by the activity of formed clinker minerals. The mechanism, kinetics and thermodynamics of the process of thermal decomposition of strontium carbonate are discussed in Chapter 4.2.

The equilibrium composition in the mixture of ternary system $CaO - SrO - Al_2O_3$, which consists of $CaCO_3$, $SrCO_3$ and Al_2O_3 in the molar ratio ½ : ½ : 1 is shown in Fig.2. Increasing amount of CaO and SrO in the raw material (Fig.3) supports the formation of C_2SrA and SrA in the reaction mixture, respectively.

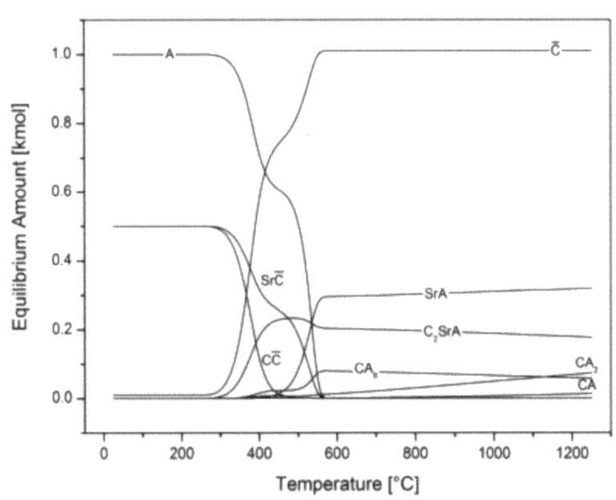

Figure 2. Influence of temperature on the equilibrium composition in the ternary system.

It was observed that increasing amount of glassy phase led to greenish tinge of fired clinker while the material treated to the temperature at which the glassy phase could not be formed was white. On the other hand, increasing amount of glassy phase increased the demand on milling of the clinker to strontium aluminate cement.

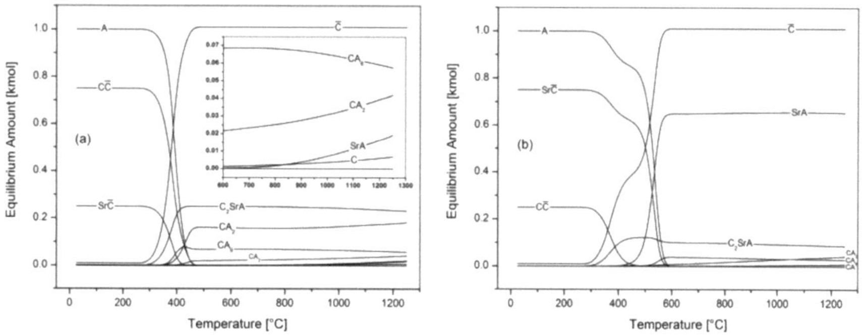

Figure 3. Temperature dependence of equilibrium composition in the systems with higher content of CaO (a) and SrO (b).

1. Thermal treatment of alumina

The α-Al_2O_3 modification of alumina can be used for the synthesis of strontium aluminate clinker. On the other hand, the most stable and well crystalline phases are also less reactive. That often means a long time synthesis which must be performed at high temperatures. Due the thermodynamic reasons, it is better to use **aluminium hydroxides** ($Al(OH)_3$ or $Al_2O_3{\cdot}3H_2O$), **oxyhydroxides** ($AlO(OH)$ or $Al_2O_3{\cdot}H_2O$) or the **transitional alumina oxides** (Fig.4) for the preparation of strontium aluminate as the main clinker mineral for strontium aluminate cements.

Individual $Al(OH)_3$ polymorphs differ in the terms of stacking sequences of layers which are bound together by weak hydrogen bonds. The following main types of aluminium hydroxides and oxohydroxides are [325-335]:

- **Gibbsite (hydrargyllite**, Fig.5(a)) is the monoclinic γ-$Al(OH)_3$ ($P2_1/n$, a=8.684 Å, b=5.078 Å, c=9.736 Å, β=94.54°). The crystals mostly show tabular pseudohexagonal habit. The structure of gibbsite is often drawn as sheets of *hcp* layers with open packing between successive sheets. In the lateral extension of hexagonal closed packed sheets each Al^{3+}ion is coordinated by six hydroxyl (OH) groups. Each OH group is coordinated by two Al^{3+}ions with one vacant octahedral site. Six octahedra, each sharing two edges, yield $[Al_6(OH)_{12}]^{6+}$ ring. Gibbsite shows a perfect cleavage parallel to the basal plane (001). The layers are arranged in the AA-BB-AA sequence. Gibbsite was firstly described in 1820 [336,337].
- **Bayerite** (Fig.5(b)) is the monoclinic α-$Al(OH)_3$ ($P2_1/a$, a=5.0626 Å, b=8.6719 Å, c=9.4254 Å, β=90.26°). It mostly occurs in very fine fibers in radiating hemispherical aggregates and sometimes forms flaky tabular crystals with the size of about 0.1 mm. The crystal lattice of bayerite is composed of layers of hydroxyl groups similar to those in gibbsite. These layers are arranged in the AB-AB-AB sequence, so that hydroxyl groups of the third layer lie in

the depressions between hydroxyl positions of the second layer. Bayerite was firstly described in 1925 [338].

- **Boehmite** is rhombic γ-AlO(OH) (*Amam*, *a*=3.6936 Å, *b*=12.214 Å, *c*=2.8679 Å) which has the same structure as γ-FeO(OH). The structure of boehmite consists of double layers of oxygen octahedra partially filled with Al^{3+}cations. The stacking arrangement of the three oxygen layers is such that the double octahedral layer is in cubic closed packing. Within the double layer one can distinguish two different types of oxygen. Each oxygen atom in between the double layer is shared by four other octahedra, while oxygen atoms on the outer side are only shared by two octahedra. These outer oxygen atoms are hydrogen-bound to two other similarly coordinated oxygen atoms in the neighboring double layers above and below. The stacking of the layers is such that the hydroxyl groups of one layer are located over the depression of hydroxyl groups in the adjacent layer.

- **Pseudoboehmite** (**gelatinous boehmite**) contains higher amount of water than boehmite (γ-AlO(OH)$\cdot nH_2O$, where n ranges from 0.08 to 0.62) [339].

- **Doyleite** is triclinic $Al(OH)_3$ ($P\bar{1}$or P1,*a*=5.002 Å, *b*=5.175 Å, *c*=4.98 Å, α=97.5°, β=118.6, γ=104.74° and Z=2) polymorph of gibbsite, bayerite and nordstrandite. Doyleite was firstly described in 1985 [340].

- **Nordstrandite** (Fig.5(c)) is triclinic polymorph of $Al(OH)_3$ (P1, *a*=6.148 Å, *b*=6.936 Å, *c*=5.074 Å, α=79.76°, β=99.06, γ=83.3°) synthesized and identified by van Nordstrand at al. in 1956 [341][1]. This polymorph occurs in nature and can be prepared by the crystallization of alumina gels at high pH and the presence of chelating agent.

- **Diaspore** is orthorhombic α-AlO(OH) (*PBnm*, *a*=4.379 Å, *b*=9.421 Å, *c*=2.844 Å and Z=4) which is rare and occurs in the metamorphic and sedimentary bauxite ores.

- **Amorphous gel forms** precipitate from aluminium salts. According to applied conditions, various crystalline forms mentioned above can be prepared.

Some selected properties of $Al(OH)_3$ polymorphs including the point of zero charge are listed in Table 1. The experimental methods used to determine the PZC/IEP are described using the following abbreviations [343]:

- CIP (common intersection point) of potentiometric titration curves obtained at three or more ionic strengths or equivalent methods;

- Intersection (intersection point) of potentiometric titration curves obtained at two ionic strengths);

- pH (natural pH of the dispersion), e.g., mass titration and potentiometric titration at one electrolyte concentration;

- IEP (isoelectric point) obtained by means of electrophoresis, electroosmosis, or electroacoustic method.

1 Van Nordstrad proposed the name Bayerite II for this polymorph of Al(OH)3.

Phase	pH of PZC/ IEP	d001	Density	Al_2O_3	H_2O
		[Å]	[g·cm^{-3}]	[%]	
Gibbsite	8.1 – 9.8 (10) [342-344]	4.85	2.34	65.4	36.4
Bayerite	5.4 – 9.7 [347] 8.1 [345] 9.2 [344]	4.79	2.53		
Nordstrandite	?	4.72	2.42		
Doyleite	...	...	2.48		
Boehmite	7.3 – 7.5 [343] 8.9 [343] 9.4 [344]	6.12	3.03	85.0	15.0
Pseudoboehmite	9.2 [346,344]	...		77.7-84.0	16-22.3
Diaspore	7.9 [343] 6.9 – 8.8 [347]	...	3.38	85.0	15.0
Amorphous	9.4-9.5 [348]	...	...	...	...

Table 1. Properties of strontium aluminate polymorphs.

The thermal treatment of these compounds leads to the dehydration and subsequent formation of the most stable modification of α-Al_2O_3. The process is complicated due to the formation of different **transitional aluminas**. The industrial importance of transitional aluminas is the application of many phases as catalysts or carriers (support) for catalysts for petrochemical and fine chemical processes [349,350].

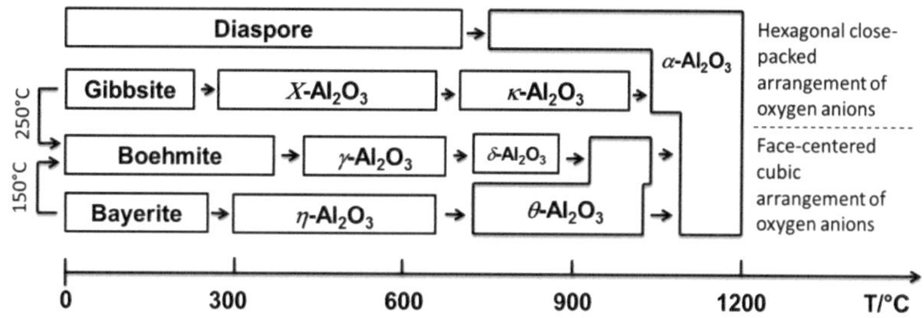

Figure 4. Temperature transformation of hydroxides or oxohydroxides to corundum via the formation of transitional alumina phases [325].

Significant effects of pH value on the composition, structure, morphology, and phase transformation of precipitated aluminium hydroxide can be observed. During the increase of pH value from 5 to 11, the aluminum hydroxides precipitating from the solution vary from amorphous aluminum hydroxide through γ-AlOOH to α-Al(OH)$_3$; and the corresponding morphology of the aluminum hydroxide particles changes from the ultrafine floccules through 50 nm blowballs up to irregular agglomerates with the average diameter of 150 nm. The aluminum hydroxides precipitating at different pH values have different transformation sequences toward α-Al_2O_3 and different formation temperatures of α-Al_2O_3 [351]:

- **Amorphous aluminum hydroxide** precipitating at the pH of 5-6 is transformed to α-Al_2O_3 at 950 °C via the following transformation sequence:

Amorphous hydroxide → amorphous alumina → α-Al_2O_3

- **Boehmite** precipitating at pH=7 transforms to α-Al_2O_3 at 950 °C via the following path:

γ-$AlOOH \rightarrow \gamma$-$Al_2O_3 \rightarrow \alpha$-$Al_2O_3$

- **Bayerite** precipitating at the pH values between 8 and 11, is transformed to α-Al_2O_3 at 1000 °C via the transformation sequence:

α-$Al(OH)_3 \rightarrow \gamma$-$Al_2O_3 \rightarrow \varepsilon$-$Al_2O_3 \rightarrow \eta$-$Al_2O_3 \rightarrow \alpha$-$Al_2O_3$

The temperature relationships between transitional alumina phases are complicated. The course of the process depends on the type of initial hydroxide or oxohydroxide due to their structural similarities, experimental conditions and content of admixtures (Fig.4). Some uncertainty still remains about the numbers and transition sequence between these phases. The course of the phase transitions seems to be given by the requirement to attain the closest structural similarity between original and newly formed phase. The structural similarity between the structures of original and newly formed phase is termed as **topotaxy**. The final product of the process is α-Al_2O_3 in all cases.

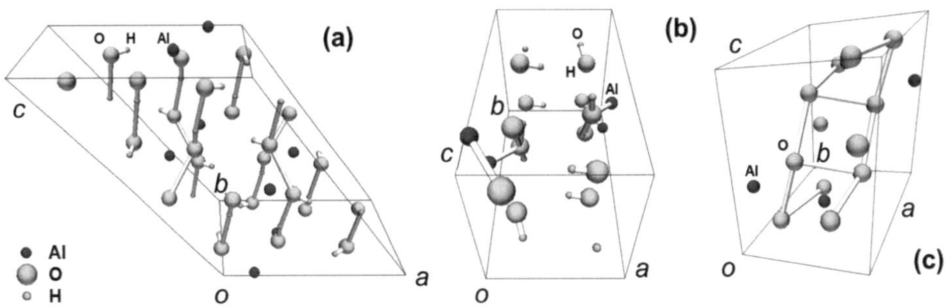

Figure 5. The structure of gibbsite (a), bayerite (b) and nordstrandite (c).

Two general groups of transitional alumina phases can be recognized [325]:

1. **Transitional alumina with face-cantered cubic arrangement of oxygen anions** including cubic γ-Al_2O_3 (*gamma*) and η-Al_2O_3 (*eta*), monoclinic θ-Al_2O_3 (*theta*) and tetragonal (or orthorhombic) δ-Al_2O_3 (*delta*) polymorphs.

2. **Transitional alumina with hexagonal close packed arrangement** where trigonal polymorph of α-Al_2O_3 (*alpha, corundum*), orthorhombic κ-Al_2O_3 (*kappa*) and hexagonal χ-Al_2O_3 (*chi*) are included.

Another classification of transitional alumina phases defines:

1. **Low-temperature transition alumina phases** (γ-and η-Al_2O_3);

2. **High-temperature transition alumina phases** (δ-and η-Al_2O_3).

The beta modifications of alumina including β-and β''-alumina (three blocks structure) are not polymorphs of Al_2O_3, but they are binary or more complex solid solutions of alumina with different metal ions. The beta alumina group of oxides has closely-packed slabs (spinel block) and loosely-packed layers (conduction planes) with mobile ions. The spinel block consists of four layers of oxygen ions with aluminium ions in both octahedral and tetrahedral interstices.

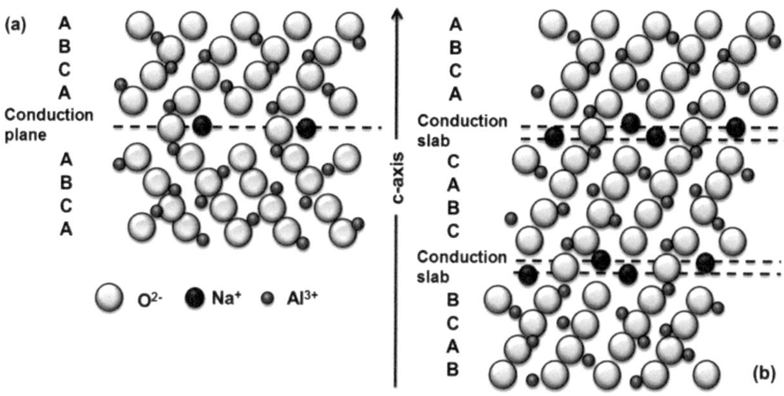

Figure 6. Stacking sequence of $\beta'$and β'' alumina in the unit cell shown in the (11 $\bar{2}$ 0) projection [**352**].

Two neighboring spinel blocks are bound via the conduction plane. There are two different crystal structures (Fig.6):

a. **Hexagonalβ-alumina** ($P6_3/mmc$; a=0.559 nm, c=2.261nm) where the conduction plane is sandwiched in between two spinel blocks (two-block structure). The ideal formula of β-alumina is Na_2O $11Al_2O_3$.

b. **Rhombohedralβ''**-alumina (R3m; a=0.560 nm, c=3.395 nm) where two conduction slabs are separated by three spinel blocks. Since they accommodate higher amount of sodium ions, the structure of β''-Al_2O_3 shows higher conductivity than β-Al_2O_3. The phase is stable up to the temperature of 1600 °C. The ideal formula of β''-alumina is Na_2O $(Al_2O_3)_{6+x}$.

These materials can be used as solid electrolytes (BASE), sodium heat engine or alkali metal thermoelectric converters for direct thermoelectric energy conversion, gas sensors, galvanic cells for the measurement of thermodynamic data [353-358].

From this viewpoint the following phases can be recognized:

- AM – β *and/ or*β''-alumina where AM denotes Alkali Metals (Li, Na, K, Rb…). Sodium β-or β''-alumina (Na_2O_{1+x} $11Al_2O_3$ where x=0.25 – 0.55) exhibits high sodium conductivity at relatively low temperatures (300 – 350 °C). The most convectional technique for the production is solid-state synthesis based on the thermal treatment of Na_2CO_3 and α-Al_2O_3

mixture. Ion exchange in molten salt is further used for the preparation of other ion-alumina phases and their mixtures [359-361].

- AEM – β *and/ or*β''-alumina where AE denotes Alkaline Earth Elements=Ca, Sr, Ba. The structure consists of two spinel block and an intermediate layer. Large cations such as Ba^{2+}occupy the nine-fold coordination sites in intermediate layer of b-alumina structure. Small cations, such as Sr^{2+}or $Ca^{2+}$are located in the twelve-fold coordination sites of MP structure in aluminum oxide [362].
- REE – β *and/ or*β''-alumina where REE denotes Rare Earth Elements such as Gd^{3+}, Eu^{3+}, Nd^{3+} ... [363]
- M – β *and/ or*β''-alumina where Ag^{+}, Pb^{2+}mobile ions were used.

1.1. Influence of the source of alumina

It can be concluded that there are several ways to produce strontium aluminate clinker from raw material bases:

1. The mixture of strontium carbonate with alumina;
2. The mixture of strontium carbonate with transitional aluminas;
3. The mixture of strontium carbonate with aluminium hydroxides, oxohydroxides or gels;
4. The mixture of strontium carbonate with bauxite;
5. The advanced techniques.

Using alumina is promising way for the preparation of high purity product, where the course of solid state reaction (Eq.43), the phase composition and the properties of product can be easily controlled. Moreover, transition aluminas, hydroxides, oxohydroxides or gels have an advantage in higher reactivity.

Using bauxite may increase the content of Fe in the strontium aluminate cement which decreases the thermal stability of materials especially after the reduction conditions. In principle the same production process as for alumina cement can be used.

The activated aluminas, i.e. acidic, neutral and basic aluminas (no definite chemical compositions; made by adding various amounts of water to activated aluminas) are used extensively as adsorbents because of their affinity for water and other polar molecules; and as catalysts because of their large surface area and appropriate pore structure. As adsorbents, they are used for drying gases and liquids; and in adsorption chromatography. The catalytic properties may be attributed to the presence of surface active sites (primarily OH^{-}, O^{2-}, and Al^{3+}ions). Such catalytic applications include the sulphur recovery from H_2S (Clauss catalysis); the dehydration of alcohols, the isomerization of olefins; and as the catalyst support in petroleum refining [91].

Fig.7 shows the temperature dependence of the synthesis of molar unit of strontium aluminate phase using different sources of alumina. These results clearly indicate that γ-Al_2O_3 is the most reactive oxide form for the synthesis of strontium aluminate while thermodynamically most stable α-Al_2O_3 reacts less willingly. Aluminum hydroxides and oxide-hydroxides are even more reactive, but they are the subject of dehydroxylation and phase transformation before sufficient temperature is reached. Therefore we need to identify aluminium hydroxide or oxide-hydroxide which form γ-Al_2O_3 within the temperature interval where the formation of strontium aluminate proceeds.

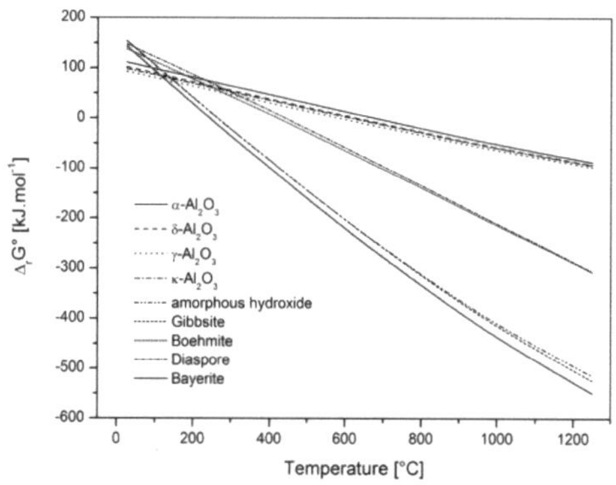

Figure 7. Temperature dependence of $\Delta_r G°$ for the formation of structure unit of $SrAl_2O_4$ from different sources of alumina.

Diaspore (α-AlO(OH)) is directly transformed to the lesser reactive form of α-Al_2O_3 (Fig.4) within the temperature range from 350 to 500 °C (please compare Fig.8(a) and (c)). From this viewpoint, it is not appropriate to use it for the synthesis. In the temperature interval where the formation of strontium aluminate takes place, the thermal transformations of gibbsite, bayerite and boehmite lead to κ-Al_2O_3, Θ-Al_2O_3 and δ-Al_2O_3, respectively.

These transitional alumina phases are all more reactive than corundum, but the question is which of the transitional alumina phases is the most reactive or which transition sequence is optimal and will prevail the for synthesis of strontium aluminate clinker. In the assumption that the phases which require the minimum structural changes (or have the highest structural similarity) are formed predominantly, the bayerite leading to the monoclinic Θ-Al_2O_3 should be considered as the most reactive phase. On the other hand, there is another factor such as alumina source availability. From this point of view, the gibbsite as the main product of Bayer process is promising raw material.

The course of synthesis of strontium aluminate clinker can be significantly changed by the source of Al_2O_3. Regardless of raw material prices, the usage of thermodynamically most stable modification of α-Al_2O_3 (corundum[2]) as the source of aluminium oxide requires the highest temperatures and the longest time for the synthesis.

2 Corundum for ceramics purposes is often produced from bauxite via Bayer process (Chapter 2.1.2).

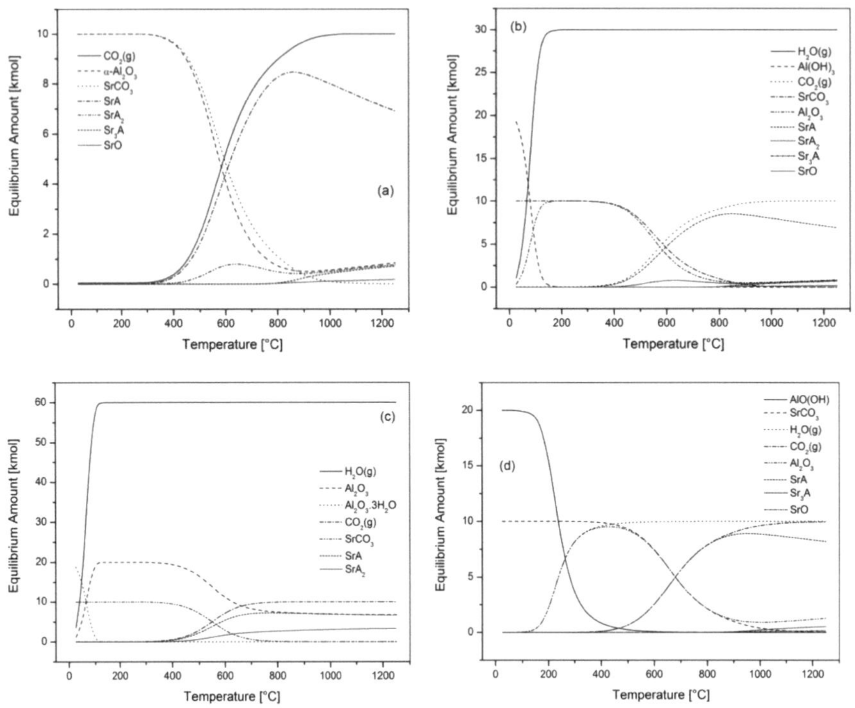

Figure 8. Formation of strontium aluminate clinker from the mixture of $SrCO_3$ with corundum (a), gibbsite, bayerite (c) and diaspore (d)

1.2. Thermal decomposition of $Al(OH)_3$

Thermal decomposition of gibbsite can be described by the reaction [364,365]:

$$2\,Al(OH)_3(s) \rightarrow Al_2O_3(s) + 3\,H_2O(g) \tag{1}$$

The published data related to the kinetic triplet, which includes the activation energy, the frequency factor and the kinetic factor of the process, are listed in Table.2.

Kinetic triplet data			**Note**		**Reference**
E_a [kJ·mol^{-1}]	**A [s^{-1}]**	**n**			
108.5	$2.93 \cdot 10^9$	~1	Gibbsite	Kissinger Equation	[365]
198	---	---	Boehmite	Staring Equation	[368]
175	$1.20 \bullet 10^9$	$m = 0{,}205$ $n = 0{,}738$	Diaspore	Šesták-Berggren	[366]

Table 2. Thermal decomposition of aluminas.

It is generally accepted that hydrothermal conditions favor the formation of boehmite from gibbsite. The pathway of the dehydration is affected by the particle size, partial pressure of water vapor and heating rate. On the contrary to the dehydration sequence in Fig.4, there are at least three ways for the dehydration of gibbsite [367]:

a. Small crystals (< 10 μm) in dry air at atmospheric pressure;

b. Large crystal (> 100 μm) in wet air under the pressure higher than atmospheric;

c. Flash calcination of gibbsite.

This transformation sequences are shown in Fig.9.

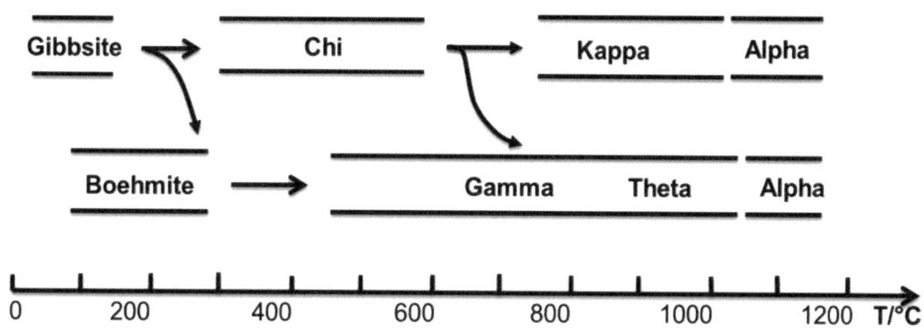

Figure 9. Influence of conditions on the reaction pathway for the dehydration of gibbsite [367].

When mechanically activated by vigorous grinding, many crystalline materials, including gibbsite, are known to lose their long-range order and become X-ray amorphous. Amorphous phase retains its water content and becomes "gel-like". The thermal dehydration of this amorphous phase broadens the characteristic gibbsite dehydroxylation endotherm and lowers its temperature to about 150 – 200°C. The dehydrated product is also X-ray amorphous and is thermally converted to α-alumina (Fig.4) either via η-alumina or via the γ-δ-η sequence [364].

The dehydration of boehmite is a topotactic process, which can be described by the following decomposition reaction [73,368]:

$$2\ AlOOH(s) \rightarrow Al_2O_3(s)+H_2O(g) \tag{2}$$

The mechanism of boehmite transformation to γ-Al_2O_3 involves the elimination of water formed by protons and hydroxyl groups and the migration of Al cations. The latter is the rate-limiting step.

2. Thermal decomposition of strontium carbonate

The thermal decomposition of strontium carbonate should be described by simple first order chemical equation:

$$\alpha\text{-SrCO}_3(s) \xrightarrow{930^{\circ}C} \beta\text{-SrCO}_3(s) \xrightarrow{T} SrO(s) + CO_2(g) \tag{3}$$

From the thermodynamic point of view, there is only simple chemical reaction in the system:

$$\sum_{i=1}^{N} \nu_i C_i = 0 \tag{4}$$

where ν_i is the stoichiometric coefficient for species C_i. According to the convection:

- The value of $\nu_i < 0$ for reactants;
- The value of $\nu_i > 0$ for products;
- The value of $\nu_i=0$ for specimens which do not participate in the reaction.

The change in the number of moles for the reaction 4 is then equal to $\Sigma \nu_i$.

Since pure condensed phases ($SrCO_3$ and SrO) which participate in reaction 3 do not form solid solution, their activities are equal to one (a_{SrCO3}=1 and a_{SrO}=1) if pure solid phase at given temperature and pressure is considered as standard state. In the other words, the activities of condensed phases do not influence the value of equilibrium constant and $\Sigma \nu_i$ includes gas species (CO_2) only.

The Gibbs energy (G) of open homogeneous system is described by the general equation:

$$G = f(T, p, n_1, n_2, \ldots n_i) \tag{5}$$

where T is the temperature, p is the pressure and n_1, n_2,...n_i denote the number of moles of chemical species of the reaction system. The expression for total differential of Eq.5 is given as the following relation:

$$dG = \left(\frac{\partial G}{\partial T}\right)_{p,n_i} dT + \left(\frac{\partial G}{\partial p}\right)_{T,n_i} dp + \sum \left(\frac{\partial G}{\partial n_i}\right)_{p,T,n_{j\neq i}} dn_i \tag{6}$$

Based on the relations:

$$\left(\frac{\partial G}{\partial T}\right)_{p,n_i} = -S\,; \quad \left(\frac{\partial G}{\partial p}\right)_{T,n_i} = V\,; \quad \left(\frac{\partial G}{\partial n_i}\right) p,T,n_{j\neq i} = \mu_i \tag{7}$$

Eq.6 can be written as:

$$dG = -S\,dT + V\,dp + \sum \mu_i dn_i \tag{8}$$

$$dG = -S\,dT + V\,dp + \sum \mu_i \nu_i\, d\xi \qquad (9)$$

where μ_i is the chemical potential, $dn_i = \nu_i\, d\xi$ and ξ is the extent of the reaction. For complete chemical reaction where ν_i mols of species i react, the value of $d\xi = 1$.

The equilibrium state is required to fulfill the following conditions:

$$dG_{T,p} = 0 \qquad (10)$$

$$\sum \nu_i \mu_i = 0 \qquad (11)$$

From the consideration mentioned above the Gibbs energy of chemical reaction reaches the minimum for the equilibrium state and should be written as:

$$\Delta_r G = \left(\frac{\partial G}{\partial \xi}\right)_{T,p} = \sum \frac{\partial G}{\partial n_i}\frac{\partial n_i}{\partial \xi} = \sum \nu_i \mu_i = 0 \qquad (12)$$

where $dn_i/d\xi = \nu_i$. The chemical potential is defined as follows:

$$\mu_i = \mu_i^{\circ} + RT \ln a_i \qquad (13)$$

where μ_i° is the standard chemical potential and a_i is the activity. For the standard state, the value of $a_i = 1$ and $\mu_i = \mu_i^{\circ}$. From the combination of Eqs.12 and 13 follows:

$$\Delta_r G = \sum \nu_i \mu_i^{\circ} + RT \sum \nu_i \ln a_i = 0 \qquad (14)$$

$$\Delta_r G = \Delta_r G^{\circ} + RT \sum \nu_i \ln a_i = 0 \qquad (15)$$

$$\Delta_r G^{\circ} = -RT \sum \nu_i \ln a_i \qquad (16)$$

$$\Delta_r G^{\circ} = -RT \ln \prod a_i^{\nu_i} = -RT \ln K \qquad (17)$$

$$K = \exp\left(\frac{-\Delta_r G}{RT}\right) \qquad (18)$$

where K is the equilibrium constant of reaction 4.

From the definition of Gibbs energy:

$$G = H - TS \tag{19}$$

where H and S are enthalpy and entropy, the law for reaction Gibbs energy can then be written as:

$$\Delta_r G^\circ = \Delta_r H^\circ - T\Delta_r S^\circ \tag{20}$$

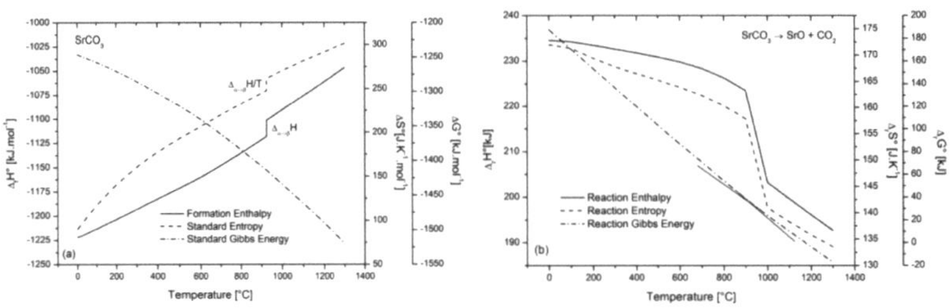

Figure 10. Temperature dependence of the thermodynamic potentials for species (a) and reaction (b).

The temperature dependence of $\Delta_r G^\circ$ (Fig.10) can be calculated using the following equations:

$$\Delta_r H^\circ(T_2) = \Delta_r H^\circ(T_1) + \int_{T_1}^{T_{\alpha\to\beta}} \Delta c_p^\alpha \, dT + \Delta_{\alpha\to\beta} H^\circ + \int_{T_{\alpha\to\beta}}^{T_2} \Delta c_p^\beta \, dT \tag{21}$$

$$\Delta_r S^\circ(T_2) = \Delta_r S^\circ(T_1) + \int_{T_1}^{T_{\alpha\to\beta}} \frac{\Delta c_p^\alpha}{T} dT + \frac{\Delta_{\alpha\to\beta} H^\circ}{T_{\alpha\to\beta}} + \int_{T_{\alpha\to\beta}}^{T_2} \frac{\Delta c_p^\beta}{T} dT \tag{22}$$

where $\Delta_r H^\circ$ and $\Delta_r S^\circ$ are the standard reaction enthalpy and entropy, respectively, $\Delta_{\alpha\to\beta} H^\circ$ and $\Delta_{\alpha\to\beta} H^\circ / T = \Delta_{\alpha\to\beta} S^\circ$ are the enthalpy and entropy related to the $\alpha\to\beta$ phase transformation and Δc_p is the isobaric heat capacity:

$$\Delta c_p = \left(\frac{\partial \Delta_r H^\circ}{\partial T}\right)_p = \sum \nu_i c_{p,i} \tag{23}$$

The temperature dependence of heat capacity is given by law:

$$c_{p,i} = A + BT + \frac{C}{T^2} + DT^2 \tag{24}$$

Where *A*, *B*, *C* and *D* are the constants for given species.

Another option for the calculation of dependence of $\Delta_r G°$ is to use the equation which was derived from Eq.20:

$$\left(\frac{\partial \Delta_r G^\circ / T}{\partial T} \right)_p = -\frac{\Delta_r H^\circ}{T^2} \tag{25}$$

The activity of the ideal mixture of gasses is given by the following relation:

$$a_i = \frac{P_i}{P^\circ} = \frac{P\,x_i}{P^\circ} = P^* x_i \tag{26}$$

where P^*, $P°$, P and P_i are the relative pressure, the standard pressure, the pressure and the partial pressure which is given by the Dalton law ($P_i = P \bullet x_i$) and $x_i = n_i / \Sigma n_j$. Therefore, the value of K should be expressed as follows:

$$K = \prod_i x_i^{\nu_i} P^{*\sum_i \nu_i} = \prod_i n_i^{\nu_i} \left(\frac{P^*}{\sum_j n_j} \right)^{\sum_i \nu_i} \tag{27}$$

The thermodynamic equilibrium constant of the process 3 is then:

$$K = n_{CO_2} \frac{P^*}{n_{CO_2}} = P^*_{CO_2} \tag{28}$$

Therefore the following can be derived:

$$\Delta_r G^\circ = -RT \ln K = -RT \ln p^*_{CO_2} \tag{29}$$

In order to explain the influence of temperature and relative pressure of carbon dioxide on the thermal stability of $SrCO_3$ as well as other carbonates that are important for the chemistry of strontium aluminate cement (Fig.11) was constructed. For this purpose, the following type of reaction is supposed:

$$\mathrm{MeO{+}CO_2 \rightarrow MeCO_3}\text{, where Me=Ca, Mg, Sr, Ba, Fe ...} \tag{30}$$

All reactions are balanced for 1 mol of CO_2 hence the standard Gibbs energy for the reaction is given by the formula:

$$\Delta_r G^\circ = RT \ln P^*_{CO_2} \tag{31}$$

The value of Gibbs energy which is necessary for one mol of CO_2 to expand from standard to the equilibrium pressure related to given temperature is:

$$\Delta G = -RT \ln \frac{P}{P^\circ} = -RT \ln P^* = -(R \ln P^*)\,T \tag{32}$$

The lines representing the constant relative pressure form the $P^*_{CO_2}$ scale of diagram in Fig.11.

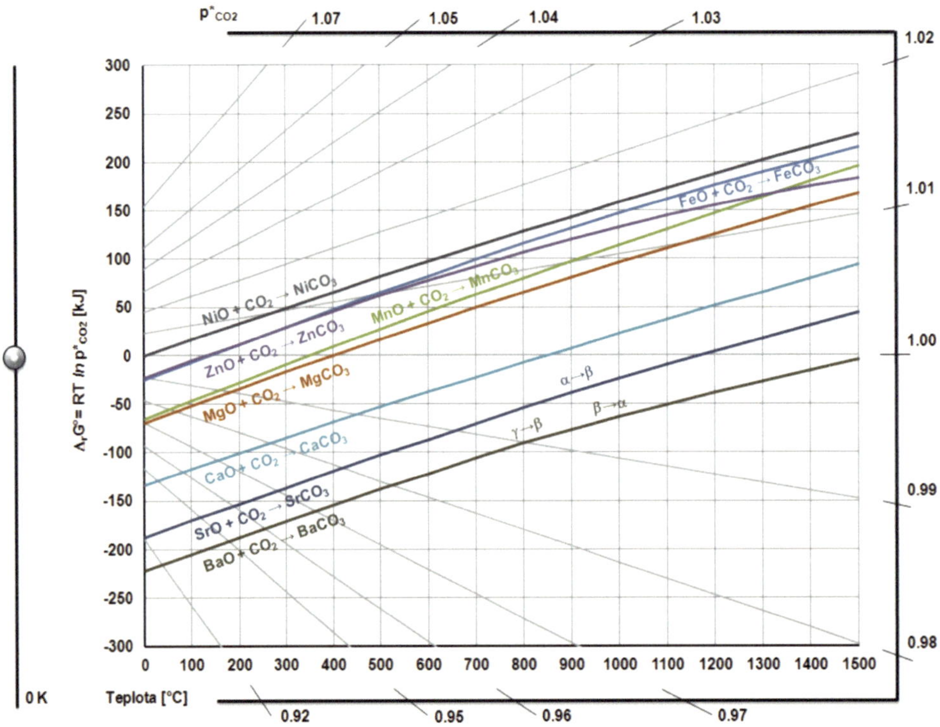

Figure 11. Thermal stability of carbonates.

The temperature of spontaneous thermal decomposition of carbonate can be found as the cross point of the temperature dependence of $\Delta_r G^\circ$ with the line for $P^*_{CO_2}=1$. For strontium carbonate we can read the temperature of 1175 °C. The change in line direction means the phase transformation. Strontium carbonate shows the transformation of orthorhombic (α) to hexagonal (β) modification at 925 - 935 °C. From the diagram we can read the value which is necessary for the thermal decomposition of $SrCO_3$, below the temperature of phase transformation. The extrapolation of line between the point on the $0K$ axis and the point on the temperature dependence of $\Delta_r G°$ for 925 °C to the axis of relative pressure of CO_2 provides the value of 0.9965. Therefore the partial pressure of CO_2 lower than 100.97 kPa is needed. On the contrary, the equilibrium temperature of thermal decomposition can be increased by increasing the value of $P^*_{CO_2}$. For example, for the temperature of 1400 °C the attainment of $P^*_{CO_2}$ of 1.002, i.e. the partial pressure of 101,53 kPa is required.

In some cases the cross point between the temperature dependences of $\Delta_r G°$ for two carbonates occurrs which means the chemical reaction. For example, at the temperature of 1330 °C there is an equilibrium of two reactions under $P^*_{CO_2} = 1.013$:

$$MnO+CO_2 \rightarrow MnCO_3 \quad (a) \tag{33}$$

$$ZnO+CO_2 \rightarrow ZnCO_3 \quad (b) \tag{34}$$

Therefore the following reaction can be supposed:

$$ZnO+MnCO_3 \rightarrow MnO+ZnCO_3 \quad (c) \tag{35}$$

$$\Delta_r G°(c)=\Delta_r G°(b)-\Delta_r G° \quad (a) \tag{36}$$

The value of $\Delta_r G°(c) > 0$ below the equilibrium temperature (reaction 35 shows opposite direction), while $\Delta_r G°(c) < 0$ if the temperature is higher than 1330 °C.

The reaction 3 shows that the phase transformation of orthorhombic α-$SrCO_3$ to the hexagonal structure of β-$SrCO_3$ proceeds prior to the thermal decomposition of strontium carbonate. This process shows sharp endothermic effect which is well visible at DTA (Fig.12). The equilibrium of both phases means that:

$$\mu^{\alpha} = \mu^{\beta} \Rightarrow d\mu^{\alpha} = d\mu^{\beta} \Rightarrow dg^{\alpha} = dg^{\beta} \Rightarrow dg = 0 \tag{37}$$

where:

$$dg^{\alpha} = -s^{\alpha}\, dT + v^{\alpha}\, dp \tag{38}$$

$$dg^{\beta} = -s^{\beta}\, dT + v^{\beta}\, dp \tag{39}$$

From that the Clapeyron law can be derived:

$$-s^{\alpha}\, dT + v^{\alpha}\, dp = -s^{\beta}\, dT + v^{\beta}\, dp \tag{40}$$

$$(s^{\beta} - s^{\alpha})\, dT = \frac{\Delta_{\alpha\to\beta} h}{T}\, dT = (v^{\beta} - v^{\alpha})\, dp \tag{41}$$

$$\frac{dp}{dT} = \frac{\Delta_{\alpha\to\beta} h}{T\, \Delta v} \tag{42}$$

2.1. Kinetics of thermal decomposition of $SrCO_3$

The behaviour of pure strontium carbonate during the thermal treatment is discussed in this chapter. The mechanism, the kinetics and the thermodynamic stability of activated complex of the process of thermal decomposition were assessed by non-isothermal TG-DTA and EGA assessment using the model-free Kissinger kinetic approach (Eq.112 in Chapter 1). 30 mg of strontium carbonate were heated to the temperature of 1200 °C using the heating rate from 5 to 25 °C min^{-1}. The kinetics of the process was evaluated from the shift of DTA peak with the heating rate[3]. It should be pointed that the peak temperature is higher than the temperature of $\alpha\to\beta$ transformation, therefore the kinetics results are related to the thermal decomposition of high temperature (hexagonal) polymorph of $SrCO_3$.

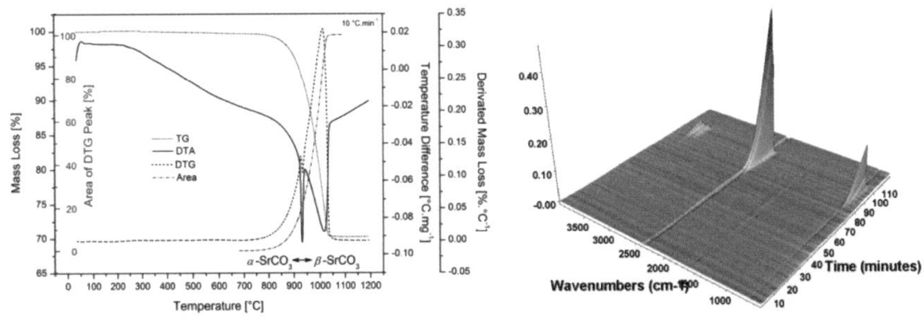

Figure 12. Typical plot of simultaneous TG-DTA and EGA analysis of strontium carbonate.

The slope of Kissinger plot (Fig.13(a)) enables to calculate the activation energy for the thermal decomposition of strontium carbonate (Eq.3) to be 223.7 kJ mol^{-1}. The kinetic factor was calculated according to the relation in Eq.113 in Chapter 1. The mechanism of the process was

3 Other options are to use the peak temperature of DTG or the EGA peak.

determined from the value of kinetic factor (n=2.73 ± 0.03) obtained via the extrapolation of the dependence of n on Θ to the nearly isothermal conditions ($\Theta \rightarrow 0$). The frequency factor can be then calculated from the intercept of Kissinger plot with y-axis to be 1.13 10^7 s^{-1}.

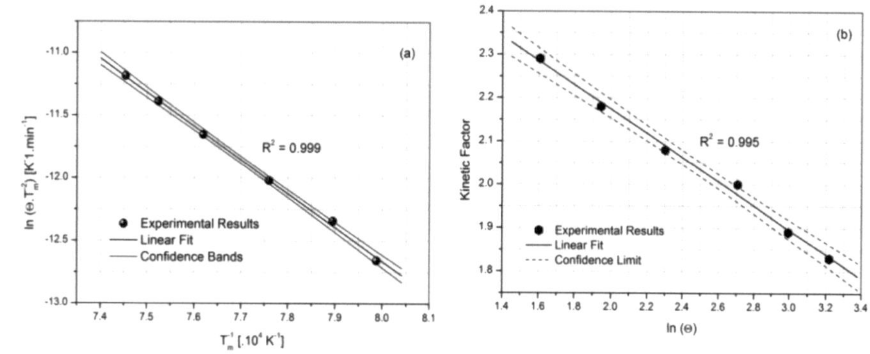

Figure 13. Kissinger plot (a) and extrapolation of kinetic factor to nearly isothermal conditions of the process (b).

The kinetics should be determined via the model fitting procedure using EGA data[4] as well. The total Gram-Schmidt reconstruction data show sharp peak related to the thermal decomposition of strontium carbonate. The peak was integrated and used to calculate the degree of conversion.

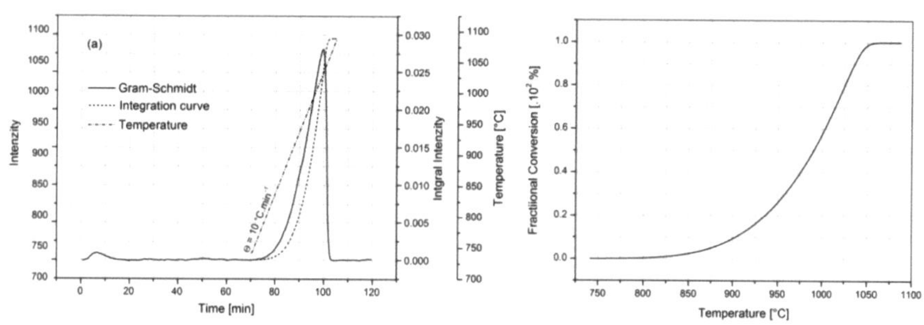

Figure 14. Using EGA to evaluate the kinetics of thermal decomposition of $SrCO_3$.

The rate determining step of the thermal decomposition of $SrCO_3$ is recognized from the slope of the dependence of $ln\ (g(\alpha)/T^2)$ on reciprocal temperature (Fig.15) as the rate of reaction on the reaction interface of contracted sphere (shrinking core, R_3: $1\text{-}(1\text{-}\alpha)^{1/3}$). The activation energy required for decarbonation of $SrCO_3$ has the value of 238.6 kJ.mol^{-1}. The frequency factor that was calculated from the intercept is 2.32 10^6 s^{-1}. Since usual error in the assessment of activation

4 Another and more often applied option is to use TG data in order to evaluate the kinetics of the process. On the other hand, there is one great advantage of EGA, as it is possible to monitor the required product (e.g. H_2O, CO_2, organics such as hydrocarbons, acetone, etc.) according to the selected wavelength.

energy usually reaches few percents of the determined value, the good agreement with the model free and model fitting method has been achieved.

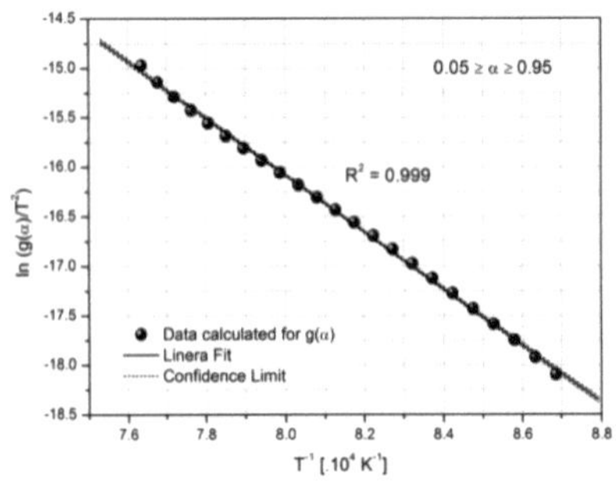

Figure 15. Determination of the activation energy for the most probable mechanism of the process.

These results are also in agreement with experimental as well as calculated[5] results published in literature, namely 233 and 238.7 kJ mol^{-1} [369], respectively.

2.2. Calculation of activation energy from thermodynamic data

The thermodynamic data required for the calculation of theoretical value of activation energy and the values of calculated equimolar (E_a^{Te}) and isobaric activation energy (E_a^{Ti}) for the process of thermal decomposition of strontium carbonate are listed in Table 3.

Parameter	Species	Temperature [K]				
		1000	1100	1200	1300	1400
$\Delta_f H°$ [kJ•mol^{-1}]	$SrCO_3$(s)	-1144.481	-1131.506	-1100.920	-1086.320	-1071.720
	SrO(g)	11.895	15.638	19.403	23.202	27.050
	SrO(s)	-556.539	-551.283	-546.023	-540.759	-535.482
	CO_2(g)	-360.097	-354.601	-349.006	-343.328	-337.580
$\Delta_r H°$	Eq.121 in Chapter 1	796.297	792.542	771.317	766.194	761.190
$\Delta_c H°$	SrO(g→s)	-568.435	-566.920	-565.426	-563.961	-562.532
$\Delta_{rc} H°$	Eq.121 in Chapter 1	512.062	509.083	488.604	484.214	479.924
E_a^{Te}	Eq.121 in	256	255	244	242	240
E_a^{Ti}	Chapter 1	512	509	489	484	480

Table 3. Thermodynamic data for the calculation of theoretical value of activation energy.

5 The method for the calculation of theoretical value of activation energy is described in Chapter 1.6.5. and the example in the next Chapter.

The schematic energy curve in Fig.16 illustrates the method used for the estimation of activation energy for the thermal decomposition of strontium carbonate according to congruent dissociative vaporization mechanism.

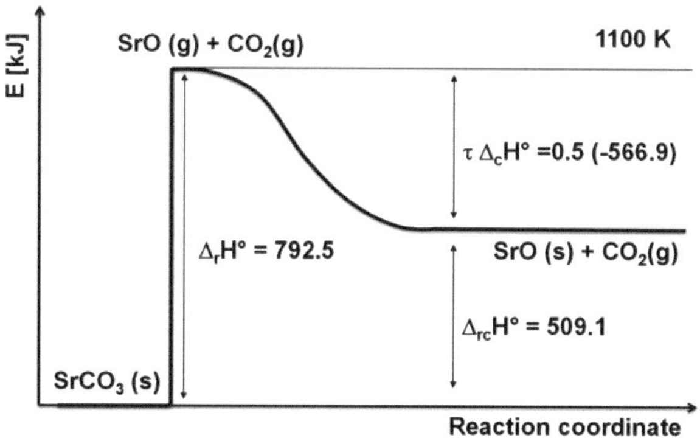

Figure 16. Schematic representation of energy curve for the thermal decomposition of $SrCO_3$.

The temperature dependence of equimolar activation energy on the heating rate is shown in Fig.17. The transitions between individual carbonate polymorphs lead to the step change in the value of activation energy.

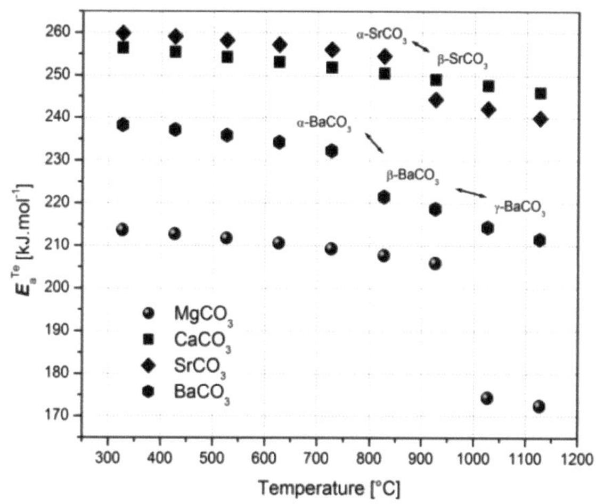

Figure 17. Dependence of E_a^{Te} on temperature.

2.3. Thermodynamics of thermal decomposition

The thermodynamic parameters of activated complex were calculated from Eqs.117 – 120 in Chapter 1 and are listed in Table 4.

T	ΔH#	ΔS#	ΔG#	K#	v	k
[K]	[J·mol⁻¹]	[J·K⁻¹·mol⁻¹]	[J·mol⁻¹]			[s⁻¹]
295.15	221.24	-123.99	251.86	$8.41\cdot10^{-45}$	$6.22\cdot10^{12}$	$5.23\cdot10^{-32}$
1000	215.42	-128.48	343.34	$1.16\cdot10^{-18}$	$2.09\cdot10^{13}$	$2.42\cdot10^{-5}$
1100	214.59	-129.12	356.38	$1.19\cdot10^{-17}$	$2.29\cdot10^{13}$	$2.73\cdot10^{-4}$
1200	213.76	-129.76	369.42	$8.30\cdot10^{-17}$	$2.50\cdot10^{13}$	$2.08\cdot10^{-3}$
1300	212.93	-130.40	382.46	$4.29\cdot10^{-16}$	$2.71\cdot10^{13}$	$1.16\cdot10^{-2}$
1400	212.10	-131.34	395.50	$1.75\cdot10^{-15}$	$2.92\cdot10^{13}$	$5.11\cdot10^{-2}$

Table 4. Thermodynamic data of activated complex.

The plot of *ln k* vs $1/T$ of course gives the Arrhenius plot with the same activation energy (Fig. 18) as was calculated by Kissinger method. Therefore non-isothermal data can be used to calculate the isothermal rate constant (k) of the investigated process.

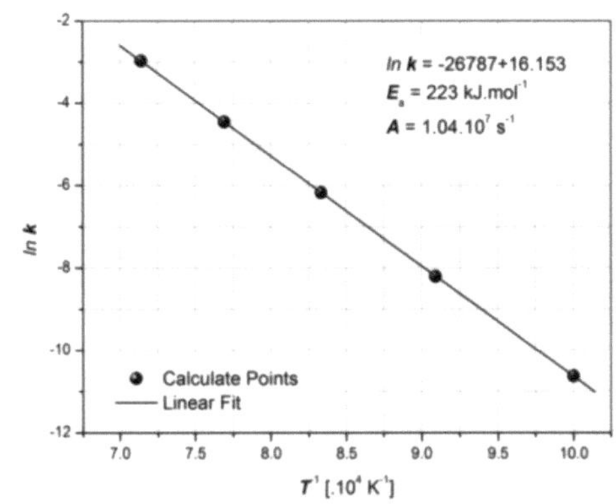

Figure 18. Arrhenius plot calculated from the data of activated complex.

3. Formation of strontium aluminate

During the early stages of sintering, mono-strontium aluminate is formed by the solid-state reaction between strontia and alumina in the powder compact:

$$SrCO_3 + Al_2O_3 \rightarrow SrAl_2O_4 + CO_2(g) \tag{43}$$

Hexaaluminate must be formed by further reaction of strontium aluminate (SrA) and alumina. That can proceed by two distinct mechanisms as is shown in Fig.19.

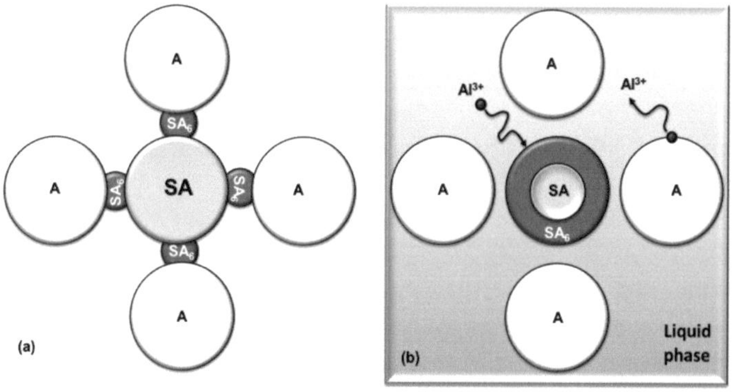

Figure 19. Mechanism of formation of strontium heaxaaluminate [371].

One possibility is that SrA_6 nucleates at the interfaces between alumina and SrA particles and the reaction proceeds by solid-state diffusion through the reactant phase. However, if the surfaces of SrA_6 and alumina grains are already wet by a liquid phase, the transformation to SrA_6 would, by necessity, proceed via the solution-precipitation reaction. The reaction by solid-state diffusion results in the formation of equiaxed SrA_6 grains, while the solution-precipitation favors the development of plate like grains. It is proposed that localized melting take place as a result of low temperature eutectic reaction in the SrO-Al_2O_3 system and it is the eutectic liquid which plays a dominant role in affecting the SrA_6 reaction mechanism (similar model is proposed by An et al. [370] for the alumina–calcia system) [371].

The difference in the final microstructure lies in the extent to which the solid-state reaction occurs between SrA and SrA_6 prior to wetting by liquid phase. This depends on a variety of factors including the particle size, the packing density and the uniform dispersion (mixing) of alumina–strontia powders. Just a slight difference in the above factors could have an appreciable effect on the subsequent wetting behaviour and the microstructure development. SrA_6 may be formed by the process [371]:

- Reaction between Sr and A:

$$\mathrm{Sr+6\ A \rightarrow SrA_6} \tag{44}$$

- Reaction between SrA and A:

$$\mathrm{SrA+5\ A \rightarrow SrA_6} \tag{45}$$

- Reaction between SrA_2 and A:

$$SrA_2 + 4\ A \rightarrow SrA_6 \tag{46}$$

- Reaction between SrA, SrA_2 and A:

$$SrA+SrA_2+9\,A \rightarrow 2\,SrA_6 \tag{47}$$

$$2\,SrA+SrA_2+14\,A \rightarrow 3\,SrA_6 \tag{48}$$

The phases SA and SA_2 are both the reaction intermediates and are minimized with the formation of SA_6. Similar reaction intermediates were proposed for the formation of calcium aluminates [371].

3.1. Thermal treatment of raw meal

The processes occurring during the formation of strontium aluminate clinker can be observed by HT-XRD analysis of raw meal upon heating. Fig.20 shows the changes in the phase composition of equimolar mixture of strontium aluminate and alumina. The diffraction lines of $SrCO_3$ start disappearing at the temperature of 800 °C. SrO formed via the thermal decomposition of $SrCO_3$ reacts with alumina and the diffraction line of tristrontium aluminate appears at 825 °C. For this reason the diffraction lines of alumina disappear at the same temperature

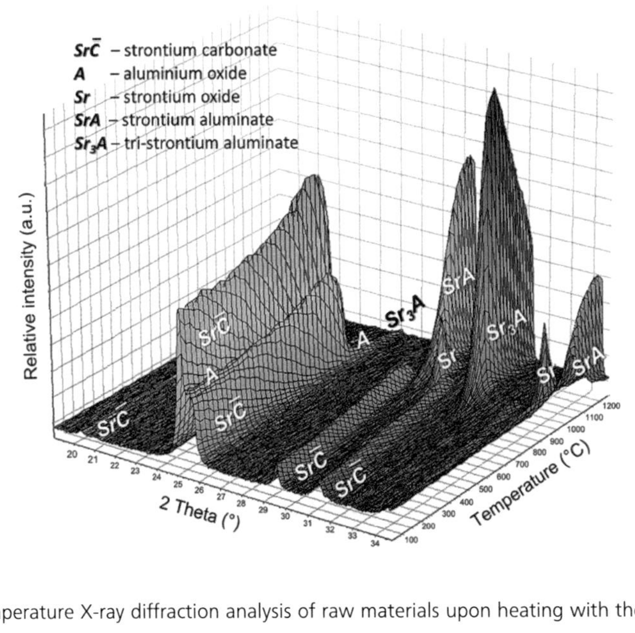

Figure 20. High-temperature X-ray diffraction analysis of raw materials upon heating with the heatinf rate of 10 °C min^{-1} [379].

The features of the main clinker phase, i.e. strontium aluminate, are recognized at the temperature of 875 °C. The process of thermal decomposition of strontium carbonate is not finished yet and its diffraction lines can be detected up to the temperature of 1000 °C. The features of free SrO gradually disappear at the temperature of 1250 °C.

The comparison of X-ray diffraction pattern of raw materials (a) and fired strontium aluminate clinker (b) is shown in Fig.21. There are two main product of thermal treatment of raw equimolar mixture of $SrCO_3$ with Al_2O_3 to the temperature of 1600 °C, namely strontium aluminate and tri-strontium aluminate. Using quantitative X-ray diffraction Rietveld analysis, the content of strontium aluminate and tri-strontium aluminate phase in prepared strontium aluminate clinker is assessed to be 96 % and 4 %, respectively.

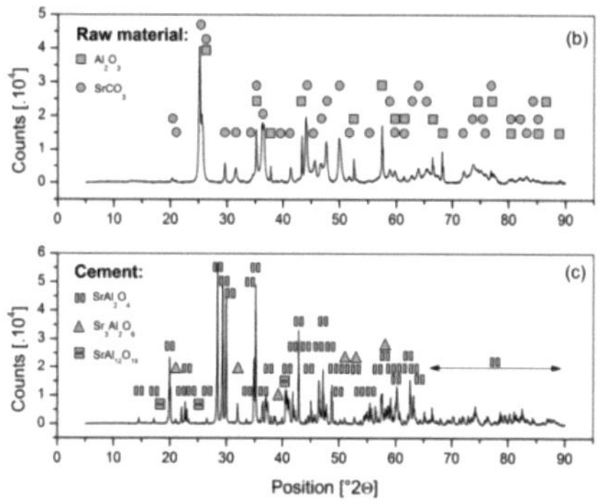

Figure 21. Initial (a) and final composition (b) of strontium aluminate clinker [379].

The comparison of infrared spectra of raw materials and product (Fig.22) shows disappearing absorption bands of planar CO_3^{2-}anion with D_3h symmetry[6] [183,186,372-374] including strong band of antisymmetric stretching at 1465 cm^{-1} ν_3, weak band of symmetric stretching at 1072 cm^{-1} ν_1, out of plane bending at 856 cm^{-1} ν_2 and antisymmetric bending vibration at 702 cm^{-1} (ν_4). There are also three combination bands $2\nu_1+\nu_2$, $\nu_1+\nu_4$ and $\nu_1+\nu_4$. Corundum in raw mixture shows four main absorption bands of stretching of (AlO_6) octahedra at 639, 591, 489 and 448 cm^{-1} [375,376].

On the contrary, the $SrAl_2O_4$ structure is built up from (AlO_4) tetrahedra with the T_d symmetry and the structural lattice channels are occupied by Sr^{2+}ions [377]. The bands appearing in the spectral regions from 900 to 780 cm^{-1} and from 650 to 550 cm^{-1} belong to antisymmetric and symmetric stretching of (AlO_4) tetrahedra. The bands related to the doublet of bending of O-Al-O are located at 446 and 418 cm^{-1}.

Simultaneous TG-DTA (a) and EGA (b) analyses of the mixture of strontium carbonate and alumina (Fig.23) show sharp exothermic peak at the temperature of 967 °C, which divides the huge endothermic effect of thermal decomposition of strontium carbonate occurring within the temperature range from 810 to 1020 °C. The CO_2 bands appear on EGA. That means, that

6 The D_3h symmetry of CO_3^{2-} anion is reduced to C_s for carbonates from group of aragonite. It leads to the splitting of degenerated modes ν_2 and ν_4 and amount of vibration modes in infrared spectrum is increasing. Furthermore, the intensity of fully symmetric stretching mode ν_1 is increasing (this mod is not IR active for carbonates from the group of calcite).

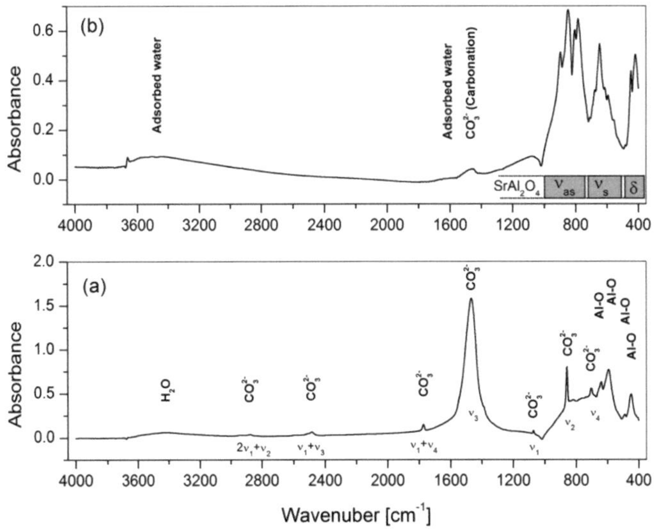

Figure 22. Initial (a) and final infrared spectra (b) of strontium aluminate clinker.

formed $SrAl_2O_4$ phase covered the surface of decomposed $SrCO_3$ particle and this layer slowed down the diffusion of CO_2 from the reaction interface to disappearing $SrCO_3$ core. Therefore, the thermal decomposition of $SrCO_3$ was suppressed. The volume changes caused by the decarbonation and recrystallization of products led to the formation of cracks, which enabled easy diffusion of CO_2 through the layer of the product. The mass of the sample was reduced by 17.7 wt. % during this process.

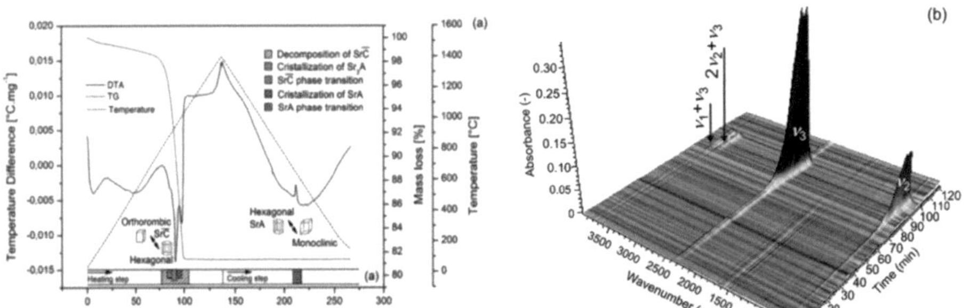

Figure 23. Thermal analysis of strontium aluminate clinker: simultaneous TG-DTA (a) and EGA (b) assessment using the heating rate of 10 °C min^{-1}.

The reaction interface (Fig.24) abundance of SrO (outer side of formed strontium aluminate layer) or Al_2O_3 (inner side) component led to the formation of tri-strontium aluminate (Sr_3A) and strontium hexaaluminate (SrA_6). The equilibrium composition of strontium aluminate (SrA) was then established via the opposite direction diffusion of $Sr^{2+}$and Al^{3+}ions. The opposite model, i.e. Al_2O_3 particle surrounded by $SrCO_3$, where the formation of strontium aluminate is limited by the diffusion of Sr^{2+}ions into the disappearing alumina core cannot explain the observed behaviour. Therefore, formed Sr_3A is actually the mesophase during the process of reaction [378].

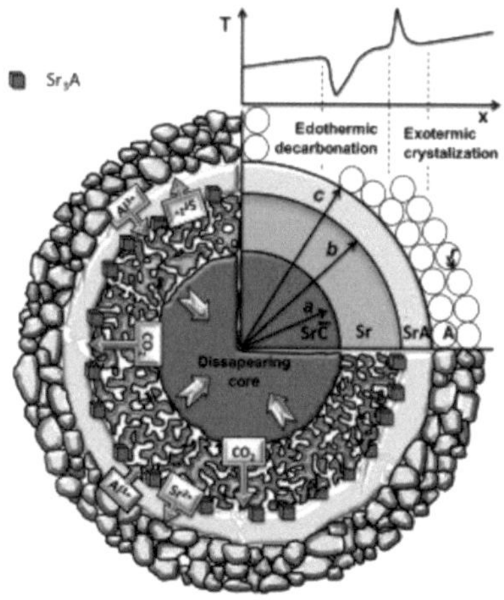

Figure 24. Mechanism of formation of strontium aluminate [379].

The cooling zone in (Fig.23) showed an exothermic effect of reversible transformation of hexagonal high-temperature modification of $SrAl_2O_4$ to low-temperature monoclinic phase at the temperature of 650 °C. During repeated cycle of heating, the exothermic transformation took place at higher temperature. The temperature hysteresis of this transformation increased linearly with the heating rate (Fig.25). Therefore, the limit value of 9.10 ± 0.07 °C should be estimated from the linear fit for $\Theta \rightarrow 0$ [379].

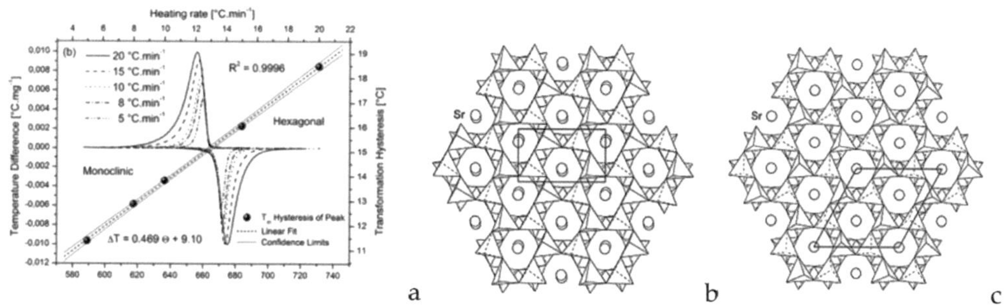

Figure 25. Temperature hysteresis in the polymorphic transformation (a) [379] between monoclinic (b) and hexgonal strontium aluminate (c) [24].

The recorded XPS (X-ray photoelectron spectroscopy) spectrum of strontium (Sr3d) in prepared strontium aluminate cement is presented in Fig.26. The results indicate that the structure of $SrAl_2O_4$ includes two binding states of strontium in the structure of strontium aluminate distributed in the ratio close to 1:2, similarly as was described for the structure $CaAl_2O_4$ where two Ca(I) and Ca(II) had octaedric coordination while Ca(III) was coordinated by six O^{2-}ions in the trigonal antiprism [380].

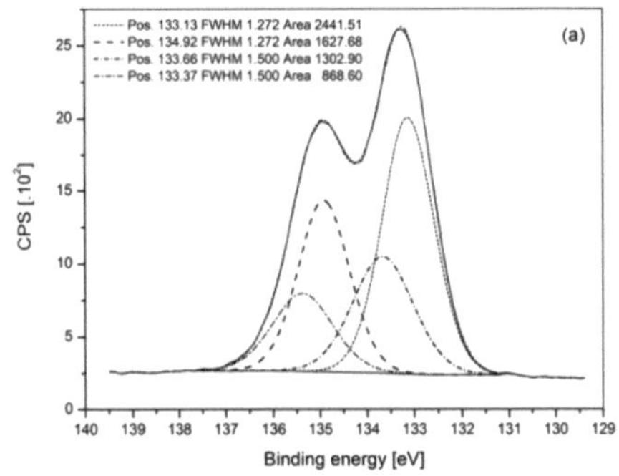

Figure 26. XPS spectrum of strontium from prepared strontium aluminate clinker.

3.2. Kinetic of formation of strontium aluminate

The Kissinger plot related to the kinetics of strontium aluminate and tri-strontium aluminate phase formed from the mixture of raw materials is shown in Fig.27.

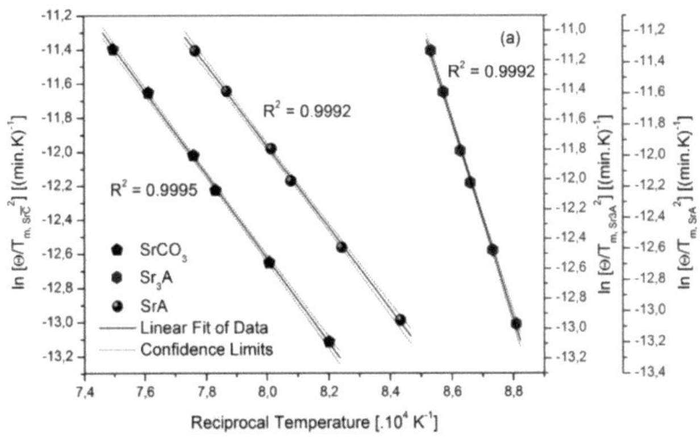

Figure 27. Kissinger plot related to the formation of SrA and Sr_3A clinker phase [379].

The apparent activation energy and the frequency factor relating to the formation of Sr_3A are 551 kJ mol^{-1} and 1.69 10^{23} s^{-1}, respectively. The value of kinetic exponents calculated according to Eq.113 in Chapter 1 is 4.0, hence the process is driven by constant nucleation rate of a new phase. SrA shows the activation energy and the frequency factor of 218 kJ·mol^{-1} and 1.63 10^7 s^{-1}, respectively. The value of kinetic exponent is equal to 5.2. Therefore, the crystallization of SrA phase is driven by increasing nucleation rate of a new phase [379].

3.3. Sintering of strontium aluminate clinker

Heating microscopy (Fig.28) showed the expansion of specimen up to 113 % of its original height. For the temperatures higher than 1050 °C the growth of specimen decreased. The initial temperatures of solid-state and liquid state sintering were determined to be 1450 and 1550 °C, respectively.

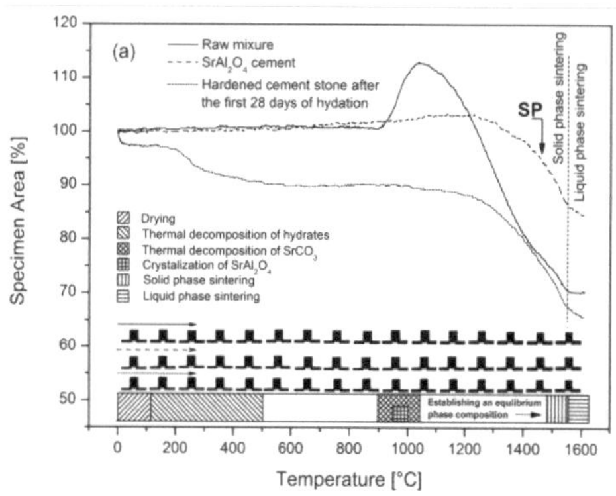

Figure 28. Behaviour of raw material specimen during heating with the heating rate of 5 °C min^{-1} [379].

The information on the mechanism and kinetics of the processes occurring during the synthesis of strontium aluminate clinker were used for the calculation of correct conditions (Fig.29) during the thermal treatment from kinetic triplet, i.e. activation energy, frequency factor and kinetic factor (Chapter 4.3.2).

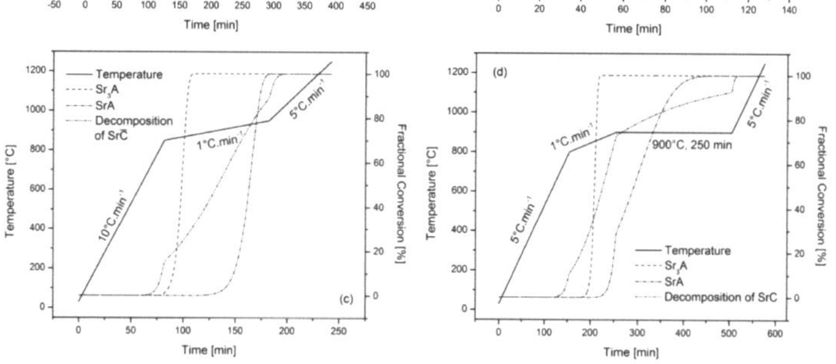

Figure 29. Formation of strontium aluminate using single ramp (a), fast ramp (b), slow heating upon reaction zone (c) and isothermal condition in reaction zone (d).

4. The influence of degree of saturation

The influence of saturation degree (*SD*, Eq.36 in Chapter 2) on the equilibrium composition of the product during the thermal treatment of strontium aluminate raw mixture consisting of $SrCO_3$ and Al_2O_3 is shown in Fig.30. Lower values of saturation of raw materials by strontium oxide support the formation of strontium dialuminate[7] while the amount of tri-strontium aluminate and strontium oxide increases with increasing SD_{SrO}. These phases significantly contribute to the heat release during mixing the clinker with water. In order to avoid the danger of overheating of mixture, the saturation degree ≥ 100 % shouldn't be used. The theoretical value, which was estimated as the highest allowed value of saturation degree for which the equilibrium amount of SrO is inconsiderable is close to 95 %.

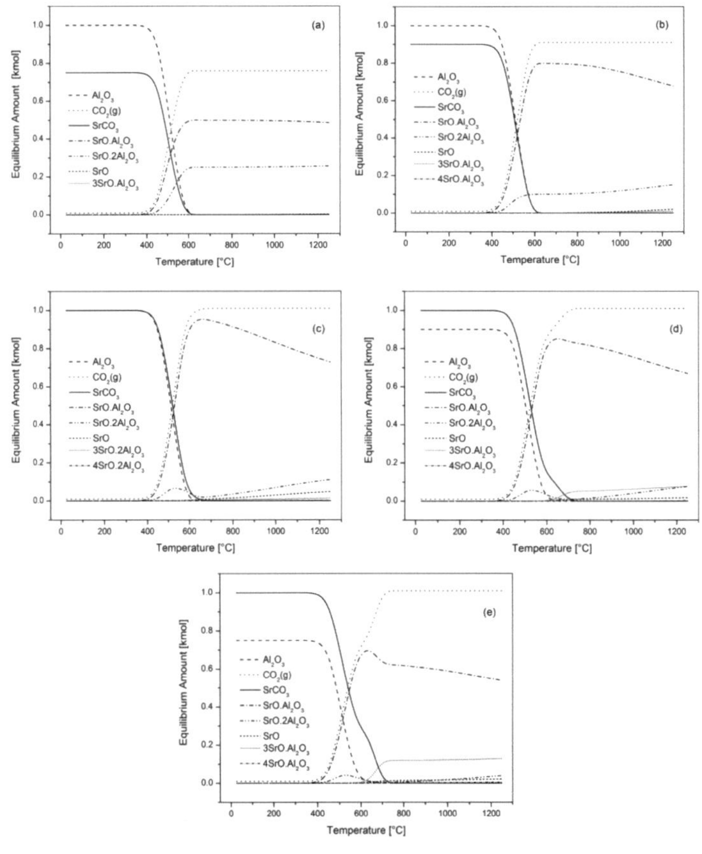

Figure 30. Influence of SD_{SrO} on the calculated equilibrium composition of reaction products: 73.8% (a), 88.6 (b), 98.4 (c), 109.4 (d) and 131.2 (e).

7 Formation SrA_6 is preferred from the formation of SrA_2 during thermal treatment of raw meal (please see discussion in Chapter 1.5).

5. Final treatment of strontium aluminate cement

Grinding and fine milling of strontium aluminate clinker (Fig.31) prepared via the calcination of pellets (a) or fine raw material powder (b) provided cement with the median of particle size of 7.52 μm (Fig.33), which was used for the hydration experiments described in the next chapter as well as for other applications described in Chapters 6-8. Using pressed pellets may influence the behaviour of raw material during the thermal treatment as was already discussed in Chapter 2.8.

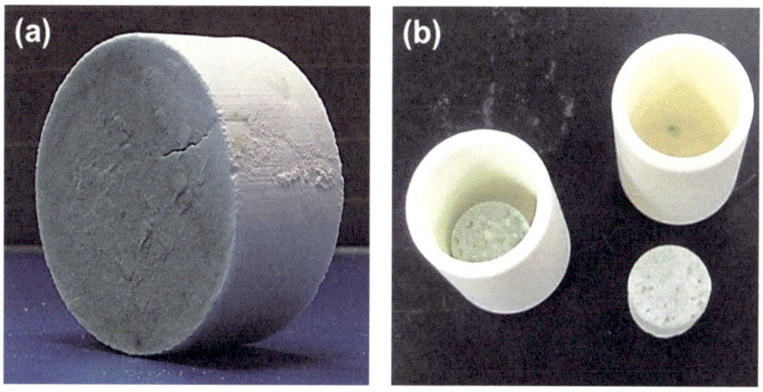

Figure 31. Strontium aluminate clinker prepared from pelletized (a) and fine powdered (b) mixture of raw materials.

The SEM picture of strontium aluminate cement (Fig.32) shows larger subhedral particles with some glass-like faces surrounded by aggregates consisting of closely associated smaller particles.

Figure 32. SEM picture of fine ground strontium aluminate cement.

The particle-size distribution (PSD) and the cumulative distribution function of prepared binder are shown in Fig.33. The multimodal particle size distribution in ground clinker should be explained by the content of other phases of different grindability (Fig.21) and of glassy phase.

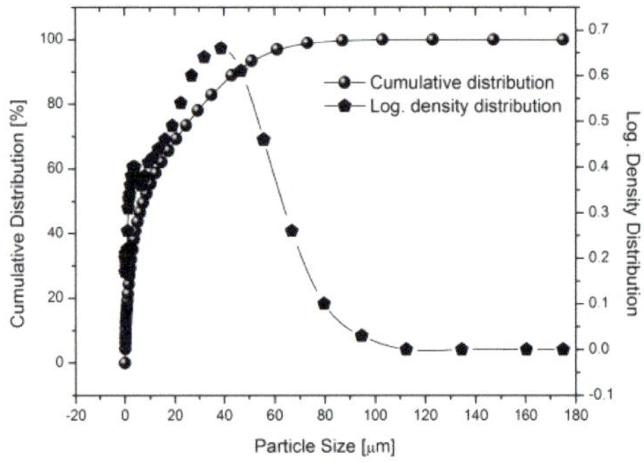

Figure 33. Particle size distribution of strontium aluminate clinker [379].

6. Clinker quality evaluation

The specific surface, the particle size distribution (Chapter 3.5 and Chapter 6.1.2), the content of major and minor clinker phases (Chapter 2.5) in cement are the main parameters which can be used for the evaluation of quality of strontium aluminate clinker similarly to other binders such as AC and PC. These parameters define the proportion of fine and coarse particles in cement. This proportion controls the water demand, the setting and hydration reactions [381-385].

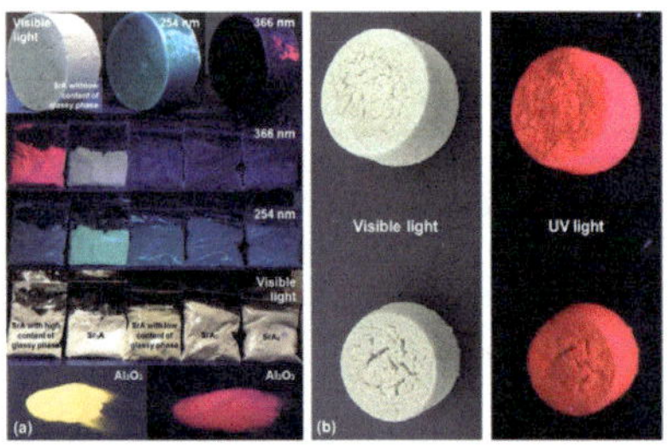

Figure 34. The clinker phase (Sr_3A, SrA, SrA_2 and SrA_6) and alumina (a) and strontium aluminate clinker with high content of glassy phase (b) under visible and UV light.

Therefore, low firing temperature or rough milling procedure can be used in order to reduce initial high reactivity of SrA cement with water. The phosphorescence of calcinate can be applied to evaluate the firing process (estimate the amount of unrecated alumina, Fig.34). Furthermore, the phosphorescence under UV light can be used for marking of cement or for preparation of cement layer or concretes of special properties.

Chapter 5

Hydration and Setting Behaviour of Strontium Aluminate Cements

Hydration of cement generally means complex changes and reactions which occur when the cement paste is prepared by mixing of anhydrous cement with water. Setting means stiffening without significant development of compressive strength which typically occurs within a few hours after mixing of cement with water. Hardening process means gradual significant development of compressive strength during curing in applied environmental conditions (storage in air or water, temperature) [263,265,270].

As was already mentioned in Chapter 2.6, cement slurries are complex reactive systems. They continuously change, physically and chemically. The microstructural and chemical evolution of cement slurry from the first minutes after mixing of cement powder with water until the beginning of setting can be observed by a variety of physicochemical methods including thermal analysis calorimetry [286-393], X-ray diffraction analysis [386,394,744], infrared Raman and Mössbauer spectroscopy [386,395,396], electron microscopy [273,275,394-399] as well as via liquid phase analysis [400,401], porosity and specific surface of cement stone measurement [402], ultrasonic pulse velocity [403-406], impedance spectroscopy measurement [407,408] and electrical conductivity/ resistivity measurement [409,410].

The knowledge of the setting characteristics of concrete is quite important in the field of concrete construction. They will help in scheduling of various stages involved in concrete construction operations such as transporting, placing, compacting and finishing of concrete. This information is a necessity when deciding whether to use a retarding admixture or accelerator or not [411].

The introduction of superplasticizers in early 1960s had great effect on the development in concrete technology. Superplasticized cement can be prepared with lower water to cement ratio, has improved workability, high strength and high durability without affecting the setting and hardening behaviour. In chemical terms, superplasticizers are organic polyelectrolytes which belong to the category of polymeric dispersants. Today, superplasticizers are considered as an integral component of High-Performance concretes [412].

ASTM C494 recognizes the following classes of superplasticizers [412]:

1. **Class F**: High-range water reducers: the most important group are sulphonated synthetic polymers such as poly-β-naphtalene sulphonate and sodium polymelamine sulphonate;
2. **Class G**: High-range water reducers and retarders.

Further sub-classifications of superplasticizers is also used and refers to the first, second and third generation of products.

The effect of superplasticizers' chemical structure on their efficiency in cement pastes was investigated by Janoska-Rekans [413]. The efficiency of industrial superplasticizers increases with the molecular weight of the superplasticizer polymer as well as with its mass fraction in considered sample. The efficiency of a superplasticizer increases with its hydrophilicity. The hydrophilic nature of the superplasticizer polymers can be measured as the ratio of ethers to esters.

General effect of admixtures such as silica fume, metakaolinite, flay ash and blast-furnace slag is the retardation of setting time of high-strength concrete [411]. Regarding limestone, the higher the limestone fines content in the mix, the shorter the initial and final setting times of concrete [414,415].

1. Analysis of hydration products

The compounds formed during 28 days of hydration of strontium aluminate clinker were investigated by X-ray diffraction analysis (Fig.1). The cubic tri-strontium aluminate hexahydrate phase (Sr_3AH_6) and the high alumina gel phase (AH_3 gel) are recognized as the main products of hydration:

$$3\,SrO\times Al_2O_3+12\,H_2O \rightarrow 3SrO\times Al_2O_3\times 6H_2O+2\,Al_2O_3\times 3H_2O \quad (1)$$

That enables to calculate the theoretic water to cement ratio for the hydration of strontium aluminate cement to be 0.36. $SrAH_{7\text{-}10}$ hydrate and gibbsite (γ-AH_3) appear after 7 and 14 days of hydration, respectively. The rate of the hydration process is significantly reduced after 14 days and the sample still contains about 20 % of unhydrated strontium aluminate phase in the 28th day of hydration (Fig.2).

The phase composition of strontium aluminate cement during the first 24 hours of hydration process is shown in Fig.3. These data confirm that tri-strontium aluminate hexahydrate is the only crystalline hydration product during early stages of hydration of strontium aluminate clinker and the initial rate of hydration is very fast as the results from calorimetric measurements show (Chapter 5.2). From this reason some important compounds improving the reactivity of strontium aluminate cement must be used in order to reach better workability (Chapter 5.4).

The infrared spectra (Fig.3) show sharp absorption band at 3664 cm^{-1}, which is related to the stretching of OH groups of Sr_3AH_6. The dehydration of this phase shows sharp endothermic peak at the temperature of 262 °C on TG-DTA and EGA (Fig.5). The test piece area decreases in about 6.0 % during this process (Fig.28 in Chapter 4). The dehydration of the gel phase shows an endothermic effect on DTA at the temperature of 198 °C, but the peak intensity and the temperature decrease with the time of hydration.

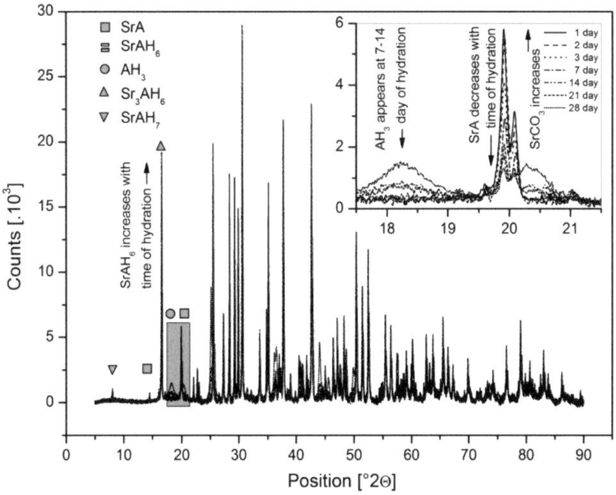

Figure 1. Hydration of strontium aluminate clinker upon the first 28th days hydration [379].

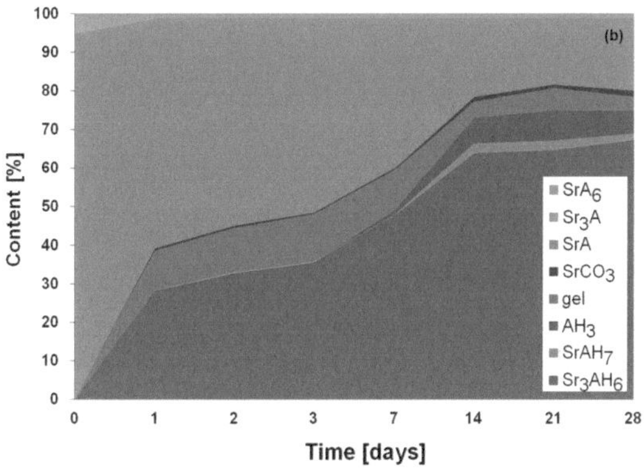

Figure 2. Course of hydration of strontium aluminate cement. The figure was constructed via the combination of XRD and TG-DTA results [379].

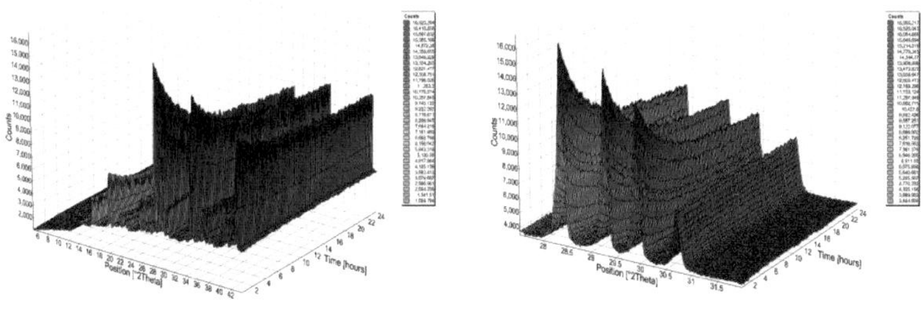

Figure 3. X-ray diffraction analysis upon early stage of hydration of strontium aluminate clinker.

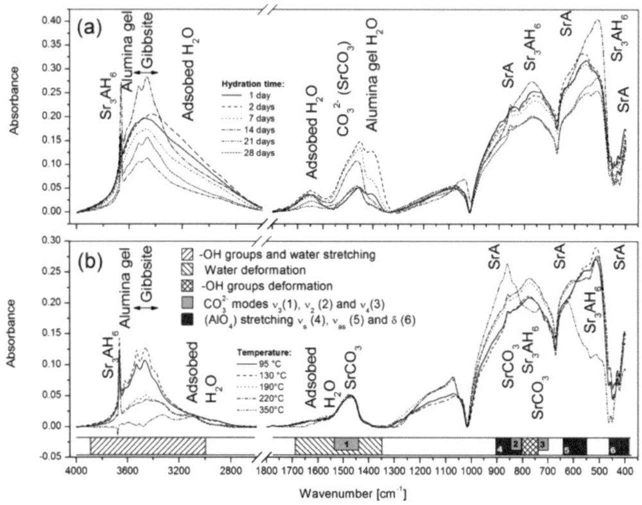

Figure 4. Infrared spectra of strontium aluminate clinker upon the first 28th days of hydration (a) and thermal treatment (b) [379].

The SEM pictures of strontium aluminate cement (**a**) and hydrated cement stone (**b-d**) during the first 28 days of hydration process are shown in Fig.6. Well-developed crystals of Sr_3AH_6 (**b**) precipitated and grew from supper-saturated liquid phase of cement slurry before setting. The hydration products usually formed during the setting do not occur in well-developed rhombic dodecahedron of Sr_3AH_6.

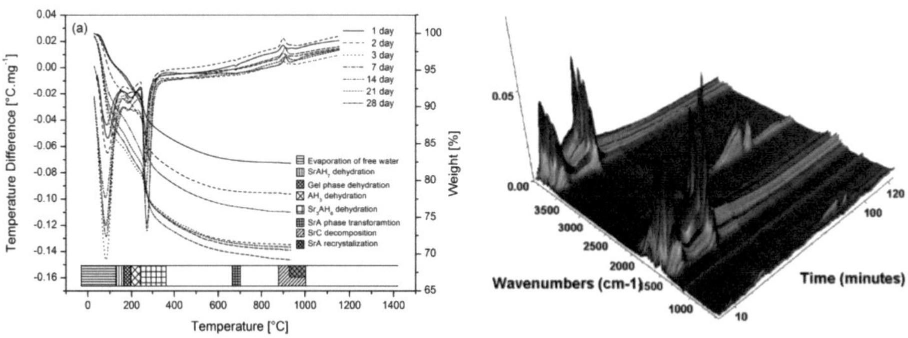

Figure 5. Thermal analysis of hydrated sample of strontium aluminate clinker performed at the heating rate of 10 °C min-1 [379].

The SEM (Fig.6) picture of hydrated cement stone after 21 and 28 days of hydration shows well-developed Sr_3AH_6 crystals surrounded by kidney-shaped aggregates of $SrAH_7$ [379].

Figure 6. SEM pictures of strontium aluminate clinker (a) and hydrated cement stone after 3 (b) and 21(c) and 28(d) days of hydration process [379].

2. Hydration and activation energy of hydration

Fig.7 shows the heat flow in the combination with the heat of hydration of strontium aluminate cement (a) and calcium aluminate cement (b, Secar 51) during 50 hours of hydration process at the temperature of 25 °C. Fast evolution of heat immediately after mixing with water is typical for strontium aluminate cement while 10 hours long induction period was measured for calcium aluminate cement. There isn't any induction period observed hence the hydration of strontium aluminate cement starts immediately after mixing of strontium aluminate cement with water and the heat flow then asymptotically decreases to the baseline. The main hydration reaction in calcium aluminate cement arises 13.5 hours after mixing with water.

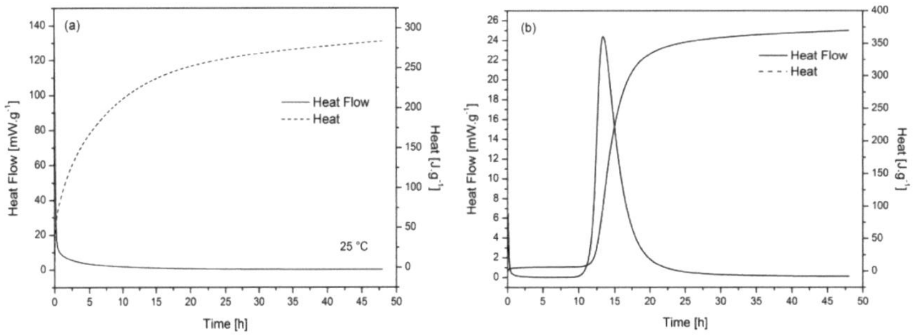

Figure 7. Heat flow and heat of hydration of strontium [379] (a) and calcium aluminate cement (b).

After the first 50 hours of hydration the heat flow of both cements already drops down to the baseline, but still produces 0.21 and 0.11 mW g^{-1} for strontium aluminate and calcium aluminate cement, respectively.

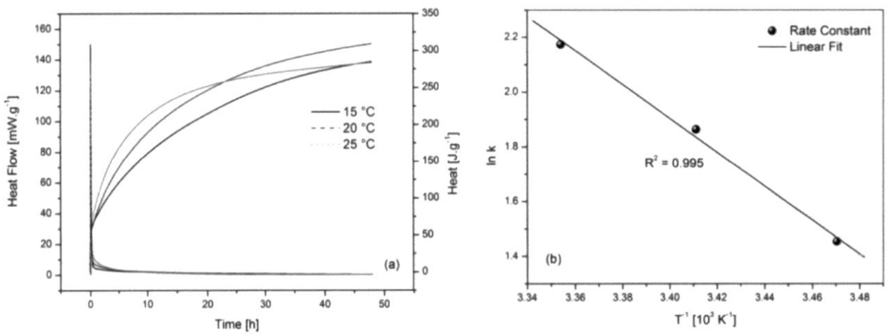

Figure 8. Arrhenius plot for the hydration of strontium aluminate cement.

The calorimetric assessments at the temperature of 15, 20 and 25 °C were applied in order to investigate the activation energy of hydration of strontium aluminate cement (Fig.8(a)). The apparent activation energy and the frequency factor of the process were calculated from the Arrhenius plot to be 51 kJ•mol^{-1} and 2.55•10^6 s^{-1} (b), respectively. The D_4 kinetic equation was evaluated as the most probable mechanism of the process of hydration.

3. Setting behaviour and influence of temperature

The measurement of storage modulus, loss modulus and damping factor (*tan* θ) were performed in order to evaluate the behaviour of strontium aluminate slurry prepared with the water to cement ratio of 0.3 at different temperatures. The results (Fig.9) indicate that the setting time of strontium aluminate clinker is significantly reduced by increasing temperature.

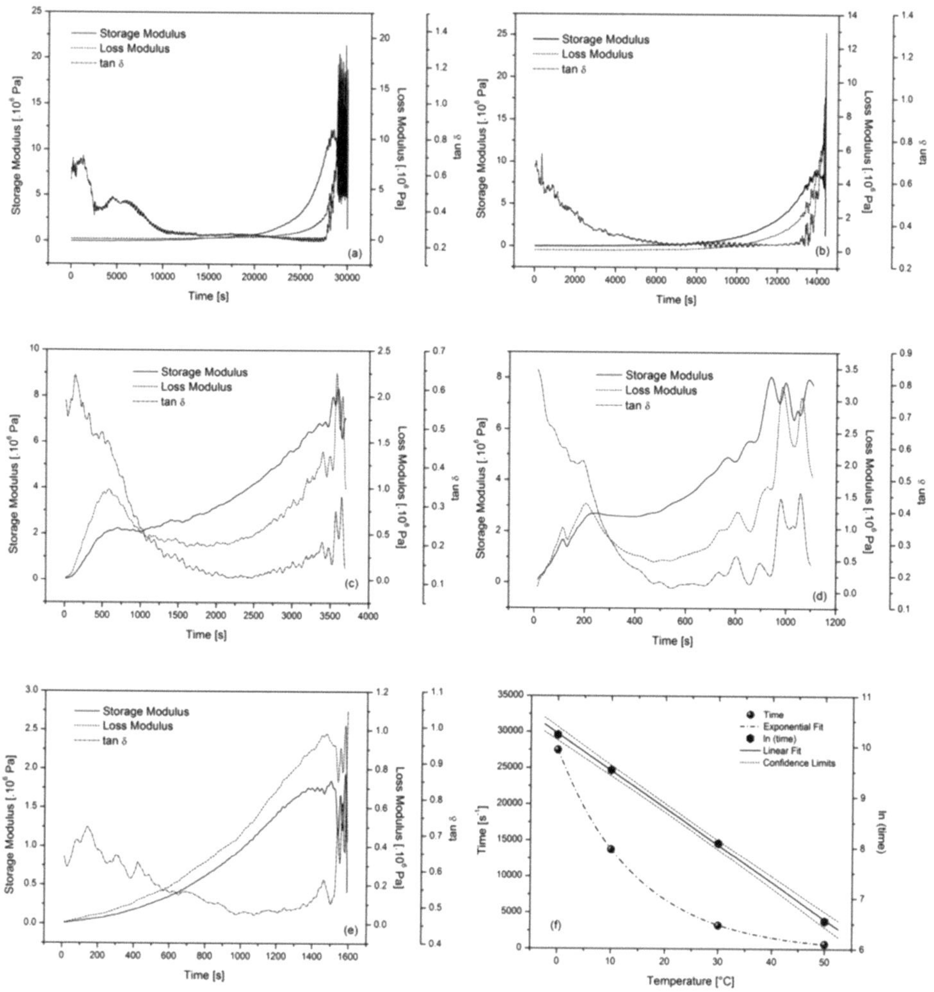

Figure 9. Tan delta for strontium aluminate clinker set at the temperature of 0 (a), 10 (b), 30 (c) and 50 °C (d) and influence of temperature on the setting time (f). The data for Secar 51 at 25 °C are shown in (e).

4. Improving mixture workability

Fast evolution of heat after mixing with water without the induction period is a disadvantage of strontium aluminate cement. In order to improve the cement workability the influence of some organic and inorganic compounds on the process of hydration was investigated by isothermal calorimetry. The formation of strontium chelates or buffer with organic acids, such as acetic and citric acid, or the formation of diffusion barrier on the surface of strontium aluminate grains via the introduction of sodium sulfate or boric acid are supposed.

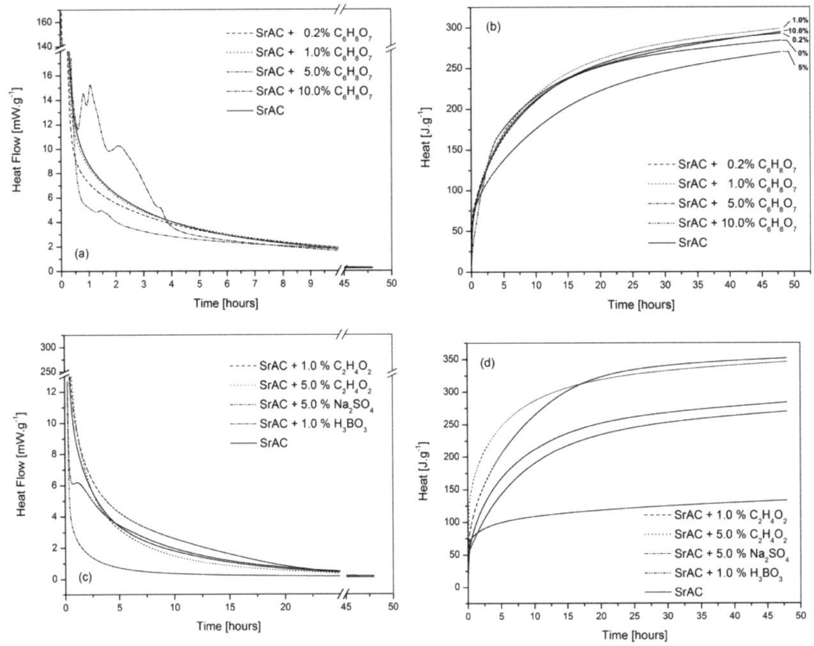

Figure 10. Calorimetry of strontium aluminate cement with acetic acid, citric acid, sodium sulfate and boric acid.

The effect of sulfates on the course of hydration of strontium aluminate cement is discussed in Chapter 8.2, but the addition of gypsum up to 5 % can improve the behaviour of cement during the hydration.

5. Chemical shrinkage

Chemical shrinkage[1] is defined as the absolute (internal) volume change of cement paste that results from the hydration of cementitious materials. Chemical shrinkage of strontium aluminate cement was evaluated using the pycnometric measuring technique described by Zhang at al. [674].

The shrinkage of cement slurry after mixing with water is shown in Fig.11. There are two linear parts, which can be extrapolated to the cross point at the time of 5.2 hours after mixing with water with corresponding volume shrinkage of 2.7 cm^3/100g.

This change in the shrinkage direction should be explained as the turn of the hydration (chemical) shrinkage to the setting shrinkage[2] (7.3 cm^3/100g). The cross point appearing at the time of 77.2 hours after mixing of SrAC with water is related to the formation of $SrAH_x$ hydrate (where x=7-10, please consult with Fig.2) [379]. The total volume shrinkage of cement slurry reaches about 10 cm^3/100g after 13 days of hydration.

1 The deeper discussion on the chemical shrinkage can be found in Chapter 8.

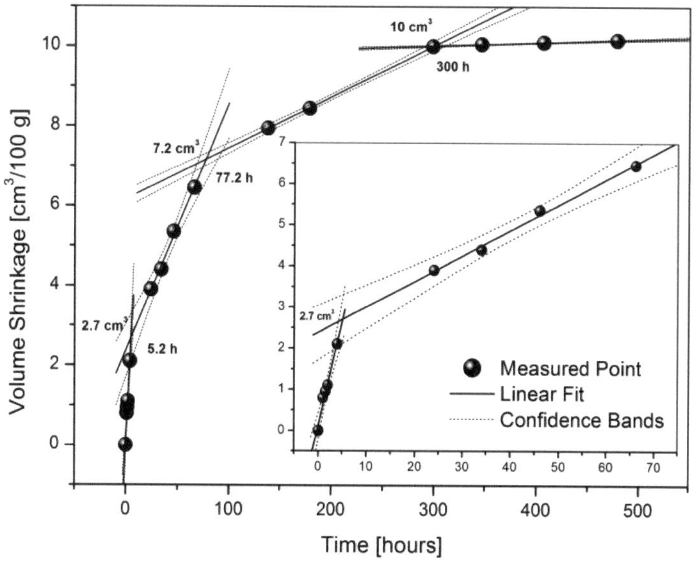

Figure 11. Early age volume shrinkage of strontium aluminate cement slurry.

6. Hydration of other phases in SrO-Al_2O_3 system

The tristrontium aluminate (Sr_3A) phase is more reactive than tricalcium aluminate (Ca_3A). There are two possible reasons for increased reactivity of tristrontium aluminate [8,416,417]:

1. Increased size of Sr^{2+}cations increases the size of channels within the Al_6O_8 rings allowing for more rapid diffusion of water.

2. Considerable overbonding at three Sr^{2+}sites occurs suggesting that there is considerable strain in $Sr_3Al_2O_6$.

The isothermal calorimetric experiments on the hydration of other strontium aluminate phases such as tri-strontium aluminate, strontium dialuminate and strontium hexaaluminate compared to strontium aluminate at the temperature of 25 °C are shown in Fig.12. The calorimetric data show that cements based on tri-strontium aluminate phase have short induction periods. The main and second hydration effect arise at the time of 4 and 48 h after mixing with water. The main hydration effect drops down to the baseline at the time of 10 h and the second at the time of 60 hours.

2 Hydration is defined as a reaction of inorganic binder with water accompanied by setting and hardening. Setting is spontaneous conversion of cement slurry into solid body. Hardening is a process of gradual increasing of strength of cement stone after setting.

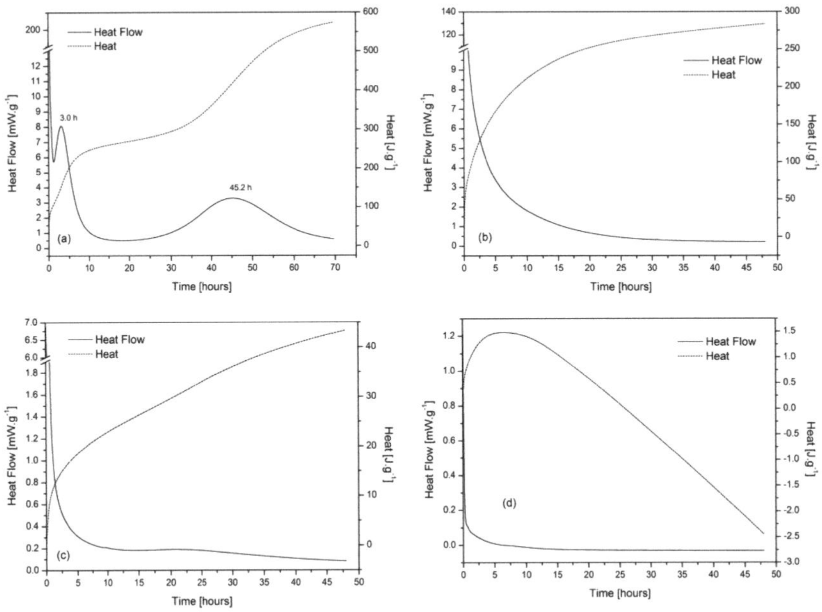

Figure 12. Heat flow and heat of hydration of tri-strontium aluminate (a), strontium aluminate (b), strontium dialuminate (c) and strontium hexaaluminate (d) at the temperature of 25 °C.

The heat of hydration decreases with decreasing strontium oxide/ aluminum oxide in the phase (Fig.13). According to the decrease of heat of hydration, the strontium aluminate phases should be ordered as follows:

$Sr_3A >> SrA >> SrA_2 > SrA_6$.

It seems to be the sigmoidal dependence of heat on hydration, where the inflexion point corresponds to the strontium aluminate phase.

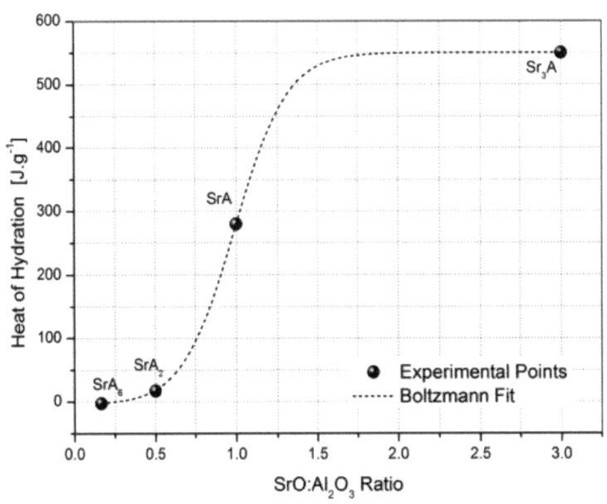

Figure 13. Hydration heat of strontium aluminate phases after the first 50 hours of hydration.

The heat flow of strontium aluminate phases is 0.58 (70 h), 0.21 (50 h) and 0.08 mW g^{-1} (50 h) for tri-strontium aluminate, strontium aluminate and strontium dialuminate phase after the first 50 hours of hydration, respectively.

Tri-strontium aluminate hexahydrate is the only crystalline hydration product that can be recognized by X-ray diffraction analysis (Fig14).

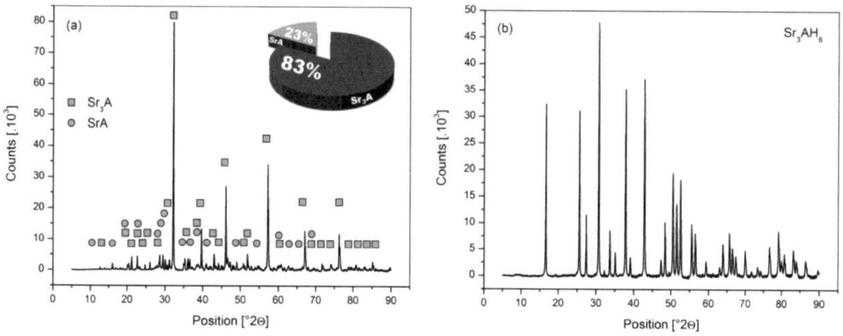

Figure 14. Ground clinker prepared from tri-strontium aluminate (a) and hydration products (b).

The results of thermal analysis of strontium aluminate clinker after the hydration are shown in Fig.15(a). The TG-DTA plot confirms that tri-strontium aluminate hexahydrate is the only hydration product of hydration of Sr_3A clinker. Please refer to Chapter 6.5 for the discussion of mechanism of thermal dehydration of tri-strontium aluminate hexahydrate.

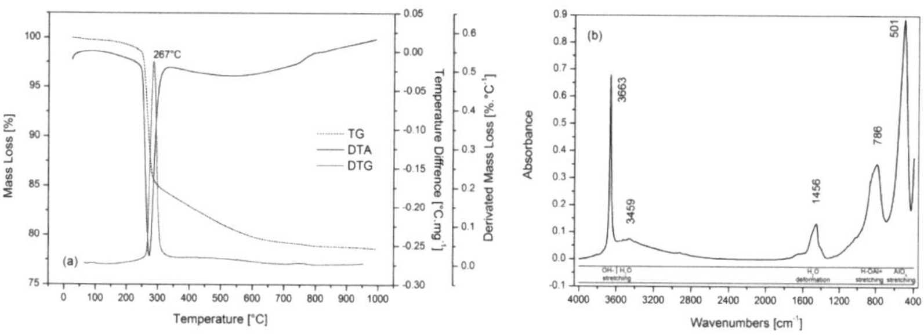

Figure 15. TG-DTA (a) and FT-IR (b) of hydrated cement stone.

The stretching of OH groups at 3663 cm^{-1} in Sr_3AH_6 is the most characteristic feature in infrared spectrum of hydrated cement stone (Fig.15(b)). The deformation mode is located at the wavenumber of 1757 cm^{-1}. The bands located at 786 and 502 cm^{-1} are most probably the composed bands of symmetric and antisymmetric stretching of =Al-OH groups [418,419,422] and the characteristic frequency of Al-O-Al stretching of AlO_6 octahedra [420,421,422], respectively.

SEM pictures of hydrated cement stone of ground tristrontium aluminate (Sr_3A) clinker are shown in Fig.16.

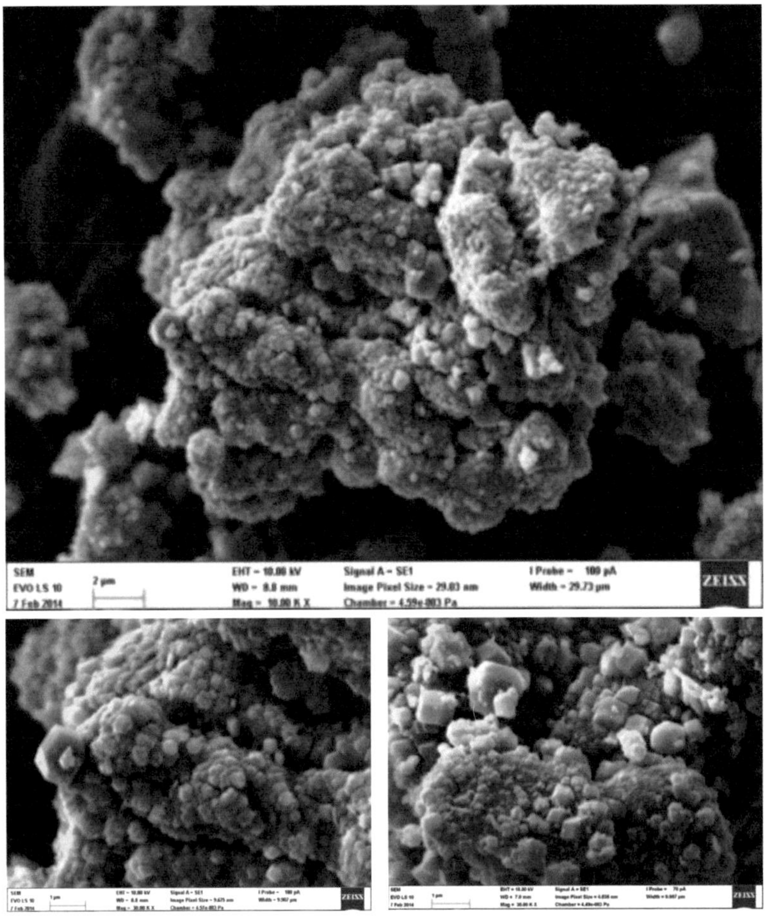

Figure 16. Tri-strontium aluminate hydrated cement stone after 72 h of hydration at 25 °C.

The hydrogarnet phase seems to be the only hydration product of Sr_3A (Fig.14), but the microphotographs show two main crystalline forms of the main product of hydration: icositetrahedron and hexagonal plates. Icositetrahedron are early forms precipitated from supersaturated solution, while hexagonal shapes are formed during later stage of hydration process.

Therefore the hydration process of tristrontium aluminate clinker should be described by the following chemical equation:

$$3SrO\times Al_2O_3 + 6\ H_2O \rightarrow 3SrO\times Al_2O_3\times 6H_2O \qquad (2)$$

which corresponds to the theoretic *w/c* ratio of about 0.21. This value is lower than theoretic water to cement ratio (Eq.1) calculated for strontium aluminate clinker (0.36).

Chapter 6

Applications of Strontium Aluminate Cements

The applications such as dense or thermal insulating refractory materials, the preparation of seawater resistance concretes and as the protection against X-ray and gamma ration are supposed for strontium aluminate cements [12,14]. The option to use strontium aluminate cement for the preparation of MDF composites as was described in Chapter 6.4 is also mentioned.

1. Refractory materials

Refractory products (**refractories**) possess the ability to resist high temperatures without physical or chemical destruction, the corresponding properties are termed as **refractoriness**. These materials can be classified according to their temperature resistance, chemical composition, porosity, etc. According to the most general viewpoint the ceramic and non-ceramic refractory materials and products can be recognized [423].

Based on the chemical composition, the following types of refractory materials can be specified:

- **Oxide ceramics:** are an extensive group of materials based on one or few oxides (SiO_2, Al_2O_3, Cr_2O_3, ZrO_2, MgO and CaO) and their stable compounds ($ZrO \bullet SiO_2$, $MgO \bullet Cr_2O_3$, $MgO \bullet Al_2O_3$, $3Al_2O_3 \bullet 2SiO_2$...). The silica refractory products, alumina-silica refractory products with different content of alumina, basic refractory products, alumina-zirconia-silica refractory products, alumina-chromic oxide-zirconia-silica refractory products, etc., are the most common examples of these materials. The alumina silicate refractories are extensively used in metallurgical, ceramic and glass industries. These materials are made primarily of refractory and kaolinitic clays, which are used both in crude and grog forms [424,425]. Spinel ($MgO \cdot Al_2O_3$) is very attractive as refractory material in heavy industries because of its high melting point, low thermal expansion, considerable hardness, high resistance to chemical attack, favorable chemical stability, and good thermal spalling [426].
- **Non-oxide ceramics:** carbides (SiC[1] [427-429], B_4C [429-431], Al_4C_3 [432],...), nitrides (Si_3N_4 [433-435], BN [436], TiN [437], AlN [438], SiAlONs [439],...) silicides ($MoSi_2$ [440,441]) can be considered as the most common types. Among the physico-chemical properties of material the kind of applied bound is an important factor for these materials.

1 The formation of SiC was firstly described by Berzelius in 1810 and 1821, but silicon carbide was latterly rediscovered by Despretz, Schützenberger and Moissan. The preparation of SiC on a large scale was managed by Acheson is 1891 who named the new substance as carborundum [429].

- **Other:** this type includes carbon or carbon containing refractory products such as Al_2O_3 – C [442,443], MgO-C [444], Al_2O_3-MgO-C [445,446], etc.

Another point of view for the classification of refractory products is the true porosity of refractory products:

- **Dense refractory products:** brick and castables of low porosity [447].
- **Insulating refractory products:** brick and castables of high porosity [447,448,498]. Insulating refractory castables are made of refractory light-weight aggregates, additives and HAC.
- **Ultra-light refractory products:** are refractory materials with true porosity higher than 75 %.

The morphology of refractory materials enables to recognize:

- **Shaped refractory products:** such as dense shaped refractory products, shaped insulating refractory products or fused cast refractory products.
- **Unshaped refractory products:** refractory castables for linings and reparations.

1.1. Regular refractory castable

The design of structures made of hydraulic–bound ceramics and subjected to high thermal stresses is of great importance today [449]. The development of refractory castables is important due to their increasing applications in metallurgical, cement and chemical industries. In last decades an increasing trend among refractories has arrived towards the replacement of bricks by castable refractories. Refractory castables can be classified on the basis of different aspects including the CaO content, the binder source, the chemical composition, the fluidity, the bulk density, etc. Their binding system plays a relevant role in different processing steps, including the workability, the dry-out and the service performance. Therefore, all efforts are made to improve the bonding agents [450-452,462].

The first reference on calcium aluminate bound refractory castables is provided by Deville [453], who prepared a refractory crucible using alumina aggregate and aluminate cement in 1856. Calcium aluminate cement and its industrial production was patented by Lafarge in 1908 [12,454] and the first industrial production was by Lafarge in 1913 and became known as "Ciment Fondu" [7,12,455]. Calcium aluminate cements are normally made by complete fusion of limestone or lime and bauxite or other aluminous materials with low content of SiO_2 at 1450 – 1600 °C. The molten clinker is tapped off continuously from the furnace, solidifies and its ground to the fines of about 3000 $cm^2 \bullet g^{-1}$. Their properties include rapid strength development, good resistance to sulfates and many other forms of chemical attack [7,12].

During the hydration of calcium aluminate as the main constituent of calcium aluminate cement, different hydrate phases can be formed. Up to 10 °C the hexagonal prisms of CAH_{10} (mono-calcium aluminate hydrate) are formed as the major product of hydration. The primary formed CAH_{10} converts to hydrogarnet (C_3AH_6), the only thermodynamically stable hydrate phase in the C-A-H system (Eq.1).

$$3\,CAH_{10} \rightarrow C_3AH_6 + 2\,AH_3 + 18\,H \quad (1)$$

$$3\,C_2AH_8 \rightarrow 2\,C_3AH_6 + H_3 + 9H \quad (2)$$

At temperatures above 10 °C, increasing amount of C_2AH_8 (di-calcium aluminate hydrate, strätlingite) is formed together with AH_3, while the content of formed CAH_{10} decreases. The conversion reactions (Eqs.1 and 2) are accelerated at high temperatures and moisture. The process is also associated with the strength loss and the formation of microcracks in the concrete structure. Above 27 °C the CAH_{10} hydrate is not formed at all and above 50 °C the C_3AH_6 is the only product of hydration. The hydration reactions of calcium aluminate can be summarized by the scheme in Fig.1. The hydration of ferrite phase is much slower than that of CA and $C_3(A,F)H_6$ is formed as the final product of hydration, with $C_2(A,F)H_x$ and $C_4(A,F)H_x$ as intermediates [7,12,415,456-460].

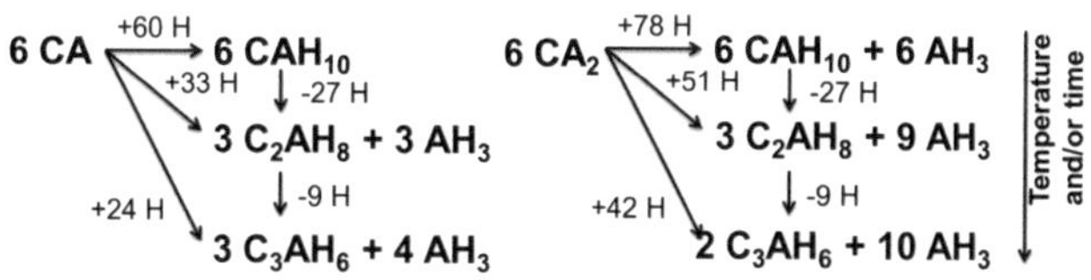

Figure 1. Hydration scheme of calcium aluminate cement [458].

Adding $CaCO_3$ to calcium aluminate cement induced the $C_3A \cdot CaCO_3 \cdot 11H$ hydrate formation (37 °C, samples immersed in water for 1 up to 30 days), inhibiting the formation/ conversion of CAH_{10} and C_2AH_8 metastable phases. Thus leading to better mechanical performance and dimensional stability. Moreover, although the mono-carbonate phase provides high mechanical strength to cement samples on the first day of curing, its decomposition combined with the C_3AH_6 and $Al(OH)_3$ generation due to continuous hydration of CA and CA_2 phases, seems to be one of the factors which lead to further improvement in compressive strength up to the 7th day [458].

CACs harden rapidly immediately after the massive precipitation of hydrates begins. Relatively high proportions of water are taken up in the hydration reactions, the theoretical w/c ratios needed for complete hydration of CA are 1.14, 0.63 and 0.46 for the formation of CAH_{10}, C_2AH_8+AH_3 and C_3AH_6+2AH_3, respectively [7,12]. The properties of the main hydrates are listed in Table 1. If the slurry of calcium aluminate cement is mixed with anhydrite (or gypsum), lime, bentonite or lithium carbonate (accelerator), the reaction is brisk and ettringite is the major product of hydration [599].

In monolithic refractories, calcium aluminate cement (CAC), one of the most widely used binders, promotes initial hardening and mechanical strength before firing. In the last few decades, the service life of alumina based refractory castables has been improved significantly by reducing the cement content. Carbon is not wetted by molten metal and does not melt,

Hydrate	**Chemical composition [%]**			**Temperature**	**Crystalline structure**	**Density**	**Decomposition [°C]**
CaO	Al_2O_3	H_2O	[°C]	[g·cm^{-3}]			
CAH_{10}	16.6	30.1	53.5	< 20	Hexagonal	1.72	120
C_2AH_8	31.3	28.4	40.3	20 - 30		1.95	170 – 195
C_3AH_6	44.4	27.0	28.6	> 30	Cubic	2.52	240 – 370
AH_3	---	65.4	34.6	> 30	Hexagonal	2.42	100

Table 1. Properties of hydrates formed in the strontium aluminate cement [460,461].

which is an excellent advantage for the refractory use. The reason for the cement content to be reduced is to prevent the formation of low melting temperature compounds in the presence of lime (CaO) [462-464].

Different types of binding systems have been developed (Table 2) throughout the years starting with hydraulic bonding, in which higher amounts of calcium aluminate cement (CAC) were used, towards coagulating binders such as colloidal silica or alumina [450,451]. In general, the refractory castables can be considered as composites, where the bonding phase (matrix) is reinforced by particles of aggregate (reinforcement) [465,480].

Nevertheless, there are some drawbacks related to the use of CAC in the systems containing microsilica and/ or magnesia, as the presence of CaO coupled with these other oxides results in the formation of low melting point phase at high temperatures. Other important concerns for cement-based castables arise during the curing and dewatering steps, which must be performed carefully in order to reduce the explosive spalling probability. In order to minimize these drawbacks, the properties of the refractory castables were improved by decreasing the cement content [111,450,466,470].

Based on the content of cement, the following types of refractory castables are defined by ASTM C401-91 (Standard Classification of Alumina and Alumina Silicate Castable Refractories):

- Normal Cement Refractory Castable (NCCs) or Conventional Cement Castables (CCs) where the content of CaO > 2.5 %.
- Low Cement Refractory Castable (LCCs) where 2.5 > CaO > 1.0 %.
- Ultra-low Cement Refractory Castable (ULCCs) where 1 > CaO > 0.2 %.
- No-Cement Refractory Castable (NLCCs) also Zero Cement Refractory Castables (ZCCs) or Cement Free Castables (ZFCCs).

Another way often applied for the classification of refractory castables is based on the density and the bonding system (see Fig.2) [455].

Bond type	1920s	1930s	1940s	1950s	1960s	1970s	1980s	1990s	2000s	2010s →
Hydraulic bond *Based on hydraulic setting and hardening*	Conventional cement bonded castables (Silicate cement, low purity CA cement)									
					Pure CA cement bonded castables (CA cement with improved purity)					
							High purity CA cement bonded castables (High purity CAC + uf.-Al_2O_3)			
							ρ-Al_2O_3 bonded castables			
								Hydratable Al_2O_3 bonded castables.		
Chemical				Phosphate bonded castables (H_3PO_4 or $Al(H_2PO_4)_3$ + MgO or CA)						
				Water glass bonded castables ($Na_2O \bullet nSiO_2 + Na_2SiF_6$)						
					Sulfate, chloride bonded castables ($Al_2(SO_4)_3$ + CA, $MgCl_2$ or $MgSO_4$)					
Polymerization					Polyphosphate bonded castables ($Na_5P_3O_{10}$ or $(NaPO_3)_6$ + MgO, CaO or CA)					
							Resin bonded castables (Phenol resin, Novalac resin + cross linker)			
Hydraulic + coagulating						Low cement castables (CAC + Clays, CAC + uf.-SiO_2)				
Coagulating						Clay bonded castables (CA + Ca-, Na-clay)				
							Ultralow cement castables (uf.-SiO_2, uf.-Al_2O_3 + CA)			
							Non-cement castables (uf.-SiO_2, uf.-Al_2O_3 + electrolyte, uf.-SiO_2 + MgO)			
							Sol bonded castables (silica sol, alumina sol + electrolytes)			
Carbon bonded								MgO – Carbon Castables		
Nano-engineered										In progress

Table 2. Progress in binding system for refractory castables [451].

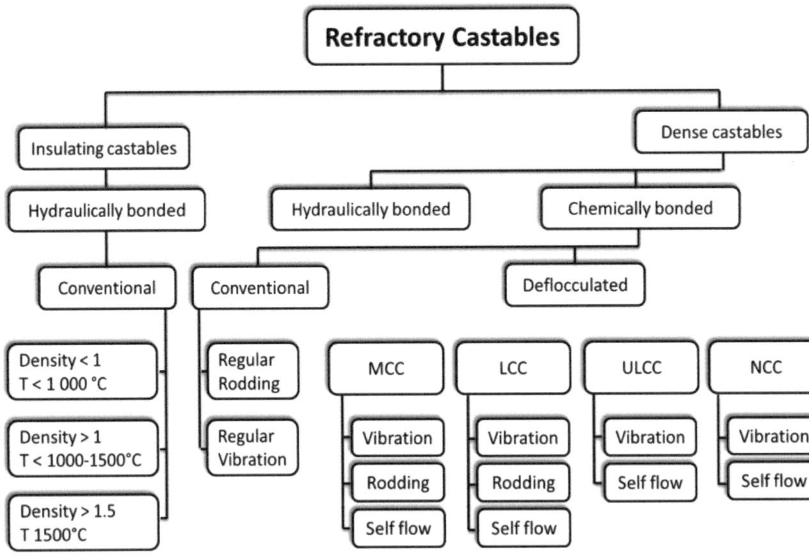

Figure 2. Types of Refractory Castables [455].

Calcium aluminate cement acts as a suspension lubricant and binder in fresh mix and after hardening in room temperature it provides the strength. Cement castables (CCs) are usually porous and open textured, due to relatively large amounts of water required for the hydration of AC. They also exhibit a characteristic drop in strength at intermediate temperatures as sluggish sintering does not allow the development of ceramic bond after breaking down the hydraulic bond. Furthermore, high content of lime in these castables favors the formation of low melting phases such as anorthite and gehlenite at elevated temperatures. These phases are known to degrade the refractoriness and corrosion resistance of conventional CCs [111,466].

Especially low-(LCC) and ultralow-cement castables (ULCC) are widely used in steel industry due to their superior rheological and physical properties. Initially, castable refractories were composed of only cement and aggregates. Then, the addition of deflocculants and fine fillers has followed with the aim to optimize and control the properties such as workability as a function of time [463]. Such castables have valuable properties, including low thermal expansion coefficient, good thermal conductivity, good thermal shock resistance, high resistance to slag and liquid metal corrosion and high strength at low and high temperatures. ULCC refractories have strong slag penetration resistance [462].

Different calcium-free binding systems such as hydratable alumina (HA), colloidal silica (CS) and colloidal alumina (CA) were developed for the refractory castables. Hydratable aluminas (HA) are amorphous mesophase transition aluminas (TA)[2] which are produced by flash

2 Transition aluminas are described in Chapter 4.1.

calcination of gibbsite in the temperature interval from 400 to 800 °C[3], which leads to high specific surface area transition phase of ρ-alumina (rho-alumina) [486,450,467-470].

The binding ability of ρ-alumina derives from its particular characteristic while undergoing the rehydration when in contact with water (or water vapor). During the hydration, a thick layer of gel is formed, a part of the gel subsequently crystallizes into traces of boehmite and major quantities of bayerite. The remaining gel phases were identified as boehmite gel or pseudo-boehmite (a poorly crystallized boehmite), together with totally amorphous gel, and may represent up to 60% of final hydrated phases, depending on the hydration temperature and pH. Interlocking bayerite crystals and gel confer the mechanical strength to green body of the refractory structure by filling pores and interfacial defects and by forming the honeycomb structures on the surface of aggregates attaching adjacent grains to each other and to the surrounding matrix [450,470,471].

The process of TA (transition aluminas) rehydration in Hydratable Alumina-bonded Refractory Castables (HAB)[4] can be expressed by the following reaction scheme [471,472]:

$$\begin{aligned} &TA + \mathrm{H_2O} \rightarrow \text{amorphous gel} \rightarrow \text{pseudoboehmite} \\ &\rightarrow \text{boehmite} \rightarrow \text{bayerite} \end{aligned} \tag{3}$$

The hydration process can be divided into four stages [458]:

1. The intensive period of preliminary hydration.
2. The induction period.
3. The acceleratory period, in which the maximum speed of evolving heat is attained. The crystallization of amorphous gel phase and the formation of bayerite as well as the changes in the morphology of grains proceed in this period.
4. The final period, where the amount of released heat decreases asymptotically.

Although hydratable aluminas outweigh the low-melting point phase generation associated with microsilica-containing CAC-bonded castables, there are still some drawbacks attributed to this binder. Compared to equivalent CAC formulations, HA-bonded castables require longer mixing time and high water contents due to high specific surface area of the binder. Moreover, after hardening, HA-containing castables form much less permeable structure in comparison with CAC bonded castables, which leads to higher explosive spalling probability during the water dry-out [450].

A great concern has been employed during recent years to refractory castables bound by nanopowders and colloidal suspensions in order to improve the refractory's bonding and densification and the behaviour during sintering. Therefore the addition of nano-particles could improve the castable properties, as long as the nano-agglomeration problem is kept under

3 The process is commonly used for the production of catalysts, carriers and absorbents.

4 Usually prepared by flash calcination of gibbsite.

control. High price of commercially available nano-powder products can also be a hindrance. Thus, nano-particles containing aqueous suspensions (colloidal binders) are preferred to nano-powders [450].

Among colloidal suspensions, the silica sol is of greatest interest for refractory castable applications. One of the reasons for its use is the possibility of mullite formation at low temperatures for alumina-rich systems. Actually, the reaction activity of nanoSiO_2 particles in silica sol is extremely high. Colloidal particles can be absorbed on the surface of active α-Al_2O_3 ones and fill in their packing gaps, which can reduce the temperature of mullite formation, with suitable densification at about 1100 °C. Another reason for applying colloidal silica to refractory castables is related to its high solid contents, ranging from 15 to 50 wt. % of silica [450].

The fact of increasing the utilization of colloidal silica as a binder has initiated a discussion concerning the possibility of adding colloidal alumina for the same purpose. Despite the colloidal silica advantages, its addition always implies that SiO_2 will be present in the final product, which would inhibit its use in various steel plant applications. This drawback was the driving force towards using colloidal alumina. At first, the addition of colloidal alumina was restricted to the contents lower than 1wt %, most likely due to processing difficulties such as high water demand and low flowability and workability. However, after overcoming these limitations, recently higher colloidal alumina solid contents (up to 4 wt. %) could be added to the refractory castables. For lower solid contents in the composition, the role of colloidal alumina as a binder may be insufficient whereas if it exceeds 4 wt. %, higher water amounts are required for mixing and the mechanical properties may be spoiled [450].

In addition to colloidal silica and alumina, there is also a report related to the utilization of mullite and spinel sols (Fig.3(a)) as bonding agents in high alumina ULCC castables [450,473,474]. These sols with low solid content (about 5 wt. %) in water suspension were prepared by the addition of silica fume and $MgCl_2$ $6H_2O$ into vigorously stirred boehmite sol (Fig.3(b) and (c)) and used as an additive for CAC containing castables in order to evaluate their effect on physical and mechanical properties, as well as on slag corrosion and thermal shock resistance. The microsilica content in a spinel bound high alumina castable must be reduced to a very low level to eliminate the detrimental glassy phases or the formation of SiO_2-Al_2O_3 melt at the temperatures close to 1250 °C [475,478].

The spinel additive from the sol–gel route contained more fine particles and showed better performance in castables than co-precipitated spinel additive, although both of them are energetically favorable to produce desired spinel phases at lower temperature but with significant volume expansion. The magnesium aluminate spinel (MA, $MgAl_2O_4$) is a very desirable phase in castables which improves the hot strength, the creep resistance, the thermal stability and restricts slag penetration of spinel containing castables [474,476,477,480].

Alumina-magnesia castables are widely known for their expansive behaviour, due to in situ formation of $MgAl_2O_4$ at temperatures higher than 1000 °C. CAC-containing alumina-magnesia castables performed better regarding the mechanical strength and thermal shock resistance. Nevertheless, they attained higher in situ expansion, due to the CA_6 formation

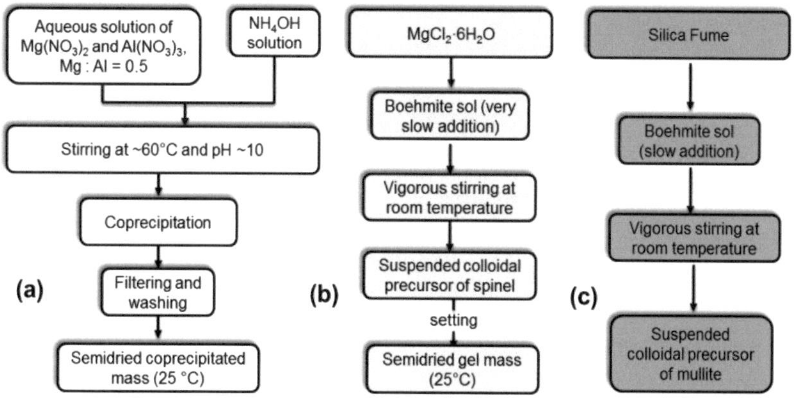

Figure 3. Preparation of co-precipitated spinel (a), sol-gel derived spinel (b) and mullite (c) [473,474].

during the thermal treatment. An efficient way to control the volumetric stability can be achieved by the addition of hydratable alumina, as its shrinkage behaviour counterbalances the in situ spinel expansion. Furthermore, this binder leads to higher creep resistance and also low apparent porosity levels, which can be useful to reduce the slag infiltration [445,478,479].

Self-flow refractory castables (SRFC) are characterized by their consistency after mixing, which allows them to flow and de-air without the application of external energy, i.e. vibration. Added water promotes the flowability of the particulate system but it has to be kept at minimum dosage (typically between 3 and 8 wt.% in conventional castables), to avoid coarse particles segregation and the increase in final porosity. In order to compose the Self-Flow Refractory Castable, the deflocculants and the particle size distribution of aggregates have to be selected carefully since the particle size distribution is one of the most important factors that affect the refractory's rheological properties [480-482].

The reliability of linings and pieces of refractories used in vessels (ovens, converters and ladles) must also be ensured because these structures are subjected to high temperatures under normal conditions. Previous studies were undertaken to model the thermo-mechanical behaviour of concrete and refractories. For these materials the temperature range of 20-800°C is of particular interest because, due to the dehydration, the cement undergoes significant microstructural transformations during the first thermal load [111,449].

These transformations considerably modify the thermo-mechanical behaviour of the material and must be taken into account for a realistic structural analysis. The chemical reactions may be described by the following equations [449]:

$$AH_3 \rightarrow AH + 2\,H \tag{4}$$

$$AH \rightarrow A + H \tag{5}$$

$$C_3AH_6 \rightarrow C_3AH_{1.5} + 4.5\,H \tag{6}$$

$$C_3AH_{1.5} \rightarrow \left(C_{12}A_7, CA, CA_2, CA_7, A\right) + 1.5\,H \tag{7}$$

The hydrogarnet phase C_3AH_6 is decomposed into two phases $C_{12}A_7H$ and CH, the chemical condensed formula of which is $C_3AH_{1.5}$. Therefore it should be written [457,483]:

$$7\,C_3AH_6 \rightarrow C_{12}A_7H + 9\,CH + 32\,H \tag{8}$$

Growing demand for refractory castables with designed properties has brought about continuous technological development involving the simultaneous understanding of particle size distribution, the use of additives improving workability as well as sintering additives and the binder performance [484].

1.2. Particle packing density

Particle packing density (PSD) was evaluated by the Andreasen model [463,480,485,486]:

$$CPFT = 100 \times \left(\frac{D}{D_L}\right)^n \quad [\%] \tag{9}$$

where ***CPFT*** denotes the cumulative percent of grains finer than ***D***, D_L is the size of maximum used grain, ***D*** means the grain size and ***n*** is the particle size distribution parameter.

Another packing model is proposed by Dinger and Funk [485,487]:

$$CPFT = 100 \times \left(\frac{D^q - D_S^q}{D_L^q - D_S^q}\right) \quad [\%] \tag{10}$$

where D_S and D_L are the smallest and largest particle sizes, respectively. The parameter ***q*** is the coefficient, which depends on the particle size distribution.

Despite the fact that water is used in rather small amount (usually 3 – 8 % in convectional refractory castables), it still remains the key ingredient that enables to keep the desired rheological characteristics of fresh castable (Fig.4). Since water fills the voids between solid particles first, improving the packing density is an obvious way to reduce the water requirement. The workability can be improved by external vibration but when it is not possible or advisable, self-flow refractory castables (SFRC) are used. SFRC "works" as a suspension of powders in which the fine matrix is the flow medium that envelops the aggregate particles, fills in the voids between them and suspends them, promoting higher flowability [486-489].

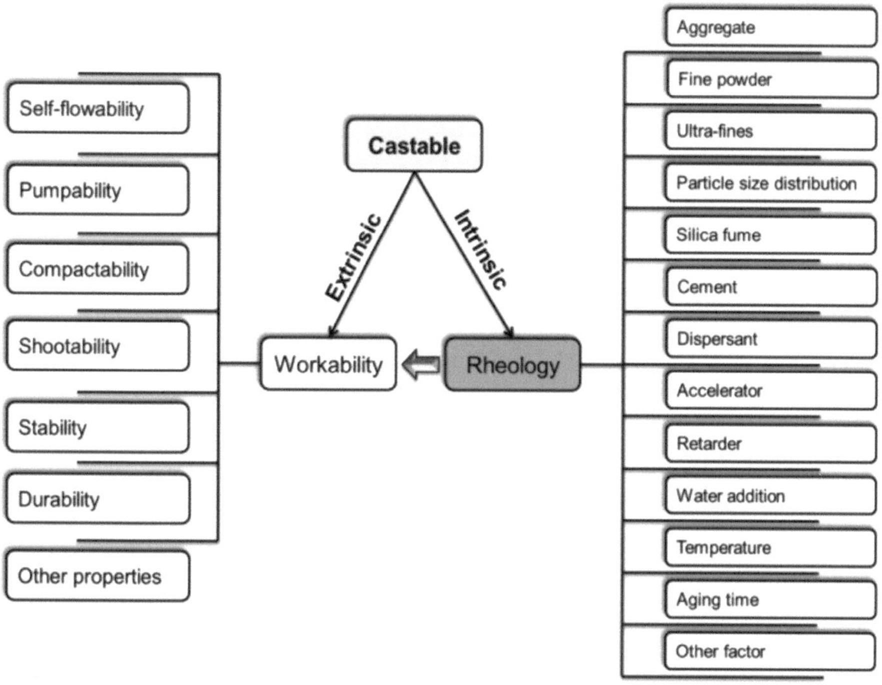

Figure 4. Factors influencing the rheology of castables [489].

1.3. Opening material

Opening material (aggregate) and filler usually improve the drying and firing shrinkage, the material mechanical properties, the thermal stability, the corrosion resistance and the thermal conductivity of refractory materials. Dense refractory castables from alumina-silica system often use corundum or calcined bauxite [110,131,463], mullite [484], refractory clays [423,490] or sand [491]. Moreover, the aggregates such as SiC [464,492,493] or zircon [494,495] are used.

1.4. Sintering additive

The use of boron-based sintering additive to speed up the refractory densification at lower temperatures can be a suitable alternative to develop novel compositions for fluid catalytic convertors (FCC) of petrochemical units [484]. Cr_2O_3 nanoparitcles are effective sintering additive for MgO based refractories due to the formation of $MgCr_2O_4$ spinel. The layer of spinel precipitates at the grain boundaries of MgO grains [496]. The addition of SiO_2, $CaCO_3$, TiO_2 and Y_2O_3 also improves the densification of refractories based on $MgCr_2O_4$ spinel [497].

1.5. SrAC bound refractory castable

Refractory castables based on strontium aluminate cement can be prepared by mixing strontium aluminate cement, grog (Chapter 6.1.3) and additives (Chapter 5.4). The increase of Al_2O_3 content in the castable by the addition of corundum or calcined bauxite (Fig.3 in Chapter 2) shifts the equilibrium composition towards the high alumina phases (Fig.1 in Chapter 1.2) such as calcium dialuminate and calcium hexaaluminate. Build line or casted panel is less sensitive to humidity, because newly formed strontium aluminate phases are much less reactive after mixing with water (Fig.12 in Chapter 5) than original strontium aluminate cement.

Optical and electron microscopy of prepared refractory castable based on strontium aluminate cement is shown in Fig.5 and Fig.6, respectively.

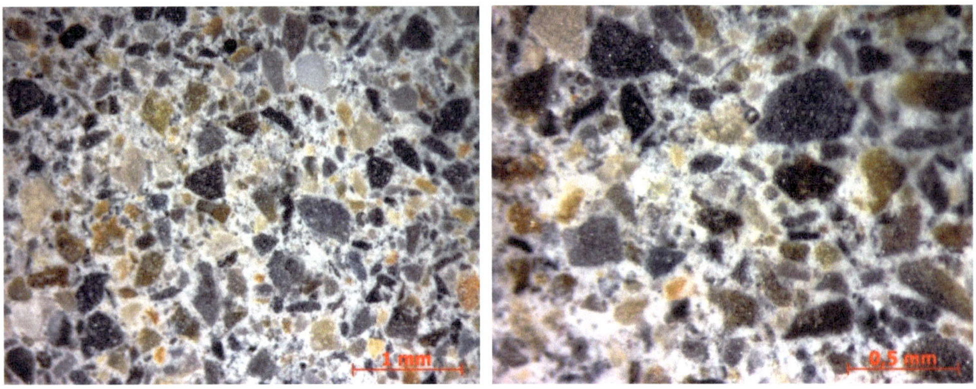

Figure 5. Optical microscopy of texture of SrAC bound refractory castable with bauxite aggregate.

The sample fired to the temperature of 1300 °C shows zero total shrinkage[5], therefore the castable has sufficient temperature stability that is necessary in order to avoid cracks formed during drying and further thermal treatment.

The texture of fired sample is shown in Fig.7. The texture of the material shows hexagonal plates, which are the pseudomorphosis of original hydrates. These hydrates are already decomposed at temperatures lower than 600 °C (please see Fig.5 in Chapter 5). Moreover, the excess of Al_2O_3 from applied aggregate (calcined bauxite) changes the equilibrium composition as was discussed above. Some sites show the initial stage of sintering process, i.e. the formation of necks between neighboring particles, and newly formed phases nucleating from the melted zone.

5 Total shrinkage consists of the change of specimen dimensions during setting, drying and thermal treatment.

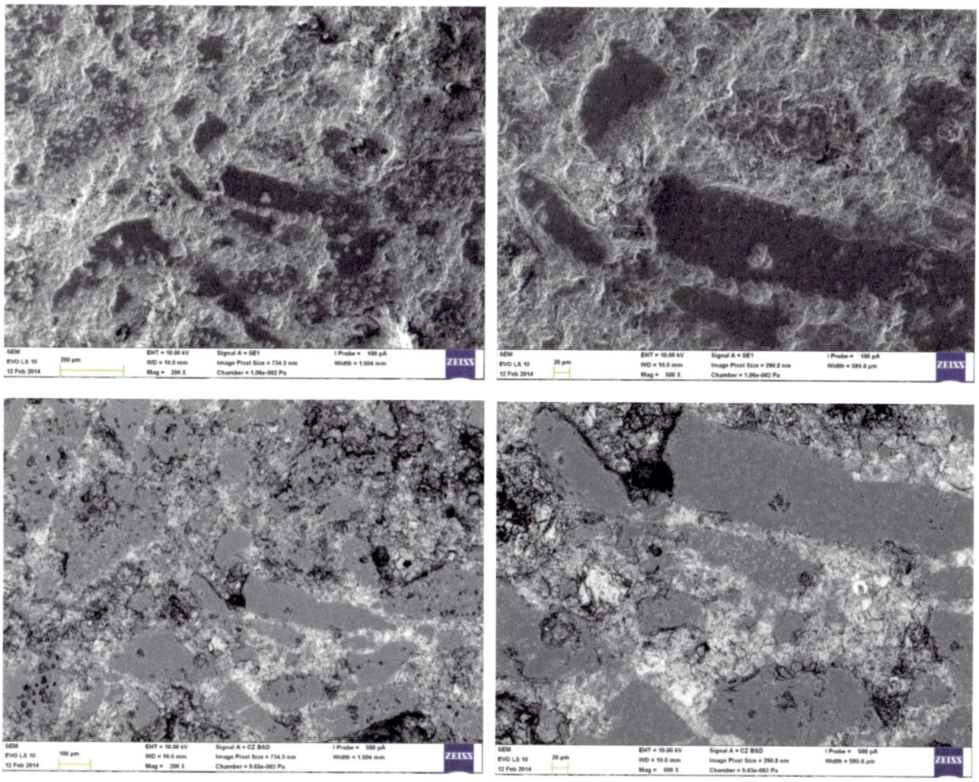

Figure 6. Electron microscopy of refractory castable before the thermal treatment.

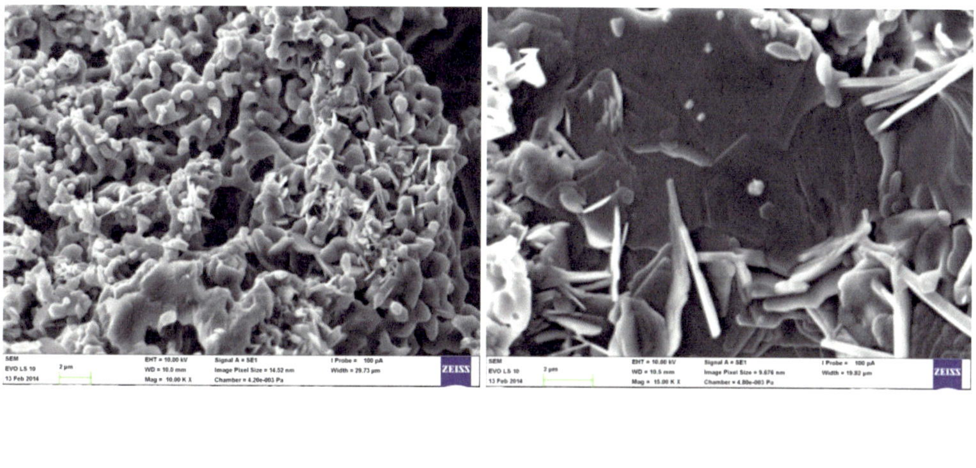

Figure 7. Electron microscopy image of refractory castable treated to the temperature of 1300°C.

The recrystallization or better sinter-crystallization takes place in the material treated to the temperature of 1400 °C (Fig.8). The process is connected with increasing dimension[6] of specimen, which is also visible in Fig.28 in Chapter 4. The first marks of crystallization of hexagonal plates from melted phase can be found in the SEM picture of the sample heated to the temperature of 1300 °C (Fig.7), but at the temperature of 1400 °C the sample crystals do not show glassy phase. Therefore, the crystallization from non-equilibrium melt takes place.

XRD analysis of sintered sample (Fig.8) confirms the formation of non-hydraulic compound ($SrO \cdot 6Al_2O_3$) due to the reaction with bauxite. The traces of quartz in bauxite cause the formation of feldspar strontian ($SrO \cdot Al_2O_3 \cdot 2SiO_2$, described in Chapter 10.4).

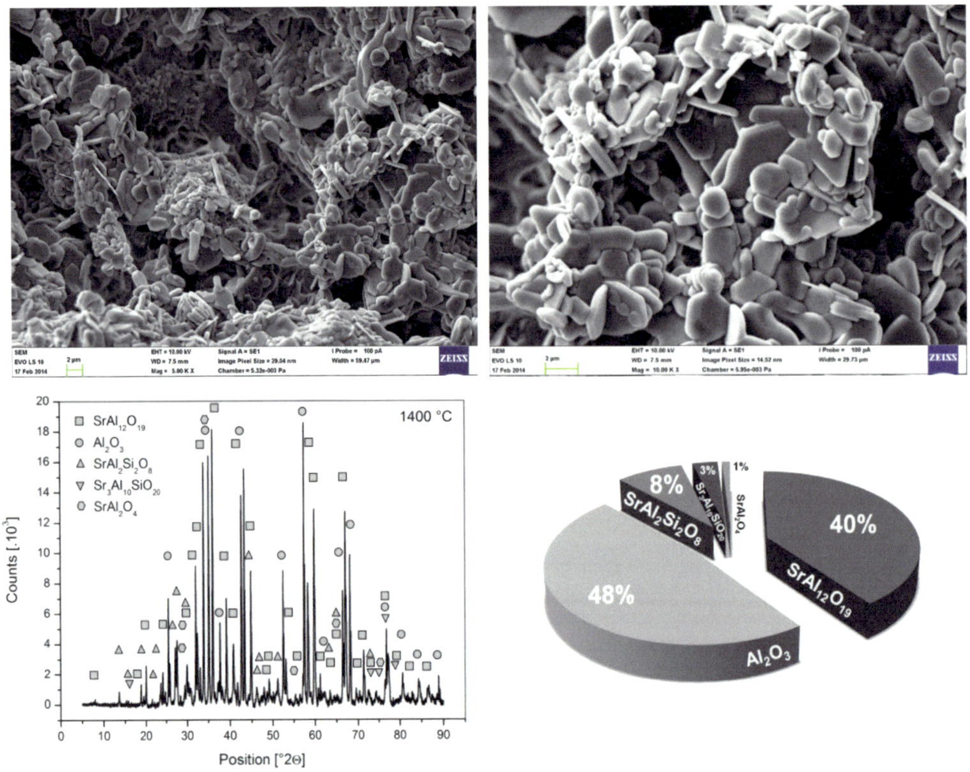

Figure 8. Electron microscopy image of refractory castable treated to the temperature of 1400°C.

Specimen heated to the temperature of 1500 °C (Fig.9) shows the volume density of 2.28 $g \cdot cm^{-3}$ and the cold crushing strength of 21.6 MPa. Measured expansion related to the thermal treatment to the temperature of 1500 °C is about 2.2 %. As the temperature increases, the small crystals are consumed during the large crystals growth and are dissolved in the melt phase.

6 This expansion should be avoided by the addition of silica bearing component into the refractory castable as was discussed in Chapter 6.2.3.

Figure 9. Electron microscopy image of refractory castable treated to the temperature of 1500°C.

After the treatment to the temperature of 1600 °C, the ceramic body of prepared refractory material consists of corundum (59 %), strontium hexaaluminate (29 %) and strontianite (12 %). Electron microscopy (Fig.10) shows that the specimen consists of large tabular pseudo-hexagonal crystals of SrA_6 surrounded by solidified glass. The sintering process reduces the dimension and porosity of the sample.

2. Insulating refractoriness

Insulating firebricks (IFBs) represent one of the refractory groups which are most commonly used for the heat insulation in industrial applications today. They are highly porous and lightweight refractories having much lower thermal conductivity and heat capacity than other

Figure 10. Electron microscopy image of refractory castable treated to the temperature of 1600°C.

refractories. Different types of insulating firebricks are manufactured mainly by using the raw materials such as diatomite, perlite, expanded vermiculite, calcium silicate, fireclay, kaolin, quartz, alumina and lightweight refractory aggregates by conventional methods. The porosity is usually created by the addition of combustible material to the raw material mixture. During firing, the combustible material burns out, and leaves large fraction of pores within the fired body [498].

Different types of pore-formers such as sawdust, foam polystyrene, fine coke, binders and organic foams, or granular materials such as hollow microspheres and bubble alumina are commonly used to obtain decreased density or to produce porous bodies in the insulating material. A considerable research is carried out recently on the reuse of pulp and paper industry wastes as a new source of raw material or pore-former in the ceramic materials due to their organic and inorganic contents [498-501].

Insulating firebricks which have highly porous structure (between 45% and 90% porosity) exhibit low thermal conductivity values. The thermal conductivity not only depends on their total porosity, but also on their pore size and shape, chemical and mineralogical composition [498]. Insulating firebricks are classified by ASTM according to the bulk density and maximum service temperature [502]. The methods for measuring the weight, dimensional measurements and bulk density of the produced bricks are defined by ASTM C134-95 [503]. Apparent porosity measurements of samples were performed by immersion into boiling water for 2 h and soaking in water for 24 h [504]. The techniques for the assessment of cold crushing strength and thermal shock resistance are specified by ASTM C133-97 [505] and ASTM C1171-96 [506], respectively.

2.1. Fillers for thermal insulating refractory castables

Figure 11. Examples of lightweight filler: cenospheres, bubble alumina and fireclay.

Lightweight fillers make the structure of refractory castables lighter and provide required thermal insulating properties via the formation of pores in structure. Some examples of these materials are shown in Fig.11.

2.2. Properties of refractory materials

Currently, we observe a noticeable change of climate, worldwide, probably associated with the global warming. In order to reduce the emission of greenhouse gases, especially of carbon dioxide, the generated energy should be reasonably managed. A very large amount of total energy produced in the world is heat. It is used in various industrial processes and buildings. The solution in saving the heat is a suitable construction of walls to ensure thermal insulation [507,508].

Most applications of ceramic materials are advantageously used as thermal-insulating or thermal-conducting materials. From the viewpoint of industrial applications, low thermal conductivity is needed for the thermal insulation, while high thermal conductivity is important for achieving the high heat release, high thermal shock resistance and temperature homogeneity throughout the material [509].

The quality control of thermal insulating materials typically focuses on the determination of thermal conductivity value (λ). For this purpose, special plate apparatuses are used, which are based on [510,511]:

a. Stationary heat transfer conditions;

b. Dynamic heat transfer conditions.

The heat flux vector ($\bar{q}$) is given by the Fourier´s law [509,512]:

$$\bar{q} = -\lambda \ grad \ T \quad or \quad \bar{q} = -\lambda \nabla T \tag{11}$$

$$\nabla = \left(\frac{\partial}{\partial x}\bar{i} + \frac{\partial}{\partial y}\bar{j} + \frac{\partial}{\partial x}\bar{k} \right) \tag{12}$$

where λ is the coefficient of thermal conductivity, ***T*** is the temperature. The vectors of the basis are denoted as *i*, *j* and *k*.

Effective thermal conductivity (ETC) is one of the key thermophysical properties used to quantify the thermal behaviour of heterogeneous materials like granular metals, ceramics, ceramic raw materials, composites, inorganic or polymeric foams, catalysts, etc. Much of the effective thermal conductivity literature is concerned with porous materials; however, the term porous itself may be the cause of a confusion. In some situations it refers to granular or particulate materials, in which the void volume may be occupied by either liquid or gaseous components; alternatively, it may refer to the material having continuous solid matrix that contains pores/bubbles, which may be isolated or interconnected. The problems may arise when a model that has been shown to work well for one type of porous material is assumed to be applicable to another type, simply because both materials are described as porous [512-514].

The Francl´s model [512,515] calculates the effective thermal conductivity of porous material (λ_e) from the thermal conductivity of solid material (λ_s) and its porosity as follows:

$$\lambda_e = \lambda_s(1-\varepsilon) \tag{13}$$

The law was shown to be effective for a certain porous material. However, if it were applied to a granular material such as sand, the thermal conductivity would be predicted with error of several orders of magnitude.

The influence of porosity (ε) and pore size (d) on the effective thermal conductivity is described as follows [516]:

$$\lambda_e = \lambda_s(1-\varepsilon) + \varepsilon \, \lambda_a + 4d\sigma T^3 \tag{14}$$

where the thermal conductivity of gas phase (Λ_a) and the radiation transport of heat at the temperature T are taken into account. σ denotes the Boltzmann constant.

The effective thermal conductivity of heterogeneous material is strongly affected by its composition and structure. Foams and porous materials can be considered as a two-phase (or two-component) systems consisting of solid skeleton and air, and the thermal conductivity can be described as the heat transfer through such complex system according to some of suggested analytical models [508,517-519].

Five structural models (Figs.12 and 13) are taken into account which can be expressed for two component system as follows [508,513,514,520-522,527]:

i. Series model:

$$\lambda_e = \frac{1}{(1-v_2)/\lambda_1 + v_2/\lambda_2} \tag{15}$$

ii. Parallel model:

$$\lambda_e = \lambda_1 (1 - v_2) + \lambda_2 v_2 \tag{16}$$

iii. ME-1 (Maxwell-Eucken model with one continuous component) model:

$$\lambda_e = \lambda_1 \frac{2\lambda_1 + \lambda_2 - 2(\lambda_1 - \lambda_2) v_2}{2\lambda_1 + \lambda_2 + (\lambda_1 - \lambda_2) v_2} \tag{17}$$

iv. ME-2 (Maxwell-Eucken model with two continuous components) model:

$$\lambda_e = \lambda_2 \frac{2\lambda_2 + \lambda_1 - 2(\lambda_2 - \lambda_1)(1 - v_2)}{2\lambda_2 + \lambda_1 + (\lambda_2 - \lambda_1)(1 - v_2)} \tag{18}$$

v. EMT (Effective Medium Theory) model, which represents isotropic materials with two phases dispersed in continuous medium:

$$(1 - v_2)\frac{\lambda_1 - \lambda_e}{\lambda_1 + 2\lambda_e} + v_2 \frac{\lambda_2 - \lambda_e}{\lambda_2 + 2\lambda_e} = 0 \tag{19}$$

The parameters λ and v are the thermal conductivity and the volume fraction. The subscripts e, 1 and 2 represent two component material and its first and second constituents, respectively.

The plot of relative thermal conductivity (Λ_e/Λ_1) for two-component material with Λ_1/Λ_2=20 calculated according to the five structural models mentioned above is shown in Fig.12.

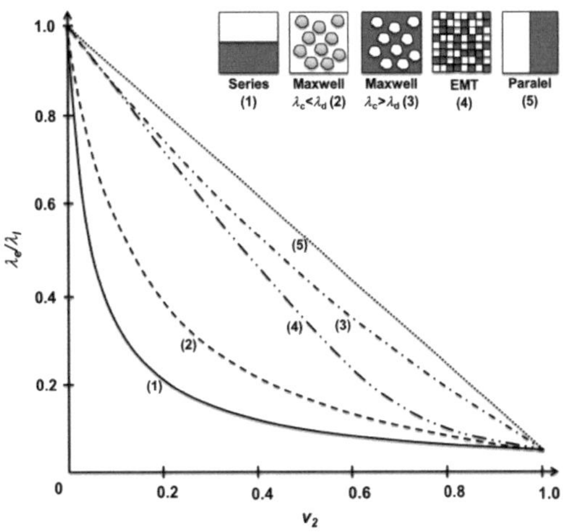

Figure 12. Relative effective thermal conductivity according to different structural models [512].

If the system consists of *i* components and is considered for *n* small spheres with the radius $\boldsymbol{R}_i$ and the thermal conductivity Λ_i, which are dispersed in uniform medium of thermal conductivity Λ_m (Fig.13),the distribution of temperature for the single small sphere under steady-state conditions is given by the Laplace´s equation [508,512]:

$$\frac{1}{r^2}\frac{\partial}{\partial r}\left(r^2\frac{\partial T}{\partial r}\right)+\frac{1}{r^2\sin\Theta}\frac{\partial}{\partial\Theta}\left(\sin\Theta\frac{\partial T}{\partial\Theta}\right)+\frac{1}{r^2\sin^2\Theta}\frac{\partial^2 T}{\partial\varphi^2}=0 \tag{20}$$

with following boundary conditions applied:

- When r=0 then $T_i \neq \infty$.
- When $r=R_i$ then $\lambda_i\frac{\partial T_i}{\partial r}=\lambda_i\frac{\partial T_m}{\partial r}$ and $\frac{\partial T_i}{\partial \Theta}=\frac{\partial T_m}{\partial \Theta}$.
- When $r \gg R_i$ then $T_m=br\cos\Theta$.

where $\boldsymbol{r}$ and $\boldsymbol{\Theta}$ are the polar radius and the polar angle, respectively, $\boldsymbol{T}$ is the temperature and $\boldsymbol{b}$ denotes the temperature gradient in the continuous medium. Assuming the symmetry of the *z*-axis so that $\boldsymbol{T}$ is independent of $\boldsymbol{\phi}$, Eq.20 has a general solution:

$$T=A+\frac{B}{r}+Cr\cos\Theta+\frac{D}{r^2}\cos\Theta \tag{21}$$

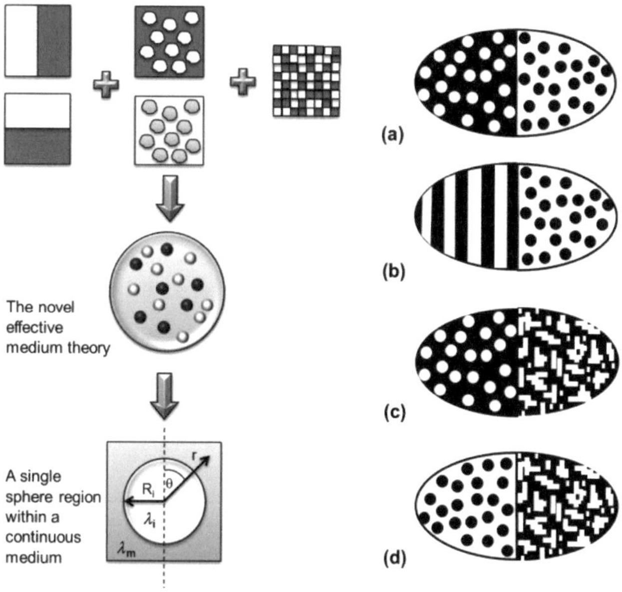

Figure 13. Structural models applied for the derivation theory for thermal conductivity of multiphase material [508] and schematic representation of tw-component material as the uniform mixture of two fundamental structural models [522]: ME-1+ME-2 (a), Parallel+ME2(b), ME-1+EMT (c) and ME-2+EMT (d).

The solution using the boundary conditions to substitute for A, B, C and D in Eq.21 and further transformation yield to the equation [508]:

$$\sum v_i \left(\frac{\lambda_i - \lambda_e}{\lambda_i + 2\lambda_m} \right) = 0 \tag{22}$$

For two-phase system (e.g. porous solid material – air) Eq.22 can be then written as:

$$(1-\varepsilon)\frac{\lambda_s - \lambda_e}{\lambda_s + 2\lambda_m} + \varepsilon \frac{\lambda_a - \lambda_e}{\lambda_a + 2\lambda_m} = 0 \tag{23}$$

where λ and ε are the thermal conductivity and the porosity and subscripts ***e***, ***a*** and ***s*** denote two-phase material, air and solid, respectively. Eq.23 represents individual structural models:

- Series model when λ_m=0;
- Parallel model when λ_m=∞;
- Maxwell-Eucken model with one continuous component when λ_m=λ_s;
- Maxwell-Eucken model with two continuous components when λ_m=λ_a;

- EMT model when $\lambda_m = \lambda_e$.

To avoid potential misapplication of the effective thermal conductivity models, the materials which may be described as porous should be divided into two classes:

1. Internal porosity materials which have bubbles/pores suspended within a continuous condensed phase (e.g. sponges, foams, honeycombs), where the optimal heat transfer pathway is through the continuous phase.
2. External porosity materials which include granular particulate materials where the optimal heat conduction pathway is through the dispersed phase.

A model that accurately predicts the effective thermal conductivity of internal porosity materials will not necessarily be applicable to external porosity materials or vice versa.

Carson at al. [512] proposed that the effective thermal conductivity of internal porosity materials was bounded above by the Maxwell–Eucken equation for the lower-conductivity material as the dispersed phase, and below by the EMT equation; and that the effective thermal conductivity of external porosity materials is bounded above by the EMT equation, and below by the Maxwell–Eucken equation for the lower conductivity material as the continuous phase.

The equation which unifies five structural models mentioned above for heterogeneous material was developed by Wang [522]:

$$\lambda_e = \frac{\sum_{i=1}^{m} \lambda_i v_i \frac{d_i \bar{\lambda}}{(d_i - 1)\bar{\lambda} + \lambda_i}}{\sum_{i=1}^{m} v_i \frac{d_i \bar{\lambda}}{(d_i - 1)\bar{\lambda} + \lambda_i}} \tag{24}$$

Each structural model can be derived by suitable choice of parameters $\boldsymbol{d}_i$ and $\bar{\lambda}$:

- Series model when d_i=1 or $\bar{\lambda} \to 0$;
- Parallel model when d_i=∞ or $\bar{\lambda} \to \lambda_i$;
- Maxwell-Eucken model when d_i=3 or $\bar{\lambda} \to \lambda_m$;
- EMT model whend_i=3 or $\bar{\lambda} \to \lambda_e$.

Parameter $\boldsymbol{d}_i$ can have a physical interpretation. Kirkpatrick [523] related similar parameter to the number of Euclidean dimensions of the system involved, while Fricke [524] and Hamilton and Crosser [525] related it to the sphericity of dispersed phase. However, it may be possible to define a parameter that combines both aspects; the component shape and the number of Euclidean dimensions. This is the topic of ongoing investigations by the authors, but it falls outside the scope of this publication. The most common approach is to use $\boldsymbol{d}_i$=3 (spherical dispersed phase) [508,522].

The model suggested by Levy [526] is based on the Maxwell–Eucken model and is based solely on algebraic operations and has no stated physical basis:

$$\lambda_e = \lambda_1 \frac{2\lambda_1 + \lambda_2 - 2(\lambda_1 - \lambda_2)F}{2\lambda_1 + \lambda_2 + (\lambda_1 - \lambda_2)F} \tag{25}$$

where

$$F = \frac{2/G - 1 + 2v_2 - \sqrt{\left(2/G - 1 + 2v_2\right)^2 - 8v_2/G}}{2} \tag{26}$$

$$G = \frac{(\lambda_1 - \lambda_2)^2}{(\lambda_1 + \lambda_2)^2 + \lambda_1\lambda_2/2} \tag{27}$$

In order to make the basic structural models more general for different structures, the additional parameter is sometimes introduced. The Krischer's approach assumed that the complex structure could be approximated by a mixture of simpler structures, where the relative amounts of each of the simpler structures was determined empirically. We define the "structure volume fractions (ε_j)" to be distinct from the component volume fractions ($\boldsymbol{v}_i$) for the $\boldsymbol{j}$-th type of structure by the relation [522]:

$$\varepsilon_j = \sum_i v_i \phi_{ij} \tag{28}$$

$$\sum_j \varepsilon_j = 1 \tag{29}$$

The "structure composition factors" ϕ_{ij}, are the measures of the fractions of material component i that is a part of structure j. Therefore the total amount of component i must be distributed between the structures:

$$\sum_j \phi_{ij} = 1 \tag{30}$$

The effective thermal conductivity of structureΛ_j, which consists of n components, is the function:

$$\Lambda_j = \Lambda_j(v_1, v_2 \ldots v_n, \phi_{1j}, \phi_{2j} \ldots \phi_{nj}, \lambda_1, \lambda_2 \ldots \lambda_m) \tag{31}$$

Since this structure can be described by one of five structural models listed above, it can be written as[7]:

$$\Lambda_j = \frac{\sum_{i=1}^{m} \lambda_i v_i \phi_{ij} \frac{d_i \bar{\lambda}}{(d_i - 1)\,\bar{\lambda} + \lambda_i}}{\sum_{i=1}^{m} v_i \phi_{ij} \frac{d_i \bar{\lambda}}{(d_i - 1)\,\bar{\lambda} + \lambda_i}} \tag{32}$$

The effective thermal conductivity of the material which is formed from *z* structures can be considered for the function:

$$\lambda_e = \lambda_e(\Lambda_1, \Lambda_2 \cdots \Lambda_z) \tag{33}$$

In order to solve Eq.32, based only on Λ_i and v_i without any empirical parameters, it is necessary to determine the expressions for ϕ_{ij} as functions of Λ_i and v_i. Therefore, the assumption of relative amounts of each structure ε_j, in total volume and of relative contributions of different Λ_j to λ_e is necessary. The infinite number of relationships between ε'_j and Λ_j could be found. It is possible to relate ε'_j to structural characteristics. However, such detailed information is often unavailable. Therefore the simplest possible relationships were adopted. They are to assume that each structure comprises an equal fraction of total volume:

$$\varepsilon_j = \frac{1}{z} \tag{34}$$

and that the thermal conductivity of each structure is equal to the effective thermal conductivity:

$$\lambda_e = \Lambda_1 = \Lambda_2 = \cdots = \Lambda_z \tag{35}$$

The solution for selected binary structure models (Fig.13) with suitable choice of d_i and $\bar{\lambda}$ according to Eq.24 can be expressed as follows [522]:

1. **ME-1+ME-2 model:**

$$\lambda_e = \frac{\lambda_1 v_1 \phi_{11} + \lambda_2 v_2 \left(\frac{1 - 2v_1\phi_{11}}{2\lambda_2}\right)\left(\frac{3\lambda_1}{2\lambda_1 + \lambda_2}\right)}{v_1 \phi_{11} + v_2 \left(\frac{1 - 2v_1\phi_{11}}{2\lambda_2}\right)\left(\frac{3\lambda_1}{2\lambda_1 + \lambda_2}\right)} \tag{36}$$

7 Please consult with Eq.24.

2. **Parallel+ME-2 model:**

$$\lambda_e = \frac{\lambda_1 v_1 \phi_{11} + \lambda_2 v_2 \phi_{21}}{v_1 \phi_{11} + v_2 \phi_{21}} = \frac{\lambda_2 v_2 \phi_{22} + \lambda_1 v_1 \phi_{12} \dfrac{3\lambda_2}{2\lambda_2 + \lambda_1}}{v_2 \phi_{22} + v_1 \phi_{12} \dfrac{3\lambda_2}{2\lambda_2 + \lambda_1}} \tag{37}$$

3. **ME-1+EMT model:**

$$\lambda_e = \frac{C + \sqrt{C^2 + 2\lambda_1 \lambda_2}}{2} \tag{38}$$

where

$$C = (2\lambda_1 - \lambda_2) v_1 \left(\frac{2v_1 + 2v_2 \phi_{21} - 1}{2v_1} \right) + (2\lambda_2 - \lambda_1) v_2 (1 - \phi_{21}) \tag{39}$$

4. **ME-2+EMT model:**

$$\lambda_e = \frac{D + \sqrt{D^2 + 2\lambda_1 \lambda_2}}{2} \tag{40}$$

where

$$D = (2\lambda_1 - \lambda_2) v_1 (1 - \phi_{11}) + (2\lambda_2 - \lambda_1) v_2 \left(\frac{2v_2 + 2v_1 \phi_{11} - 1}{2v_2} \right) \tag{41}$$

For ternary-structure models composed of EMT+ME-2+Series the following can be derived [522]:

$$\lambda_e = \frac{\lambda_1 v_1 \phi_{12} \dfrac{3K}{2K + \lambda_1} + \lambda_2 v_2 \phi_{22} \dfrac{3K}{2K + \lambda_2}}{v_1 \phi_{12} \dfrac{3K}{2K + \lambda_1} + v_2 \phi_{22} \dfrac{3K}{2K + \lambda_2}} \tag{42}$$

Where:

$$\nu_1\phi_{11} + \nu_2\phi_{21} = 1/3$$
$$\nu_1\phi_{12} + \nu_2\phi_{22} = 1/3$$
$$\nu_1\phi_{13} + \nu_2\phi_{23} = 1/3$$
$$\phi_{11} + \phi_{12} + \phi_{13} = 1$$
$$\phi_{21} + \phi_{22} + \phi_{23} = 1$$

Krischer uses the empirical approach f which is sometimes referred as the distribution factor. This parameter ranges from 0 to 1 and serves as the weighting factor that makes the basic structural model more universal for different structures. The value of this parameter must be determined by the experiment. This parameter is used in [508,522,527]:

- **Krischer model**, i.e. weighted harmonic mean between the series model and the parallel model:

$$\lambda_e = \frac{1}{\frac{1-f}{\lambda_1(1-\nu_2)+\lambda_2\nu_2} + f\left(\frac{1-\nu_2}{\lambda_1} + \frac{\nu_2}{\lambda_2}\right)} \tag{43}$$

- **Hamilton model:**

$$\lambda_e = \lambda_1 \frac{(f-1)\lambda_1 + \lambda_2 - (f-1)(\lambda_1 - \lambda_2)\nu_2}{(f-1)\lambda_1 + \lambda_2 + (\lambda_1 - \lambda_2)\nu_2} \tag{44}$$

- **Chaudhary-Bhandari model:**

$$\lambda_e = \left[\lambda_1(1-\nu_2) + \lambda_2\nu_2\right]^f \left(\frac{1-\nu_2}{\lambda_1} + \frac{\nu_2}{\lambda_2}\right)^{(1-f)} \tag{45}$$

- **Kirkpatrick model:**

$$\sum_{i=1} \nu_i \frac{\lambda_i - \lambda_e}{\lambda_i + \left(\frac{f}{2} - 1\right)\lambda_e} = 0 \tag{46}$$

2.3. Thermal insulating castables

The refractory castables filled by cenospheres (Fig.14), alumina bubbles (Fig.15) and fireclay (Fig.16) were prepared as an example of thermal insulating castables based on strontium aluminate cement. The specimens were prepared using strontium aluminate cement, bauxite and filler in the mass ratio of 1:3:1 and *w/c*=1. The applied opening material was described in Chapter 6.1.3.

The microphotographs and some properties and information on applied filler can be found in Chapter 6.2.1.

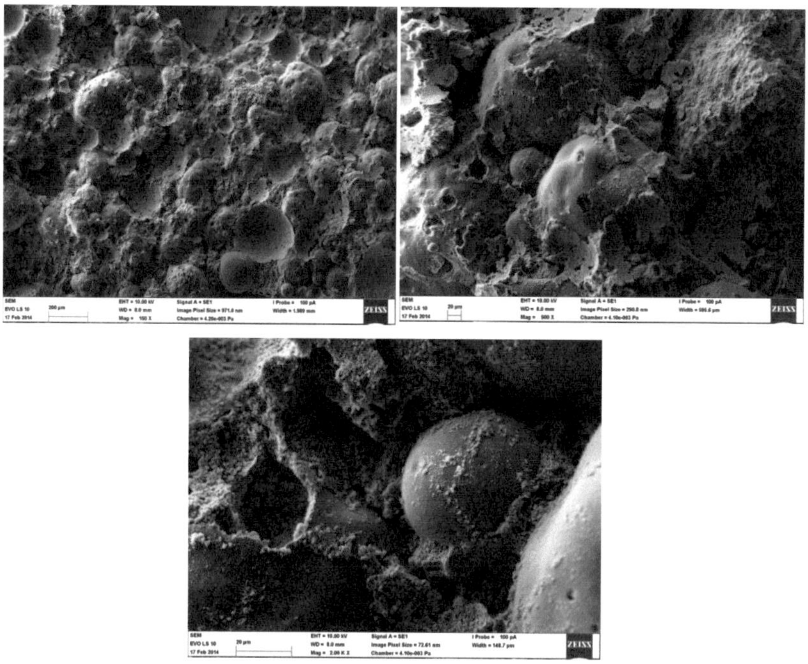

Figure 14. Thermal insulating castable based on cenospheres.

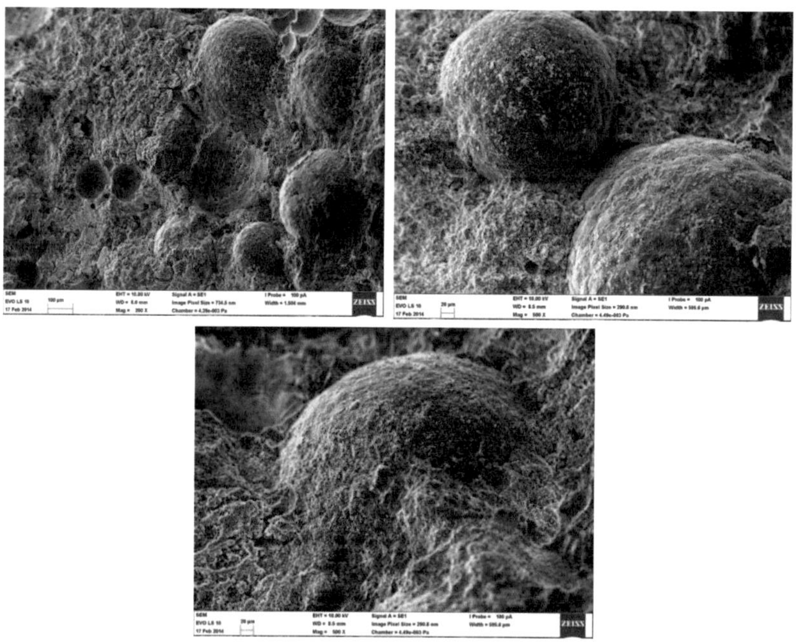

Figure 15. Thermal insulating castable filled with alumina bubbles.

Some properties of prepared samples of thermal insulating castables before and after (Fig. 17) thermal treatment are listed in Table 3.

Used opening material	Volume density [g.cm^{-3}]		Shrinkage [%]			Cold crushing strength [MPa]
	After setting	After thermal treatment	Autogenous	Thermal treatment	Total	
Cenosphere	1.28	1.65	0.3	25.0	25.3	---
Bubble alumina	2.05	1.76	0.3	+6.3*	+6.6*	16.2
Fireclay	1.96	2.25	+4.0*	11.4	15.4	24.7

* Expansion with regard to the initial volume of specimen.

Table 3. Properties of thermal insulating castable before and after thermal treatment to 1500°C.

While the dimensions of thermal insulating castables prepared with cenospheres and fireclay were reduced, the expansion was observed for the sample prepared with bubble alumina filler (Table 3).

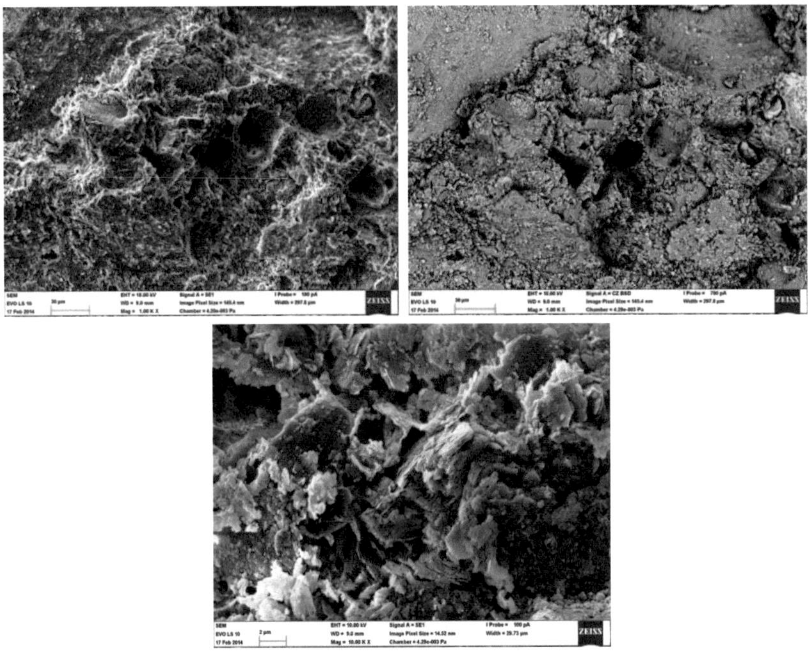

Figure 16. Thermal insulating refractory castable prepared with fireclay.

The reason for the expansion behaviour of this sample is probably the same as for dense refractory castables filled by bauxite (Chapter 6.1.5). Observed behaviour provides some really interesting possibilities to control the volume changes during the thermal treatment. Since the addition of alumina led to the expansion (recrystallization) and the addition of fireclay to the shrinkage (formation of SrO Al_2O_3 $2SiO_2$ and sintration), the proper mutual ratio of these constituents enables to prepare the dimension stable refractory.

Furthermore, the addition of fireclay enables to avoid undesirable growth of large crystals during the thermal treatment (please compare Figs.7-10). The introduction of cenospheres led to the formation of eutectic melt with iron, therefore they are not optimal for high temperatures and reductive conditions.

X-ray diffraction analysis of samples prepared with fireclay and cenospheres confirms the formation of strontianite and strontium hexaaluminate (SrA_6) in the ceramic body (Fig.17(b)). Due to missing source of SiO_2, strontianite isn't formed in the sample prepared with bubble alumina. Corundum is the main phase in all prepared samples.

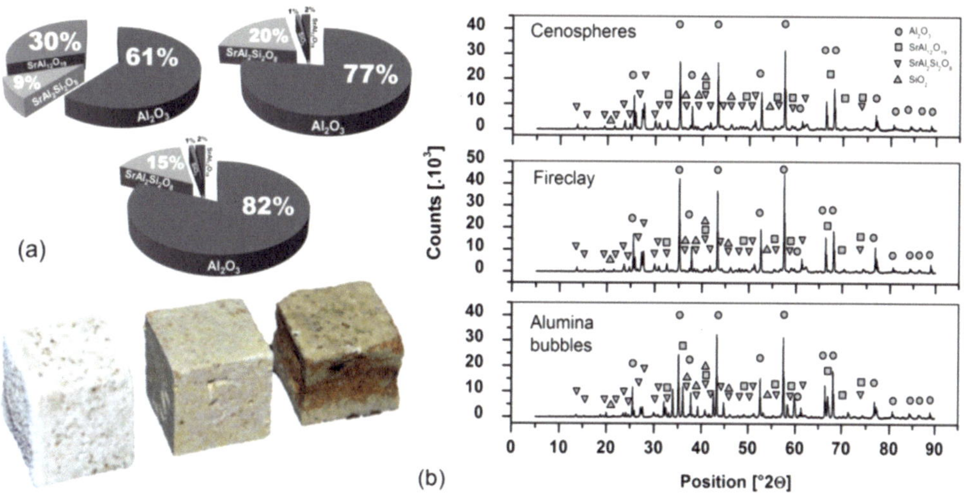

Figure 17. Specimens prepared with bubble alumina, fireclay and cenospheres treated to the temperature of 1500 °C (a) and x-ray diffraction analysis of prepared samples (b).

The microphotographs of fracture area of thermal insulating refractory castable filled with alumina bubbles after the thermal treatment is shown in Fig.18. The bubbles are surrounded by hexagonal crystals of recrystallized strontium aluminate matrix of binding phase. The inner side of alumina balls shows well-developed crystals of alumina.

Thermal insulating castable prepared with fireclay and fired to the temperature of 1500 °C is shown in Fig.19. The specimen texture shows large bauxite grains covered by matrix of recrystallized strontium aluminate phase where the artefacts of layered aggregate of fireclay filler can be seen.

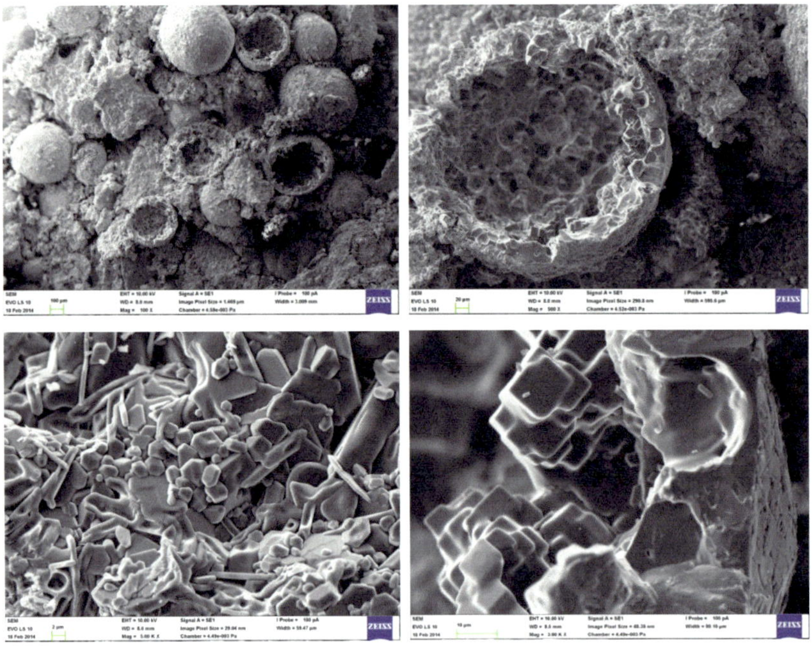

Figure 18. Thermal insulating refractory castables with bubble alumina treated to the temperature of 1500 °C.

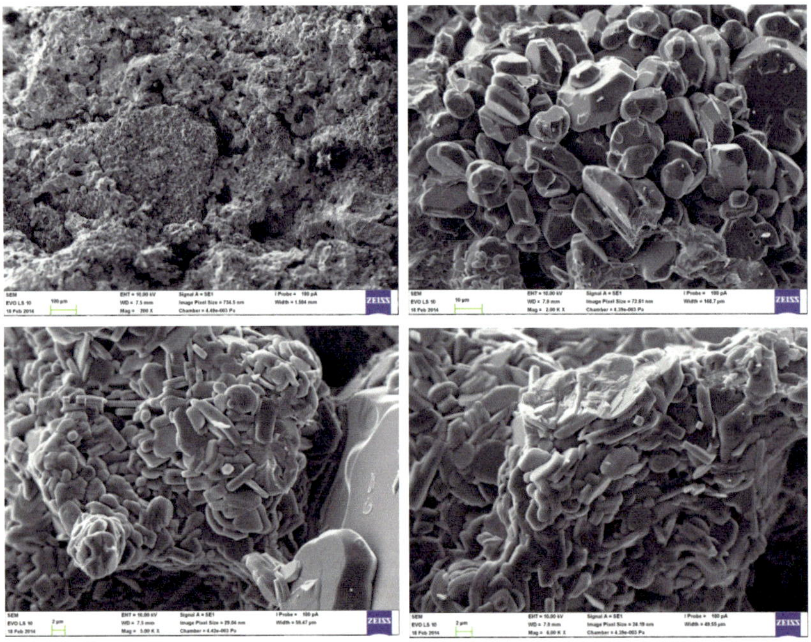

Figure 19. Thermal insulating refractory castables with fireclay treated to the temperature of 1500 °C.

Large amount of melt phase formed during the thermal treatment of specimen prepared with cenospheres led to extensive sintering, firing shrinkage and deformation of the shape. Cenospheres were completely dissolved in formed eutectic melt and there is no artefact in the body of specimen after the thermal treatment. The SEM analysis (Fig.20) shows the strontium aluminate crystals surrounded by glassy phase.

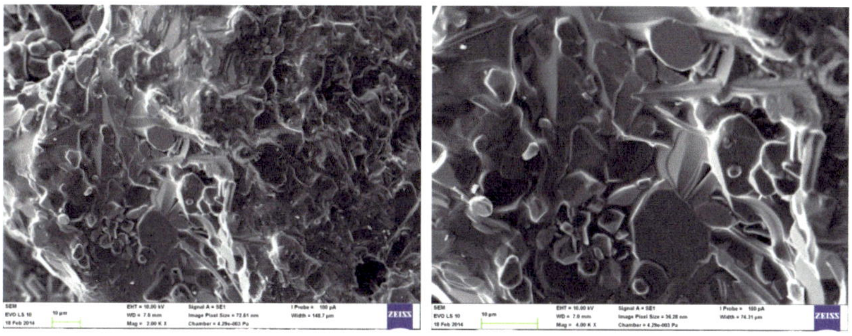

Figure 20. Thermal insulating refractory castables with cenospheres treated to the temperature of 1500 °C.

3. Ceramic Foams

3.1. Properties and processing of ceramic foams

Foams can be defined as dispersion systems based on the bubble gas phase, usually air, in continuous matrix of liquids-"aqueous and non-aqueous liquid foams", or solids "solid foams" (e.g. pumice as well as steel, glass and concrete foams). The diameter of bubbles is usually of several millimeters, but the spherical shape is not common as the weight of foam leads to mutual deformation and to the formation of irregular polyhedral cells. The structure of foams is often compared to honeycombs. The walls of bubbles make always the same angle of 120° hence maximum four bubbles can be in contact at one point. This structure provides foams with certain mechanical strength and elasticity [528-531].

Liquid foams are usually prepared via mechanical dispersion of gas in liquid in the presence of surfactants. These compounds reduce the surface tension via the formation of films on the interface as they contain both the hydrophobic and the hydrophilic groups. Foams are then always formed under dynamic conditions so the adsorption equilibrium of surfactant is usually not achieved. Applied kind and concentration of surfactant determine the time-stability and the structure of foams. Higher viscosity of surface layer, its flexibility and hydration of polar groups of surfactant decelerate the water flow-out from bubble walls and thus are the factors contributing to the time-stability of foams. This fact may have a significant effect on the magnitude of forces stabilizing the foam and foam films [530,532-536].

The research on the structure, behaviour and properties of foams is motivated by both, the preparation of foams and the prevention from undesirable formation of foam. For technical applications (usually filters, carriers of catalysts, biomaterials, thermal insulating materials, refractories [531,537,538]) the factors such as the stability, specific surface and porosity of foams are important. Foams were also used in wide range of industrial processes e.g. flotation [539].

As for the time-stability of foams, the control of bubble size and the preparation of monodispersed foams are intensively studied [540]. The structure and the processes of coarsening and drainage of foams [541] and their dynamics are also affected by fine particles [542-546].

Open cell (reticulated) ceramic foams, which are defined as dispersion systems based on the interconnected bubbles of gas phase in continuous ceramic matrix, can be prepared via different manufacturing techniques [531,547] including the transformation of ceramic powder slurry into the solid foam [536,548,549], the reticulated sponge method (replica technique) [550-553], the gel casting [554-557], the bubble generation methods [558] or direct foaming [559]. The replica technique is based on the pyrolysis of a polymeric sponge coated by ceramic slurry. Following high-temperature treatment leads to the ceramic bond via sintering. The gel casting method uses the mechanical work and foaming agents in order to foam the ceramic suspension.

Cellular cement is a lightweight material consisting of Portland cement (PC) [560,561], alumina cement (AC) [562,563] or blended cements [564,565] paste or mortar with homogeneous void or cell structure created by introducing air or gas in the form of small bubbles. Cement and concrete foams can be made by adding aluminium powder to the mix before adding gauging water or by injecting aqueous foaming agent to the cement slurry to give foams with the porosity of 50-90 %. Microsilica and a plasticizer were added to the cement-water slurry and polyester fibers were incorporated to improve the foam stability and strength [566-569]. Foamed concrete was patented in 1923 [570,571].

Using mixed cationic/anionic surfactants in aqueous foam systems has been of an interest for several years because they could provide good foam stability [572].Therefore the mixed surfactants used in pure aqueous foams are intensively studied [535,536,573].

In order to evaluate the maximum capillary pressure ($\boldsymbol{P}_{\mathrm{C,max}}$ [MPa]) and the effective bubble pressure for a stable bubble ($\Delta\boldsymbol{P}_{\mathrm{eff}}$ [MPa]), the following relationships can be used [543,574,575]:

$$P_{C,\max} = \frac{4 f \gamma}{R_P}\left(\cos\Theta + Z\right) \tag{47}$$

and

$$\Delta P_{eff} = \left(P_{atm} + \frac{2\gamma}{R_B} + \rho g H\right)\left[1 - f\left(1 - \cos^2\Theta\right)\right] \tag{48}$$

where f is the fraction area of the bubble's surface that is coated by particles, $\boldsymbol{\Theta}$ is the wetting angle of particles [°], $\boldsymbol{R}_{\mathrm{P}}$ is the particle's radius [m], $\boldsymbol{\gamma}$ is the surface tension [N m^{-1}] on the gas-liquid interface and $\boldsymbol{Z}$ is a parameter related to the sort of particle configuration within the film, P_{atm} is the atmospheric pressure [0.101 MPa], ρ is the foam density [kg m^{-3}], g is the gravity acceleration [m s^{-2}], H is the height of foam column [m] and $2R_B$ is the bubble diameter. The term ρgH is also known as the hydraulic pressure. The Laplace pressure due to the radius of curvature of the bubble is considered in the term of $2\gamma/R_B$. The value of $\Delta P_{C,max}$ is not the function of bubble size and Eq.48 is not a function of the particles size at the bubble's film.

The approach for the calculation of foam "**stability index**" (*SI*) proposed by Vivaldini et al. [575] in order to combine the $P_{C,max} - \Delta P_{eff}$ difference with $\Delta G_{ads}/k_BT$ ratio is given by the following equation:

$$SI = \left(P_{C,\max} - \Delta P_{eff}\right)\frac{\Delta G_{ads}}{k_B T} \tag{49}$$

where ΔG_{ads} is the adsorption free energy [J] of the particles at the bubble interface regarding their thermal energy expressed by term k_BT [J]. The symbols k_B and T denote the Boltzmann constant and the temperature, respectively. The dimension-less ratio $\Delta G_{ads}/k_BT$ (adsorption energy/ thermal energy) then describes the stability of the gas-liquid interface.

When a particle is attached to a gas-liquid interface, the overall Gibbs free energy of the system is reduced, which makes this process thermodynamically favorable. The energy decrease due to this adsorption process at a flat interface can be for $0 \le \Theta \le 90°$ mathematically described by the equation:

$$\Delta G_{ads} = \pi R_P^2 \gamma \left(1 - \cos\Theta\right)^2 \tag{50}$$

Nevertheless, the ΔG_{ads} value cannot be used to state whether the interface is stable or not. That argument can only be sustained when the energy released due to the adsorption of particles is compared to their own thermal energy, which is the origin of the Brownian motion of colloidal particles in suspension [575,576].

3.2. Ceramic foams stabilized by HAC

Since water is consumed by hydration, calcium aluminate cement can be used for the stabilization of raw ceramic foams. Moreover, setting of cement confers the mechanical properties to foam, which enable the manipulation prior to drying and thermal treatment. Some examples of foamed materials prepared using the techniques described in works [577,578] are shown in Fig.21. The ceramic bond formed during the thermal treatment enables to cut tapes or desks by cutting and shaping the original foamed large cast profile.

Large aggregates of kaolinite particles are delaminated and redistributed closer to the surface of liquid phase to form the solid framework of foam. The thickness of the bubble walls is approximately ten times smaller than the median of particle size of applied kaolinite. Foam with the thickness of walls of 100 nm can be prepared. Hence forced delamination of kaolinite aggregates occurs and the lamellar particles of kaolinite form the framework or solid replica of original liquid bubble wall. The thermal treatment of foam provides the ceramic foam of high service temperature and chemical stability, with high specific surface area, extremely low volume density and excellent thermal insulating properties [577,579].

Figure 21. Examples of refractory foam prepared with calcium aluminate cement (a). Raw and fired block of ceramic foam (b).

3.3. Ceramic foams stabilized by SrAC

Since the ceramic foam can be stabilized by calcium aluminate cement, it is possible to use also strontium aluminate cement for the same purpose. On the other hand, there are some differences in hydration and setting of both cements as was discussed in Chapter 5. The application of additives or certain modification of foam processing is necessary in order to improve the cement behaviour (Chapter 5.4).

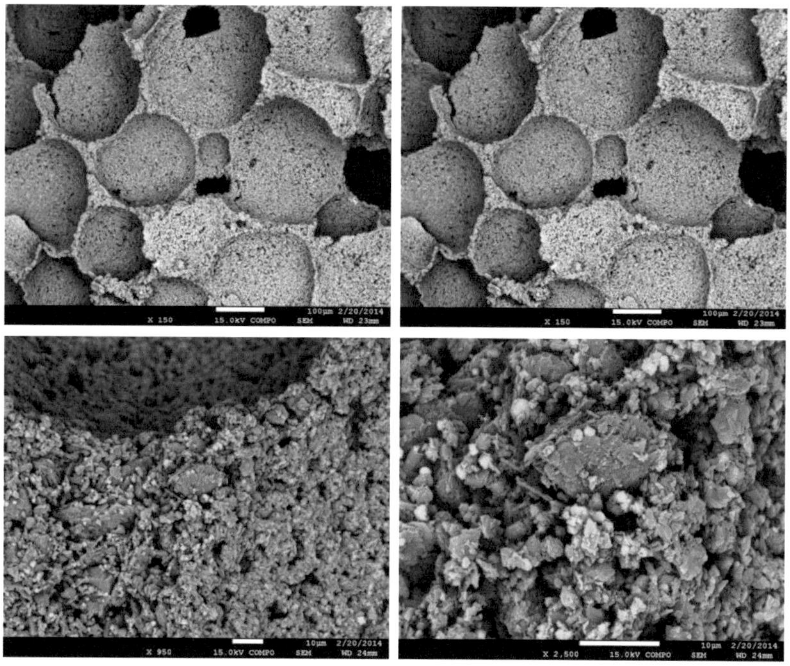

Figure 22. Raw foam before the thermal treatment.

The microphotograph of raw foam prepared via foaming the slurry prepared from strontium aluminate cement, meta-kaolinite and water in the mass ratio of 1:4:5 is shown in Fig.22. Applied method is analogical to the techniques used for the preparation of foam with calcium aluminate cement [577,578], but formed structure is quite different from the foam prepared with kaolinite. The reorganization of particles from original aggregates of kaolinite does not occur in metakaolinite. That results in the formation of thick-walled cells with more spherical shape.

The morphology of foam cells before and after the thermal treatment shows that the walls of cells in foam form an angle of ~120°. The microphotograph of foam treated to the temperature of 1500 °C is shown in Fig.23. The SEM picture shows sintered foam, where the side walls of bubbles contain large amount of needle-like crystals of mullite.

The results of thermal analysis (Fig.24(a)) show that drying, the thermal decomposition of hydrates and the pyrolysis of surfactant takes place up to the temperature of 300 °C. Large surface of foamed structure supports the carbonation. Therefore the content of carbonates in foamed samples is usually higher than that in dense materials. The thermal decomposition of $SrCO_3$ and removing of residual water from metakaolinite leads to the decrease of the sample mas by 3.4 % within the temperature range from 550 to 900 °C. SrO formed during the thermal decomposition of strontium carbonate is a reactive species which reacts with both, metakaolinite and other strontium aluminates, to strontium aluminate and Sr-gehlenite (strontiogehlenite, $SrAl_2SiO_7$ [580,581]).

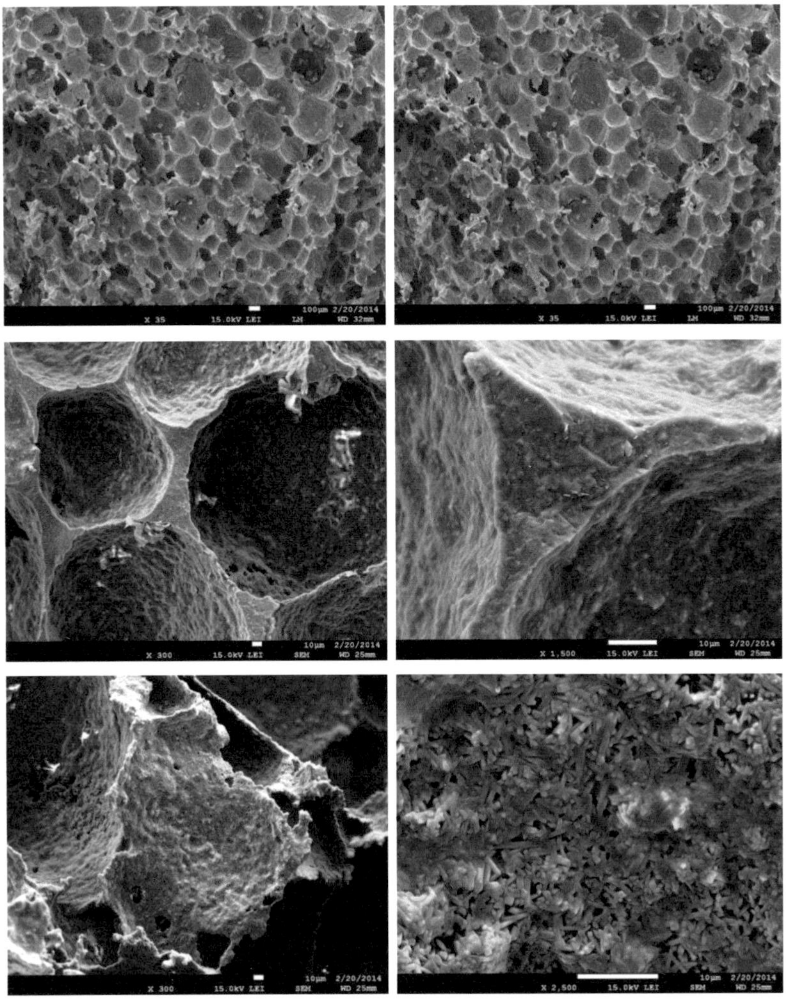

Figure 23. Foam after the thermal treatment to 1500 °C.

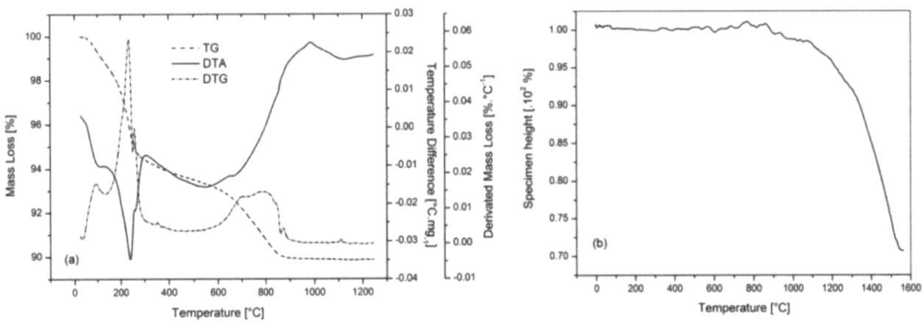

Figure 24. TG-DTA (a) and heating microscopy (b) of raw foam.

The abundance of metakaolinite leads to the formation of usual and well described products such as Al-Si spinel phase and mullite [582,583]. Strontium bearing aluminosilicates are further transformed to Sr-celsian with increasing temperature [584-586,893]. The temperature, which is required for the preparation of sintered sample was determined by heating microscopy (Fig. 24(b) and Fig.25). The sintering begins at the temperature of 1350 °C.

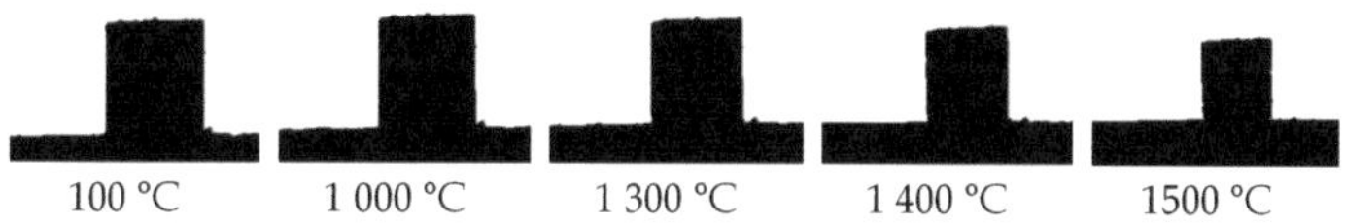

Figure 25. The behaviour of specimen of foam during heating to the temperature of 1550 °C at the rate of 5 °C min^{-1}.

Infrared spectrum of foam before the thermal treatment (Fig.26(a)) shows the features of metakaolinite [587], strontium carbonate, hydrogarnet, gibbsite, surfactant (C-H stretching in -CH_3 and -CH_2 groups in the region from 2975 to 2825 cm^{-1}) as well as the stretching (3440 cm^{-1}) and bending bands (1460 cm^{-1}) of absorbed water. The spectral features of $SrCO_3$, gibbsite and Sr_3AH_6 were already described in Chapter 5.1. Excessive carbonation results from high surface of foam.

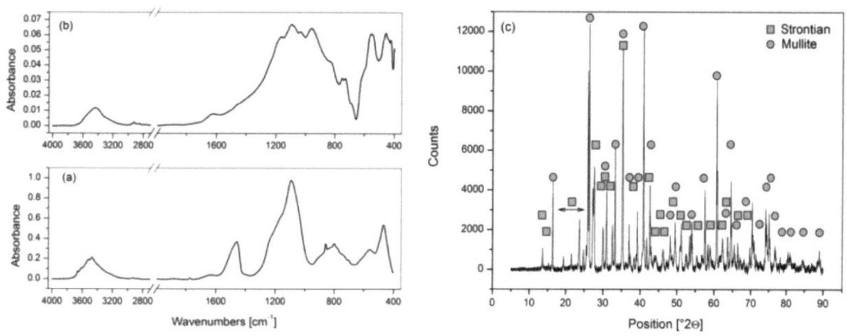

Figure 26. Infrared spectrum of raw foam (a), foam after the thermal treatment (b) and XRD of foam after the thermal treatment to 1500 °C (c).

After the thermal treatment (Fig.26(b)), the characteristic absorption bands of mullite ($3Al_2O_3$ $2SiO_2$) [583,588,589] and Sr-celsian, the Si-O stretching (1095 cm^{-1}) and bending (455 cm^{-1}) can be recognized in the spectra. The results of x-ray diffraction analysis (Fig.26(c)) confirm mullite and Sr-celsian as the main components in the sample.

4. MDF composites

Macro-defect-free (MDF) cements were developed by the research group led by Birchall at ICI during late 1970s and early 1980s [590]. MDF cements have unusual and superior properties compared to traditional cement pastes and concretes, and are often referred to as chemically bound ceramics or MDF composites. The suggestions of applications of MDF cements are extremely wide ranging. These include roofing tiles, fire resistant doors, sewage pipes, airport bridges, window shutters, plastic moulds, printing rollers, thermal insulators, tube exhaust, oil tanks, corrosion resistant tanks, cable duct covers, electric generator propellers, electrical parts, boat decking, brake lining, body armor, pallets, toys, cryogenic vessels, signboards, cooler boxes, sound insulators and electromagnetic interference screenings [591,592].

However, after 25 years since they were first patented, the incorporation of MDF materials into commercial processes has been very limited. Despite the identification of many potential applications, MDF cements have certain limitations, in particular the loss of strength on exposure to moisture or humidity. The modifications of the MDF manufacturing process and the selection of polymer used have overcome many of these problems. The major obstacles restraining MDF cements from many applications are the economics of manufacturing on a large scale and their overall commercial viability compared to existing materials [591-593].

The manufacture of MDF cement involves high shear mixing of selected polymers and hydraulic cements at low *w/c* ratios typically between 0.08 and 0.20. The key stages of the typical MDF preparation process are shown in Fig. 27. Calcium aluminate cement (CAC), polyvinyl alcohol (PVA) and glycerol are the most applied input materials [12,592].

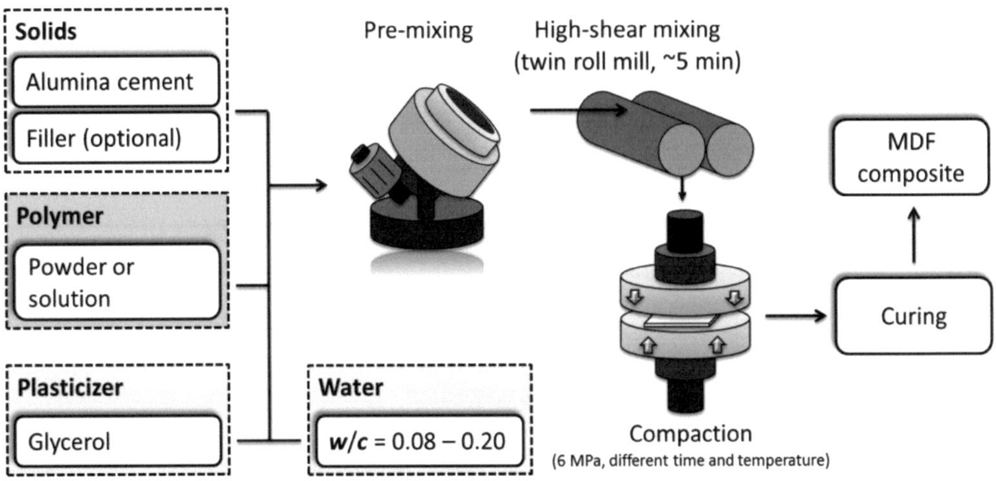

Figure 27. Typical stages of MDF manufacturing process [592].

MDF microstructure was described as close packed unhydrated cement grains within a three dimensional PVA network, with the coatings of hydrated cement on the surface, which interacts with the bulk polymer via chemical cross-linking. MDF cements comprise three distinct regions [592,594]:

a. **Unhydrated cement grains** due to the low *w/c* ratio there is a stoichiometric deficit of water that restrains all cement from full hydration;

b. **Bulk polymer phase;**

c. **Complexinter-phase region which includes hydrated cement particles and polymer chains**

A new alternative to MDF composites is the CAPR composites (Calcium Aluminate Phenol Resin), which are produced using the processing similar to that required for MDF cement manufacture (Fig.27). In CAPR composites high alumina cement (or other hydraulic cement) is mixed with a phenol resin precursor. A modifier is used to control the cross linking density and glycerol is added as a plasticizer. The process does not involve the addition of water and the cement is hydrated only by water given off by the phenol precursor during the condensation polymerization reaction that occurs in situ. This can lead to CAPR composites with an effective w/c ratio as low as 0.01 [592,595-597].

The formation of pastes in both systems is shown in Fig.28. Calcium-rich calcium aluminates, CaO, and $Ca(OH)_2$ showed high reactivity with phenol resin during roll milling while calcined alumina showed poor combining during roll milling. This indicates that Ca^{2+}ion is the reactive component whereas Al^{3+}ions control the excessive stiffness of the high alumina cement/phenol resin system.

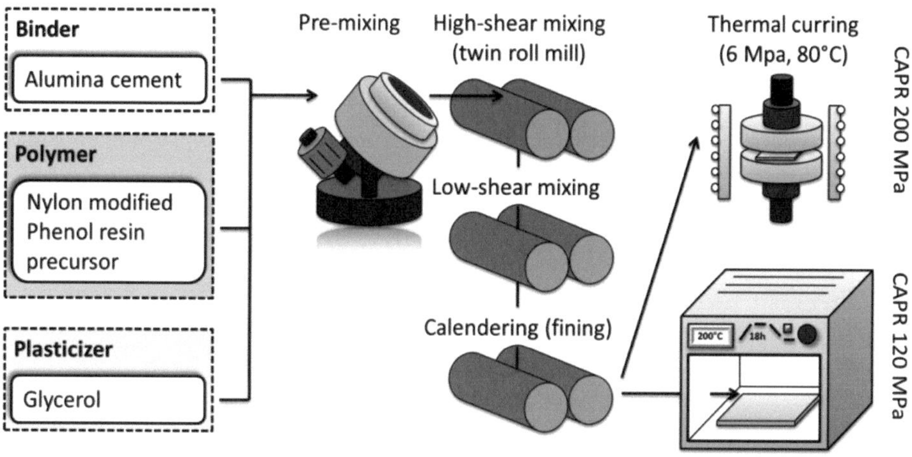

Figure 28. Flow chart of the preparation process of CAPR [595].

The formation of pastes in both systems is shown in Fig.29. Calcium-rich calcium aluminates, CaO, and $Ca(OH)_2$ showed high reactivity with phenol resin during roll milling while calcined alumina showed poor combining during roll milling. This indicates that Ca^{2+}ion is the reactive component whereas Al^{3+}ions control the excessive stiffness of the high alumina cement/phenol resin system.

Higher strength was achieved by Al^{3+}ion-rich calcium aluminate cements, and lower strength of calcium silicate based OPC indicates that the presence of Al^{3+}ions is essential for active thermal curing. On the basis of these observations, we suppose that the phenol resin precursor cross-links with Ca^{2+}ions of cement through an ionic cross-linking reaction during the roll milling and then with Al^{3+}ions through a chemical cross-linking reaction during thermal curing [598].

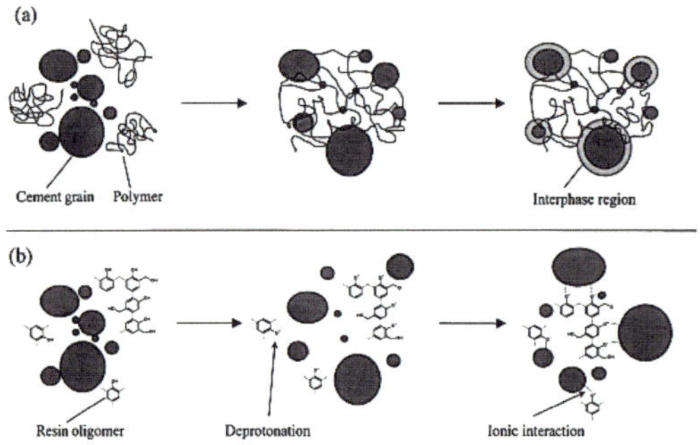

Figure 29. Mechanism of the paste formation of MDF system CA – PVA (a) and of CAPR composite (b) [597].

4.1. MDF composites based on strontium aluminate cement

MDF cement based on strontium aluminate is a promising material for the preparation of refractory products with thin wall and complicated shape. MDF composites based on strontium aluminate cement can be prepared by similar way as MDF cements based on calcium aluminate cement (Fig.27). A certain drawback is fast hydration of strontium aluminate cement, therefore the utilization of additives improving the workability of cement mixture with polymer and plasticizer is necessary.

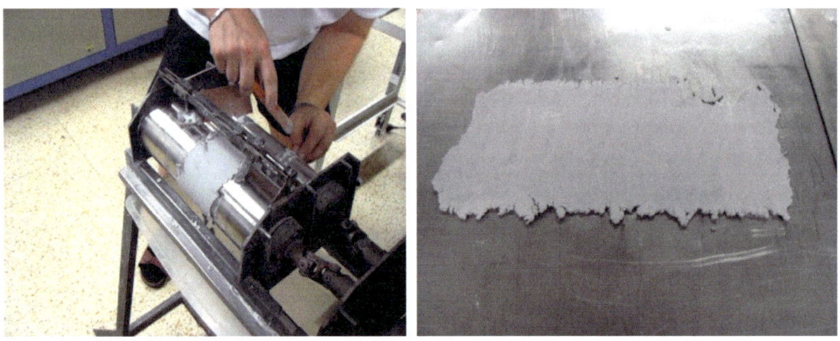

Figure 30. Preparation of strontium aluminate based MDF using twin roll mixer.

The MDF material in Fig.30 is the first prepared sample of macro defect free composite based on strontium aluminate cement. It was prepared from the mixture of SrA cement, glycerol and PVA. After premixing of MDF components the mixing on twin roll mixer was performed. The sample compacted between two steel plates was then cured at the temperature of 60 °C.

Figure 31. SEM analysis of MDF from strontium aluminate cement.

The SEM picture in the BSE-COMPO mode in Fig.31 shows the structure of prepared material where unhydrated grains of strontium aluminate cement were covered by the layer of hydration products. On the other hand the structure is not optimal due to fast hydration of strontium aluminate cement. Therefore, the additives improving cement workability are necessary in order to prepare MDF of higher quality. The other interesting options for the preparation of MDF should be the utilization of SrAC and CAC blend (Chapter 7.1.1) or the structural modification of the main clinker phase, e.g. by B_2O_3 (Chapter 7.3.2).

The preparation of shell of refractory product requires the thermal treatment to the temperatures, at which the ceramics bond is form. The behaviour of hydrated cement upon the thermal treatment is described in Chapter 6.5 and some properties of the product after the thermal treatment in Chapter 6.4.2.

4.2. Using MDF for refractory purposes

The options to use MDF composites prepared from strontium aluminate cement for refractory purposes require the information on the behaviour at high temperatures. High temperature heating microscopy performed up to the temperature of 1600 °C shows the thermal stability of MDF based on strontium aluminate cement (Fig.32).

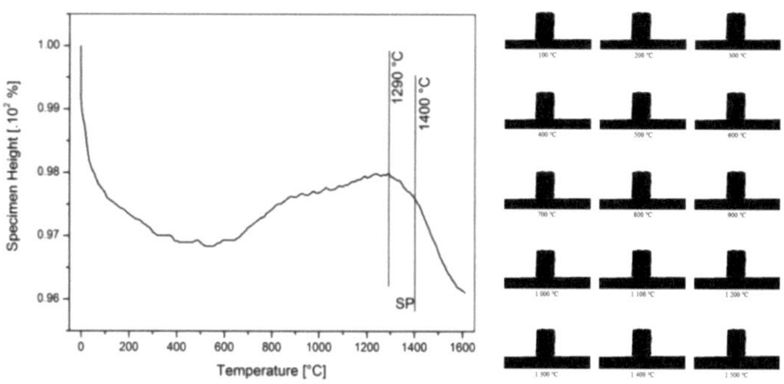

Figure 32. Heating microscopy of SrAC MDF specimen.

The initial shrinkage of specimen of about 3 % which takes place below the temperature of 600 °C is caused by the thermal decomposition of hydrates. The synthesis of strontium and tri-strontium aluminate from the product of dehydration and the thermal decomposition of strontium carbonate leads to the small expansion of 1 % within the temperature range from 650 to 900 °C. Up to the temperature of 1290 °C, the single thermal expansion of the sample takes place with the average thermal expansion coefficient of $1.073 \cdot 10^{-5}$ K^{-1}. The length contraction of sample occurs at temperatures higher than 1290 °C. The start of sintering was observed at the temperature of 1400 °C, but the change of height was still very small (about 2 %) at the temperature of 1600 °C.

The SEM image of MDF heated to the temperature of the start of sintering for 1 hour, where newly formed ceramic bond substitutes the hydraulic bond, which was lost during the thermal decomposition of hydrates (Chapter 6.5) is shown in Fig.33. The fracture area morphology shows the formation of necks (first stage sintering) between grains. In this stage, the material reaches the flexural strength of 14.6 MPa and the Young´s module of 5.2 GPa.

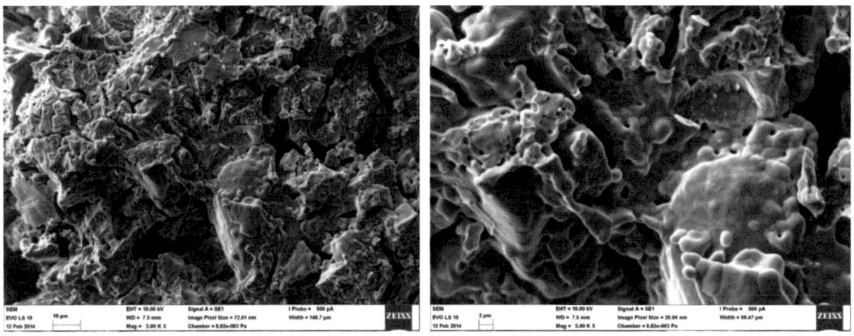

Figure 33. SEM image of fracture area of strontium aluminate MDF treated to the temperature of 1400 °C

5. Behaviour of hydrated cement stone during thermal treatment

The processes which take place during the thermal treatment of materials based on hardened strontium aluminate cement include the thermal decomposition of hydrates, repeated synthesis of strontium aluminate phases and sintering (formation of ceramic bond) in the case that material was treated to sufficient temperature. In the materials prepared with opening materials (aggregate) the newly formed phases appear. Their composition depends on the kind of applied aggregate.

The Kissinger plot related to the thermal decomposition of hydrates is shown in Fig.34. Table 4 summarizes the kinetic results of the behaviour of hardened strontium aluminate cement stone during the thermal treatment.

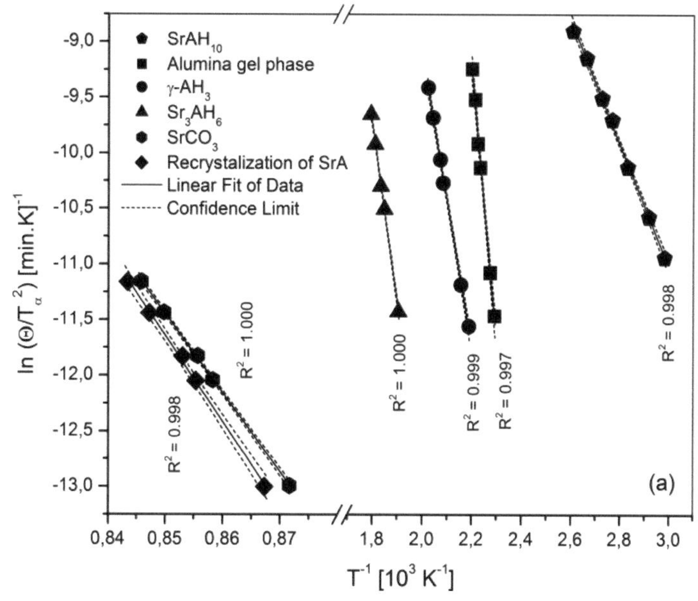

Figure 34. Kissinger plot [379].

Phase	E_a	A	a_m	$w_{1/2}$	n	The description of the mechanism of process
	[kJ•mol^{-1}]	[s^{-1}]	[%]	[°C]		
$SrAH_7$	46	$7.79{\cdot}10^5$	49.50	50.0	0.92	Growth of particles of appreciable initial volume, needles and plates of finite long dimensions.
AH_3 gel	200	$3.46{\cdot}10^{22}$	52.13	24.32	0.79	Growth of particles of appreciable initial volume, thickening of very large plates.
γ-AH_3	108	$1.45{\cdot}10^{22}$	50.48	9.14	4.18	Interface-controlled growth of a new phase, constant or increasing nucleation rate.
Sr_3AH_6	117	$3.94{\cdot}10^{10}$	49.70	17.75	2.54	Diffusion-controlled growth of a new phase, constant or increasing nucleation rate.
Sr	592	$2.28{\cdot}10^{24}$	46,11	19.31	2.31	Diffusion-controlled growth of a new phase, decreasing nucleation rate.
SrA	642	$4.30{\cdot}10^{26}$	54.25	17.95	2.48	Diffusion-controlled growth of a new phase, constant or decreasing nucleation rate.

Table 4. Mechanism and kinetics of thermal decomposition of hydration products.

The final product of the thermal treatment of hydrated cement stone is also strontium aluminate, i.e. the same compound as in the original clinker phase. Nevertheless it is formed through different reaction pathway compared to the original mixture of raw materials.

The thermal treatment of hydraulically bound refractory castables based on strontium aluminate cements leads to the thermal decomposition of hydrates. Disappearing of hydraulic bond naturally leads to decreasing strength, but bonding function is gradually adopted by ceramic bond formed with increasing temperature. The sintering of cement stone takes place at temperatures higher than 1400 °C (Fig.32) and the material becomes almost dehydrated at temperatures higher than 600 °C (Fig.35 and Fig.5 in Chapter 5). Therefore, there is a temperature range with significantly reduced strength.

The influence of temperature on to the equilibrium composition of hydrated cement stone with the composition corresponding to the 28th day of hydration process (Fig.2 in Chapter 5) is shown in Fig.35(a). The thermal treatment leads to the same compounds as the processing of strontium aluminate clinker from the mixture of raw materials (Fig.30 in Chapter 4). On the other hand, there is one significant difference, which is that the hydration process leads to better homogeneity than can be reached by the processing of raw material. That facilitates the solid state synthesis of clinker minerals.

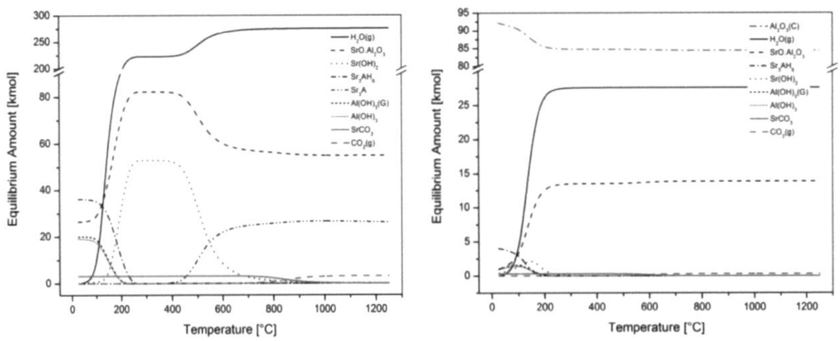

Figure 35. Effect of temperature on the equilibrium composition of hydrated cement stone (a) and hypothetical NCC (b).

The utilization of pure strontium aluminate cement without opening material (corundum, bauxite, mullite, fireclay, etc.) is not supposed, therefore Fig.35b shows the temperature evolution of equilibrium of refractory material in which the content of Al_2O_3 is increased to 90 % by the addition of alumina. These systems do not prefer the formation of tri-strontium aluminate.

The tri-strontium aluminate hexahydrate (Sr_3AH_6) is the main product of hydration of strontium aluminate cement. The thermal decomposition of analogical C_3AH_6 phase (Eqs.6 and 7), the calculation of the dependence of equilibrium composition on the temperature (Fig. 35) as well as the thermal analysis of hydrated samples (Figs.5 and 15(a) in Chapter 5) indicate the formation of strontium hydroxide during thermal decomposition of Sr_3AH_6. Assuming analogical thermals decomposition sequence as for C_3AH_6, it can be written as:

$$Sr_3AH_6 \rightarrow Sr_3AH_{1.5} + 4.5\,H \qquad (51)$$

$$Sr_3AH_{1.5} \rightarrow \textit{anhydrou sphases} + 1.5\,H \tag{52}$$

The theoretical mass loss on ignition for Sr_3AH_6 phase is 15.56+5.19=20.75 %. The mass loss of 15.36+4.98=20.34 % can be read from the first and second step of TG curve in Fig.15 in Chapter 5. That indicates that the thermal decomposition of tri-strontium aluminate hexahydrate can be described by Eqs.51 and 52. Therefore it should be written as:

$$7\,Sr_3AH_6 \rightarrow Sr_{12}A_7H{+}9SrH{+}32\,H. \tag{53}$$

Chapter 7

Blended and Multicomponent Cements Based on Strontium Aluminate

1. Blended cements

The term **blended cement** (**multicomponent cement**) is usually used to denote inorganic binder that contains a mineral additions in combination with Portland cement and often also with calcium sulfate. Blended cements usually contain single blending component as an addition, in the case that two or more blending component are applied, the term **multicomponent cement** is used. A variety of industrial by-products and natural materials quantify as mineral additions, and may be used as constituents of inorganic binders of cementitious systems. These materials may possess their own cementitious properties, or they may be latent hydraulic, pozzolanic, or even non-reactive [12,387,599-601].

1. **Hydraulically reactive materials**: these materials exhibit the cementing properties when mixed with water, without the presence of any other constituent acting as activator. Some fluidized-bed ashes or industrial slags are the examples of such materials [602-604].

2. **Latent hydraulic materials**: are also able to react hydraulically, but only in the presence of at least small amounts of suitable activator ($Ca(OH)_2$, NaOH or $CaSO_4$). They are usually glassy or amorphous and contain significant amount of CaO, SiO_2 and Al_2O_3. Blast furnace slag (Fig.1(c)) is the most widely used latent hydraulic material [505,506].

3. **Pozzolanic materials or pozzolans**: when mixed only with water, they do not exhibit cementing properties. These materials can react with calcium hydroxide in the presence of water at ambient temperatures, to form calcium silicate/aluminate hydrates. Variety of materials of natural or artificial origin, such as fly ash, microsilica (Fig.1(a)), calcined clays (b) and shales, diatomaceous earth, etc.have these properties [3,607-610].

4. **Non-reactive constituents of inorganic cements**: do not react chemically, but modify the rheology of fresh paste and some properties of hardened material.

ASTM C2 19 defines blended hydraulic cement as hydraulic cement consisting of two or more inorganic constituents (at least one of which is not Portland cement or Portland cement clinker) which separately or in combination contribute to the strength gaining properties of cement (made with or without other blendings constituents, processing additions and functional additionsadded during grinding) and are added by intergrinding or blending.

Figure 1. SEM image of microsilica (a), calcined kaolin (b) and blast furnace slag (c).

The types of blended cements covered by this specification are divided as follows [611]:

Portland blast-furnace slag cement;

Portland-pozzolan cement;

Slag cement;

Pozzolan-modified Portland cement;

Slag modified Portland cement.

1.1. Blends of strontium aluminate and aluminate cement

The effect of calcium aluminate onto the hydration of strontium aluminate was investigated by the calorimetric experiment (Fig.2) using the blend of both constituents. Isothermal calorimetric measurement shows that the main hydration effect of calcium aluminate cement[1] is delayed in time, the peak height decreases while the peak width increases with increasing content of strontium aluminate in the mixture. The effect of fast evolution of heat after mixing strontium aluminate cement with water (Chapter 5.2) increases with the content of it the sample. After the first 95 hours of hydration the hydration curves of all samples drop down to the baseline, but the cements still produces the heat flow of 0.09 (Secar 51), 0.11 (20:1), 0.09 (10:1), 0,20 (5:1), 0.08 $J \cdot g^{-1}$ (1:1).

Therefore, increasing amount of SrAC in CAC increases instantaneous heat released immediately after mixing with water and also increases the length of induction period and the width of the main hydration effect.

Hexagonal CAH_{10} hydrate as the main product of hydration was recognized by powder XRD analysis of the product of hydration (Fig.3). Furthermore, the diffraction lines of the rest of unhydrated calcium aluminate ($CaAl_2O_4$) and minor constituents of applied AC such as tetragonal gehlenite ($Ca_2Al_2SiO_7$) and calcium titanium iron oxide ($Ca_3TiFe_2O_8$), appeared.

The scanning electron microscopy of hydrated cement stone (Fig.4) shows the formation of well crystallized hexagonal product of CAH_{10} (Table 6.1). The hydrates of acicular crystal

1 Information about the hydration of AC cement can be found in Chapter 5 and 6.

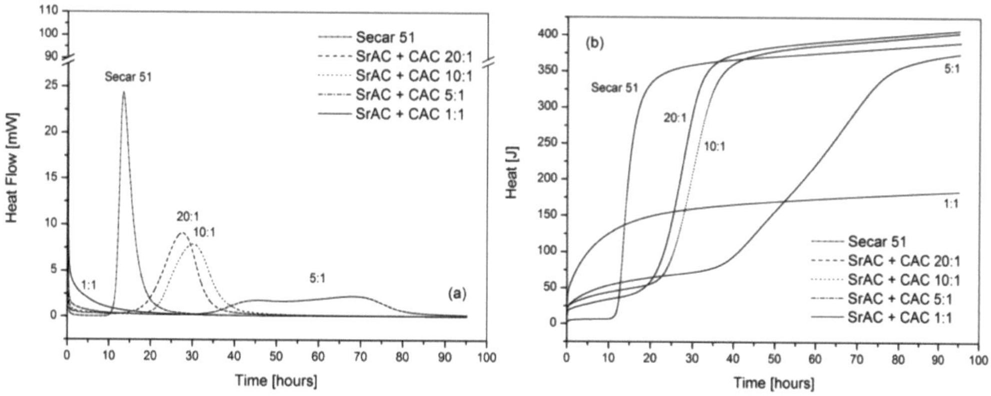

Figure 2. Heat flow (a) and heat of hydration (b) for the blend of strontium aluminate and calcium aluminate cement.

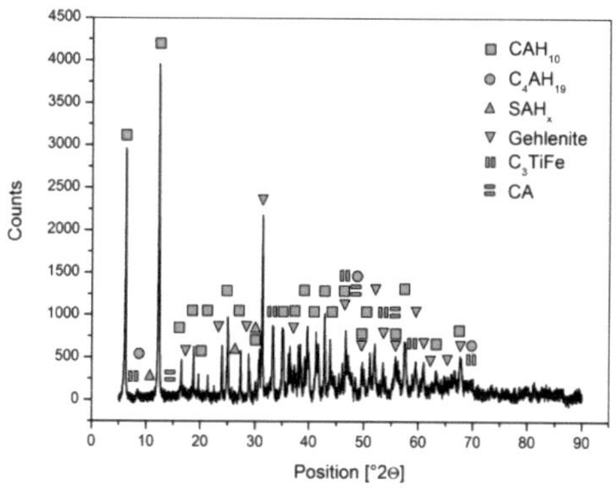

Figure 3. X-ray diffraction analysis of cement stone of SrAC and CAC blend in the mass ratio of 10:1.

character composed of radiating mass of slender, needle-like crystals with the hexagonal base were formed. These crystals are the main product of hydration in multicomponent AC cement as well (Chapter 7.4).

Thermal analysis of hydrated sample (Fig.5(a)) shows huge endothermic peak at 120 °C and smaller endothermic peat at 250 °C. The first effect is caused by the loss of molecular water of CAH_{10}. Since much water is bound very weakly, the water loss starts with drying. The peak intensity and its shape are then affected by initial drying of the specimen as well. The dehydration of crystalline CAH_{10} takes place in two steps. Firstly, interlaminar water is released and subsequent loss of water leads to the collapse of the structure. In the case that water cannot evaporate easily, the thermal decomposition of monocalcium aluminate hydrate leads to the tricalcium aluminate hexahydrate (C_3AH_6, hydrogarnet) and cubic phase of gibbsite (AH_3):

$$3\,CAH_{10} \rightarrow C_3AH_6 + 2\,AH_3 + 18\,H \qquad (1)$$

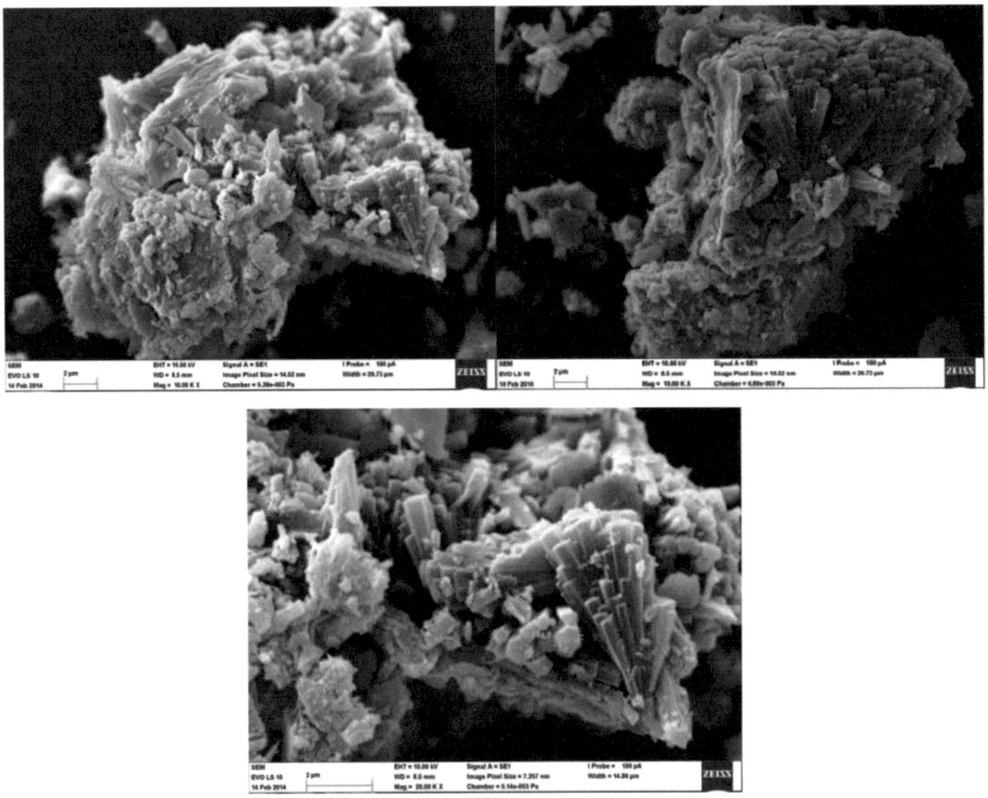

Figure 4. SEM image of hydrated cement stone of AC and SrA blend in the mass ratio of 10:1.

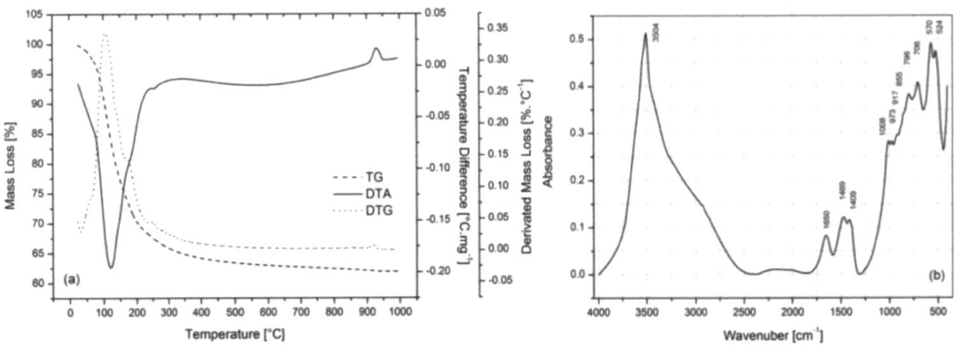

Figure 5. Thermal analysis (a) and infrared spectroscopy (b) of hydrated cement stone.

The second small peak is attributed to the dehydration of AH_3, C_3AH_6 [7,12,269,612] and Sr_3AH_6 (please see Chapter 6.5). The dehydration of hydrogarnet leads to the mayenite (dodeca-calcium heptaaluminate, $C_{12}A_7$) and $Ca(OH)_2$ [613]:

$$7\,C_3AH_6 \rightarrow C_{12}A_7 + 9\,CH + 33\,H \tag{2}$$

The thermal decomposition of calcium hydroxide takes place within the temperature range from 350 to 450 °C [613]. As the consequence of gradual dehydration of $C_{12}A_7$ (mayenite contains 1.3 % of water which corresponds to the formula $C_{12}A_7H$ [12,613]), the mass of sample decreases to the temperature of 900 °C. The effect on the right shoulder of DTA and DTG peak that appears near to the temperature of 170 °C indicates the presence of C_2AH_8 in the hydrated sample [483].

Infrared spectrum of hydrated cement stone is shown in Fig.5(b). In general the strong and broad band observed around 3400–3000 cm^{-1} corresponds to the OH stretching vibration of adsorbed water and water in formed hydrates. The maximum located at 3505 cm^{-1} belongs to the hydroxyl stretching vibration of hexagonal CAH_{10}. Water bending mode is placed at the wavenumber of about 1650 cm^{-1}. The band $\nu_3(CO_3^{2-})$ around 1410 cm^{-1} indicates the presence of both, calcium and strontium carbonate. The $\nu_2(CO_3^{2-})$ is very weak, but still recognizable part of the complex region below 1200 cm^{-1}. These features are located at 878 ($CaCO_3$) and 855 cm^{-1} ($SrCO_3$). The band at 1008 cm^{-1} was assigned to the stretching of Al-O bond.

The low-frequency rotational modes of hydroxyl groups (=Al-OH) of C_2AH_8 (920 and 705 cm^{-1}) and CAH_{10} (970 and 795 cm^{-1}) confirm the presence the both hydrates in the sample which were implied from the results of thermal analysis. The doublet band, that reaches the maximum at the wavenumber of 570 and 525 cm^{-1} is placed in the spectral region of characteristic frequency of the Al-O-Al stretching of AlO_6 octahedra [483,614].

1.2. Blends of strontium aluminate and Portland cement

Calorimetric data for the hydration of blend of strontium aluminate and Portland cement is shown in Fig.6.

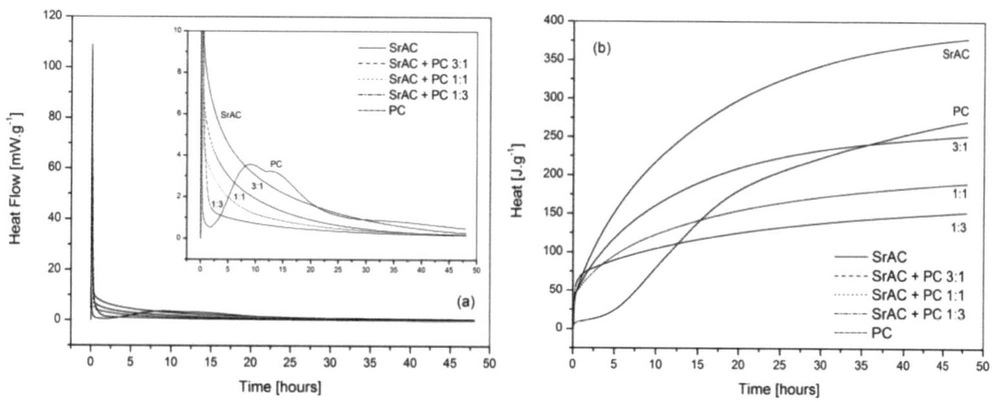

Figure 6. Heat flow (a) and heat of hydration (b) for the first 50 h of hydration of the mixture of strontium aluminate cement with Portland cement.

The heat on hydration decreases with increasing content of Portland cement in the blend. The heat flow released after mixing with water decreases more rapidly for the mixtures with higher content of Portland cement. Moreover, the comparison of calorimetric data measured for pure PC with the results of blended cement shows, that there areno measured features related to the original Portland cement.

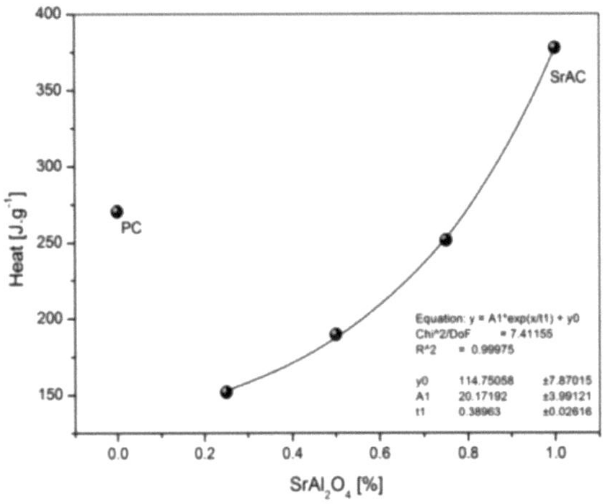

Figure 7. Influence of PC on hydration heat released by SrAC during the first 48 h of hydration.

The hydration heat released during the first 48 hours of hydration shows the highest value for strontium aluminate clinker which decreases with decreasing content of strontium aluminate clinker according to the exponential law (Fig.7). After the first 48 hour of hydration allcurves already drop down to the baseline, but the cements still produce heat flows of 0.28 (SrAC), 0.28 (3:1), 0.20 (1:1), 0.18 (1:3) and 0.50 $mW \cdot g^{-1}$ (PC). Thus the value of heat flow decreases with the content of Portland cement in the blend.

Observed decrease of heat released during the hydration of blends with increasing content of the Portland cement should be explained by the layer of hydration product of PC and ettringite formed on the surface of SrAC grains. Higher content of Portland cement also means increasing amount of gypsum in the sample and hence higher amount of ettringite, which can be formed. That often pronounces the possibility to use strontium aluminate cement as the expansion or shrinkage compensation additive as is discussed in Chapter 8.

After the calorimetric experiment, the hydration was stopped by the same way as was described in Chapter 8.3 in order to carry out the XRD, TG-DTA and IR analyses of hydrated cement stone.The SEM analysis of cement stone after the first 48 h of hydration shows the effect of Portland cement on the morphology of hydrated cement stone. The most significant features are higher amount of formed ettringite in the blends with higher content of Portland cement, i.e. higher amount of gypsum (Fig.10 and Fig.11), while only rare AFt crystals can be found in SEM images of the sample with only 25 % of Portland cement (Fig.8).

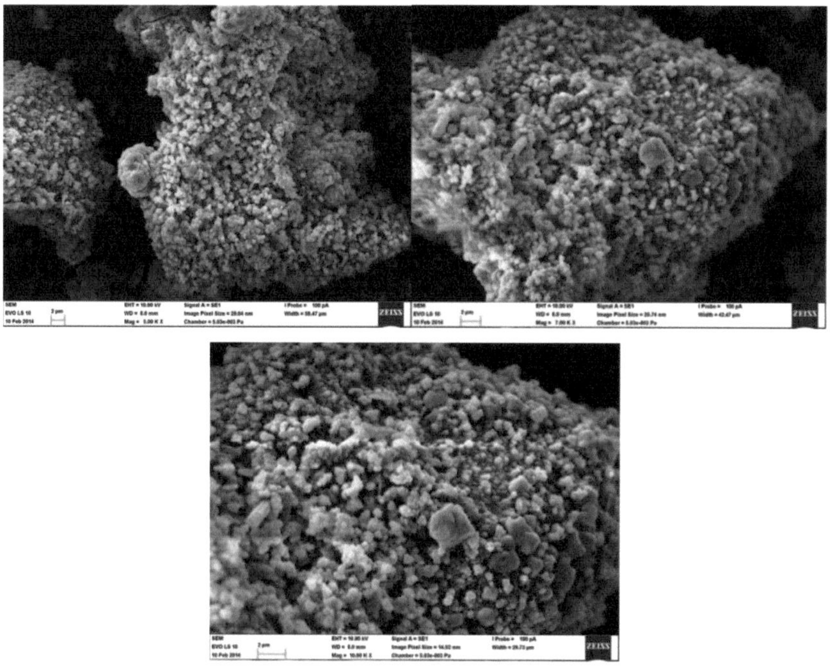

Figure 8. The hydrated cement stone prepared from the mixture of SrAC and PC (3:1).

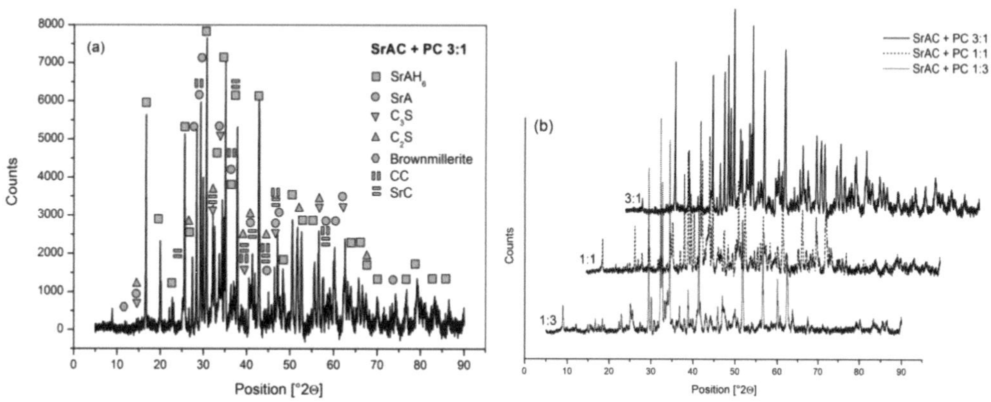

Figure 9. X-ray diffraction analysis of the blend of strontium aluminate cement with Portland cement.

Large amount of formed ettringite in the blend with the highest content of Portland cement was also recognized in XRD pattern of hydrated cement stone (Fig.9). It is an interesting fact, that the features of ettringite not only disappear from the diffraction pattern, but are substituted by growing diffraction lines (18.35, 17.43, 15.43, 25.04 and 22.34° on 2Θ scale) of new phase. It could be supposed, that is formed with increasing content of strontium aluminate cement in the blend.

Figure 10. The hydrated cement stone prepared from the mixture of SrAC and PC (1:1).

Infrared spectrum of hydrated cement stone shows that the intensity of band related to the stretching of OH groups of tri-strontium aluminate hexahydrate decreases with decreasing content of strontium aluminate clinker in the blend. On the contrary the features of hydration products of Portland cement such as the amounts of portlandite, ettringite and CSH gel increase. These results are in agreement with X-ray diffraction analysis (Fig.9), electron microscopy (Figs.8-11) as well as thermal analysis (Fig.12).

Increasing intensity of endothermic effect related to the thermal decomposition of hydrogarnet (Sr_3AH_6, please compare with Fig.5(a) in Chapter 5) at 275 °C and decreasing intensity of the first endothermic peak and its shift to lower temperatures are one of the most expressive features of increasing content of hydration products of PC. The blend of strontium aluminate and Portland cement in the mass ratio of 1:1 shows the highest carbonation as well as portlandite.

The infrared spectrum and thermal analysis show the features of both, calcium carbonate and strontium carbonate. The intensity of DTG peak of calcium (680°C) and strontium carbonate (855 °C) reflects the composition of the blend. The solid state synthesis of strontium aluminate (Eq.43 in Chapter 4) turns the effect on DTA to exothermic, while sole thermal decomposition of calcite is naturally the endothermic process.

Figure 11. The hydrated cement stone prepared from the mixture of SrAC and PC (1:3).

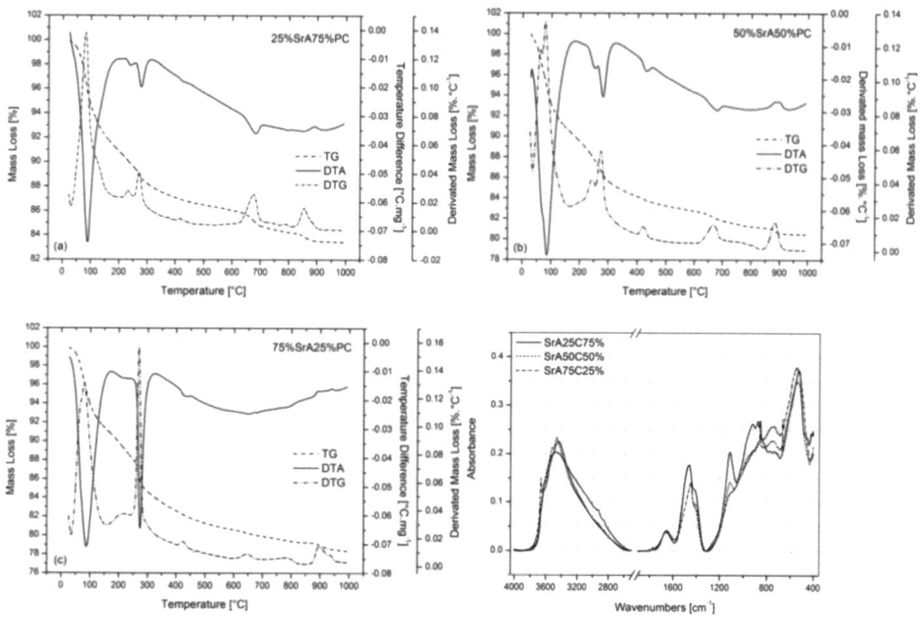

Figure 12. Thermal analysis (a-c) and infrared spectroscopy (d) of hydrated SrA – PC blend.

2. Multicomponent cements

The addition of strontium aluminate cement into calcium aluminate cement increases the time of hydration period and slows-down the rate of heat release (Chapter 7.1.1). It may be useful in order to avoid possible overheating of large constructions caused by setting calcium aluminate cement. The possibility of the preparation of multicomponent cements with admixtures (Fig.1) of pozzolanic materials (microsilica (MS) and metakaolinite (MK)) and latent hydraulic materials such as blast furnace slag (BS) are discussed in this chapter.

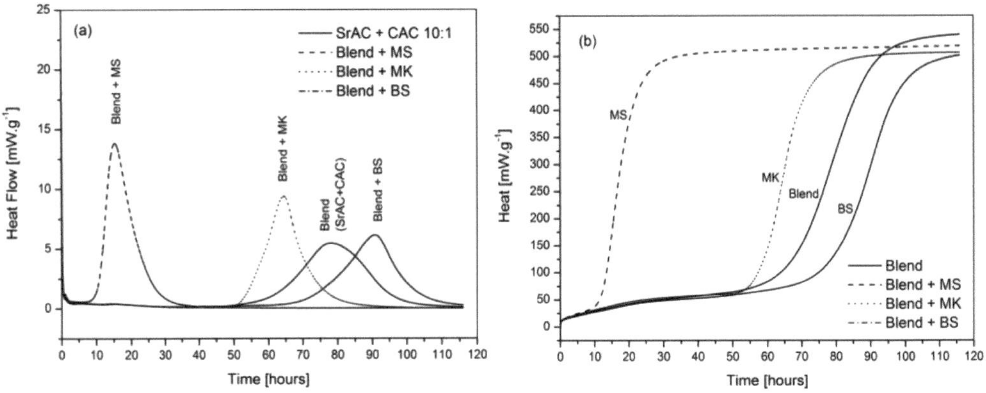

Figure 13. Heat flow (a) and heat of hydration (b) released from the multicomponent cement during the first 128 h of hydration from the blend of CAC and SrC (1:10) with microsilica (MS), metakaolinite (MK) and blast furnace slag (BS).

Isothermal calorimetry performed with the blends of strontium-, calcium aluminate cement containing the materials mentioned above is shown in Fig.13. The main hydration effect of the mixture SrAC and CAC without mineral admixture (NA) started after 50 hours from mixing with water, the heat flow from hydrated sample reached the maximum after 78 hours and was asymptotically falling to the baseline till the 110th hour of hydration. The heat flow from hydrated sample was 0.11 mW g^{-1} after the 115th hour of hydration.

The highest heat flow was measured for microsilica. Microsilica also significantly accelerated the occurrence of the main hydration effect, which began at 7.5 h after mixing with water. The main hydration effect reached the maximum at the time of 15 h and was falling down to the baseline till the 40th hour of hydration. The heat flow from hydrated sample was 0.04 mW g^{-1} after the 115th hour of hydration.

Metakaolinite also increased the heat flow release from the sample and the maximum rate of hydration (64 h). Although the main hydration effect started almost at the same temperature as for the mixture without mineral admixture, it was dropping down to the baseline much faster. Total hydration heat was almost the same for all multicomponent cements (Fig.13 (b)), however, metakaolinite accelerated the hydration of the SrAC-CAC blend. The heat flow from hydrated sample was 0.02 mW g^{-1} after the 115th hour of hydration.

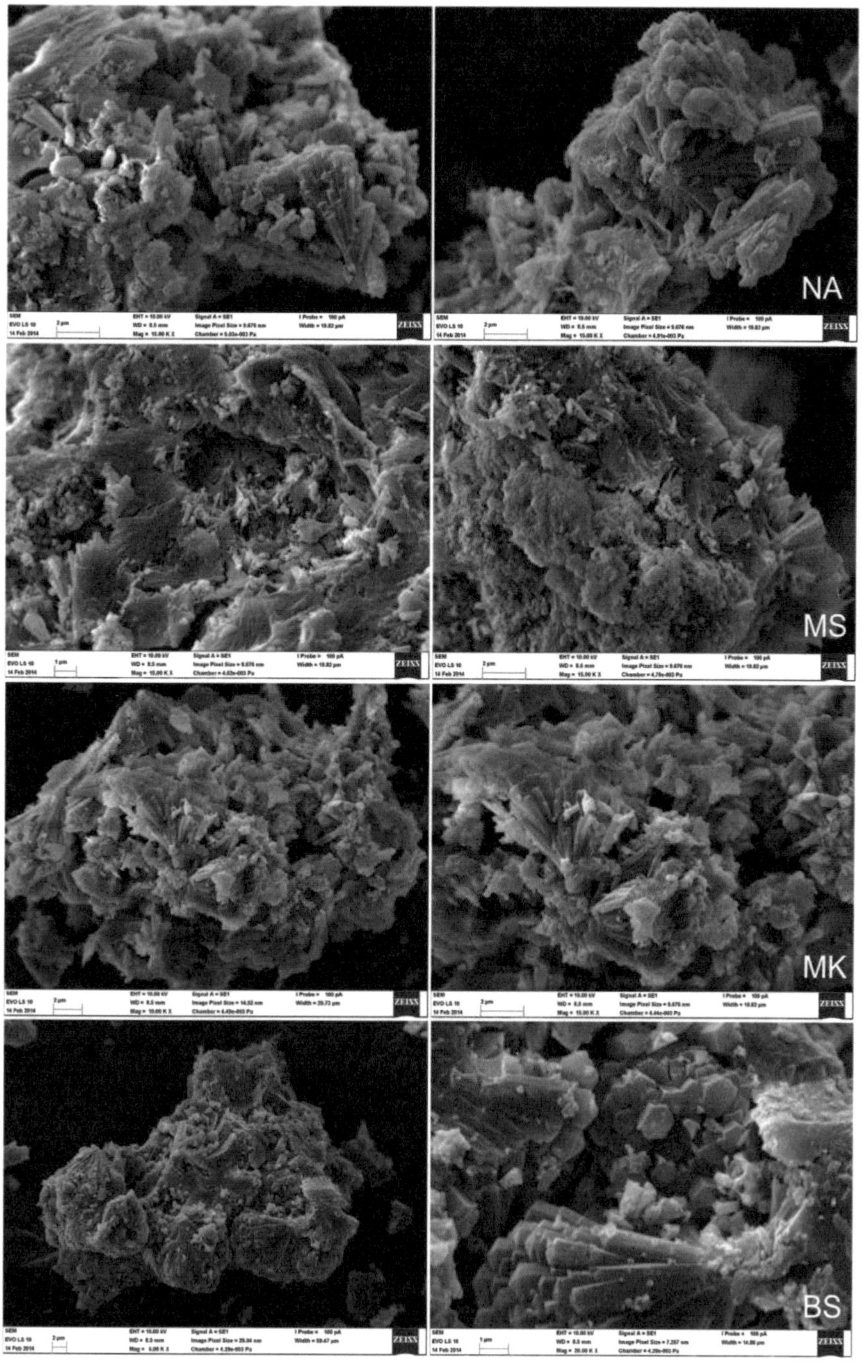

Figure 14. SEM images of hydrated cement stones consisting of CAC+SrA blend (10:1) with 10 % of microsilica (MS), metakaolinite (MK) and blast furnace slag (BS).

The main hydration effect of multicomponent cement with blast furnace slag arises 60 hours after mixing with water. The maximum of the heat flow peak was slightly higher, but it was delayed in comparison with the blend without mineral admixture. The peak was dropping to the baseline till the time of 115 hours after mixing with water. The heat flow from hydrated sample was of 0.27 mW g^{-1} after the 115th hour of hydration.

The effect of mineral admixture on the hydration of multicomponent cements should be expressed as follows:

- Heat flow peak height: MS >> MK >> BS > NA.
- Length of induction period: BS > NA > MK >> MS.

Electron microscopy of hydrated cement stones shows the formation of CSH_{10}, the shape of which is well developed in the sample containing blast furnace slag (BS) and in the blend of calcium-and strontium aluminate cement without mineral admixture (NA).

3. Effect of substitutions on the properties of strontium aluminate cements

The ionic substitutions in the structure of strontium aluminate as the main hydraulic phase of SrAC can induce some modifications in the structure, reactions and properties. As was mentioned in Chapter 1, Sr can partially replace Ca in numerous metal oxides. Therefore, it is often possible to synthesize single-phase mixed oxides, which change the reactivity with water and the setting behaviour of clinker phases or can be used for the preparation of multicomponent cements (described in Chapter 7.2).

There is a complex distribution of Ca and Sr over available sites in $Ca_{3x}Sr_xAl_2O_6$ (where $0 \leq x \leq 3$), but this distribution is not present in hydrated materials where only single Ca/Sr site exists. Therefore, the reaction proceeds by the dissolution of oxide and by the precipitation of corresponding hydroxide at crucial concentration. Since the solubilities of Ca and Sr compounds are likely to be different (in general Ca salts are more soluble), the precipitation of Sr hydroxide will commence prior to that of Ca hydroxide. In the other words, the hydration reaction is not a topotactic reaction. Alternatively, since Ca and Sr occupy different positions in the $Ca_{3x}Sr_xAl_2O_6$ lattice, it is possible that the reaction at Sr-rich phase is favored, preferentially leaching this phase into the solution and favoring the precipitation of Sr-rich hydroxide [8,615].

The investigation of deuterated samples shows that the hydration of mixed calcium-strontium aluminates of the $Ca_{3x}Sr_xAl_2O_6$ type produces a mixture of isostructural (space group of) cubic hydrogarnet phases. The hydrogarnet structure consists of four asymmetric units with the disordered Ca/Sr cations occupying one site within a distorted cube. These cations are displaced so that there are four short and four somewhat longer M-O distances. Al has approximately octahedral environment. In general, the Al-O distance (and the lattice size) increases slightly as the Sr content increases. The difference is however relatively small, the distance increases from 1.910 to 1.945 Å [8].

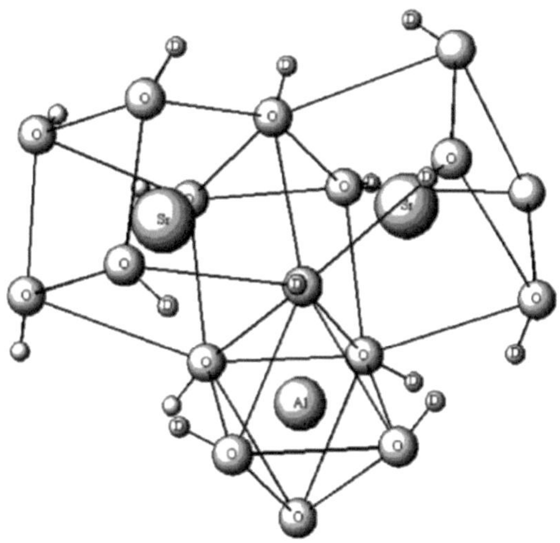

Figure 15. Structure of AlO_6 octahedra and dodecahedra according to Prodjosantoso [8].

The formation of two or three new phases from the initial single-phase oxides, suggests that these new phases may be a miscibility gap in the $Ca_3Al_2(O_4H_4)_3$-$Sr_3Al_2(O_4H_4)_3$ series. Once formed, mixed metal hydroxides seem to be stable and there is no suggestion of Ca^{2+}/Sr^{2+}ions exchange [8,416,615].

Strontium can also replace calcium in the tetra-calcium aluminoferrite ($Ca_4Fe_2Al_2O_{10}$, C_4AF), which is one of the five major components (~10 %) of Ordinary Portland Cement (OPC) and Calcium Aluminate Cement (CAC). The compound is commonly named according to naturally occurring mineral brownmillerite of similar composition or as ferrite. However, a continuous solid solution $Ca_2Al_xFe_{x-1}O_5$ exists, where $0 \leq \leq 0.7$ and both $Al^{3+}$and Fe^{3+}cations occupy the tetrahedral and octahedral sites. Therefore C_4AF is only a point in this series of solid solution for x=0.5. The end member C_2A, where x=1, can be prepared only at increased pressures of 2500 MPa. The series $x < 0.7$ is not quite isostructural, and the space group changes from to near to $x > 0.33$ [7,12,616-618].

Ordered vacancy perovskite-related type materials such as brownmillerites can be expressed by a general formula $A_2B_2O_5$ or $A_2BB'X_5$ (Fig.16(a)), where B=transition metal cations with octahedral co-ordination and B′=transition metal cations with tetrahedral coordination. The materials adopting the brownmillerite structure consist of a single B cations or of two different B cations, which are commonly disordered over both octahedral and tetrahedral sites or may be ordered according to the preference of transition metal cations for a particular co-ordination environment [619]. The oxygen nonstoichiometry of brownmillerite also changes with the oxygen partial pressure and the temperature [618,620-622].

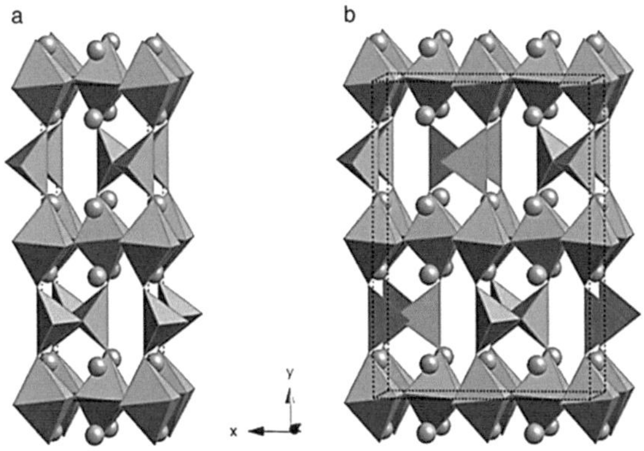

Figure 16. The conventional chain-ordered Ibm2 structure (a) and the Pbma supercell (b) of $Sr_2Fe_2O_5$ [622].

The brownmillerite structure can be explained by removing ½ of the oxygen atoms from the layer of the cubic perovskite-type structure, consisting of octahedra, creating vacancies along the [110] direction. Corner-sharing BO_6 octahedra alternates with layers of $B'O_4$ tetrahedra arranged in chains parallel to (100) direction. The crystal structure can be described as a sequence of B′–BB′–B–B′–B parallel to (010) direction. Cations A are required for the charge compensation and they occupy voids between $BO^{6-}$and $B'O^{4-}$polyhedra. Cations A are coordinated by 8 oxygen neighbours forming a distorted bicapped trigonal prism. The presence of 'channels' of oxygen vacancies between these chains allows some brownmillerites and their derivatives to display ionic conductivities comparable to the stabilized zirconia systems [618,622,623].

The displacement of tetrahedral B and O ions from their ideal positions upon the formation of brownmillerite from the parent perovskite gives rise to two opposite orientations for the tetrahedral chains, which can be assigned 'left' (L) and 'right' (R) [621,624].

The symmetry of brownmillerite is thus determined by the distribution of L-and R-chains in the structure, with three commonly recognized configurations [622]:

i. Pure L (or R), resulting in the *Ibm2* space group (Fig.16(a));

ii. Complete L/R disorder (Icmm);

iii. Strict alternation of L and R layers along b direction (Pcmn).

The brownmillerite structure type of mixed ionic/electronic conductor $Sr_2Fe_2O_5$ has been well-established since 1970s [625]. Although the diffraction studies of $Sr_2Fe_2O_5$ have clearly established a body-centered unit cell, the assignment of either the Icmm or models to describe the tetrahedral chain arrangement has varied among experiments [626]. Electron diffraction (ED) images revealed the evidence for short-range $Sr_2Fe_2O_5$ superstructures defined by complex ordering patterns of L and R chains (Fig.17) both within and between the tetrahedral layers [627]. Single crystal neutron diffraction revealed a supercell (Fig.16(b)) with a doubled axis (space group), compared to the conventional model containing disordered left-

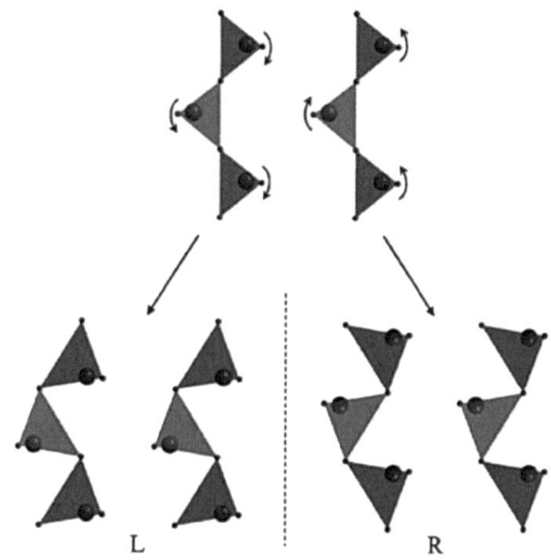

Figure 17. Cooperative twist yields to left or right-handed chains [624].

and right-handed tetrahedral chains. Therefore, the model conventionally used to describe the local chain order of $Sr_2Fe_2O_5$ is incorrect [622].

Transition brownmillerite phases $Ca_{1-x}Sr_xFe_{2.5+\delta}$ where $0.3 \le x \le 0.7$ are studied predominantly due to their promising applications in electrochemical devices at high temperature such as ceramic membranes for oxygen separation, electrodes of solid oxide fuel cells (SOFCs), electro- and photocatalysts, battery electrodes and sensor materials, etc. [620,622,628,629,630]. The transition brownmillerite phases with the composition of $La_{1-x}Sr_xCo_{0.8}Fe_{0.2}O_{3-\delta}$ can be prepared under low partial pressures of oxygen, where the value of > 0.7 [629].

The $LaACuGaO_5$, where A=Ca, Sr, brownmillerite system consists of alternating layers of corner-sharing CuO_6 octahedra and GaO_4 tetrahedra stacked along the long axis, with La and A cations statistically occupying the 10-fold cavities within the structure [619,631,632]. Anion-vacancy ordered 'triple layer' structure, related to the brownmillerite structure by the insertion of an additional layer of octahedra has the composition $LaCa_2Fe_2GaO_8$ [633].

Mn-containing brownmillerites ($A_2MnBO_{5+\delta}$, where A=Ca, Sr and B=Ga, Al) may be perspective compounds as low-field CMR materials. The system $Sr_2MnGaO_{5+\delta}$ (Fig.18) consists of alternating MnO_2 and $GaO_{1+\delta}$ layers ($-MnO_2-AO-GaO-AO-MnO_2-AO-GaO-MnO_2-$) [619,634-638].

Perovskite oxide with the general formula of ABO_3 is a kind of frequently encountered structure in inorganic chemistry. Its cubic crystal structure is composed of a three-dimensional framework of corner-sharing BO_6 octahedra. $SrFeO_x$ (SFO), where $2.5 \le x \le 3.0$, were extensively studied as ferromagnet and antiferromagnet and due to their photocatalytic activity [620,639-641].

Brownmillerite-and perovskite-type compounds exhibit also some serious disadvantages including the thermodynamic instability under large oxygen chemical potential gradients,

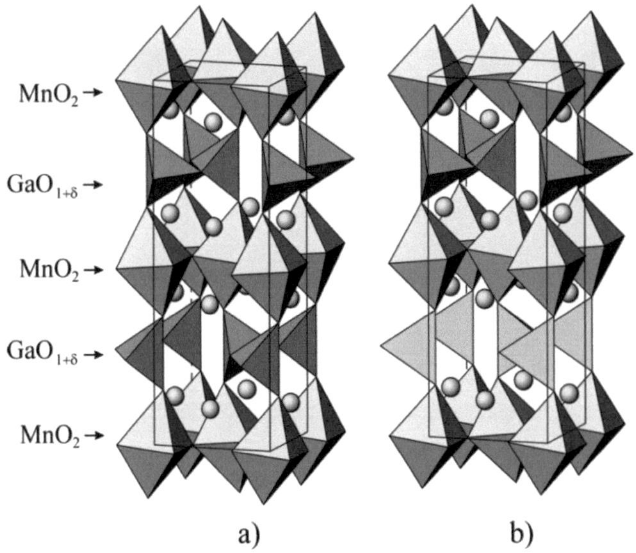

Figure 18. Crystal structure of Sr_2MnGaO_5 brownmillerites with Ibm2 (a) and Pnma (b) space symmetry. Cations are drawn as spheres [638].

poor thermomechanical properties associated with very high chemical and thermal expansion and the reactivity with CO_2 [642].The relation between the orthorhombic brownmillerite and the cubic perovskite-type structure is given as: $a \sim \sqrt{a_{per}}$, $b \sim 4a_{per}$, $c \sim \sqrt{a_{per}}$, where a_{per} is the cell dimension of the cubic perovskite [618].

The protonic conduction was discovered for the brownmillerite-type $Ba_2In_2O_5$ in 1995 by Zhang and Smyth [643] that is related to the formation of $Ba_2In_2O_5$ H_2O hydrate in a humidified atmosphere. Protons bound to oxygen in hydrate cross-links between oxygen [644-646]. In order to improve the proton conductivity it is necessary to stabilize the hydrate compound at high temperature (573 K). The dehydration temperature increases~50 K due to the substitution of In by Sc, Lu and Y. Since the substitution of In by $M^{3+}$was not possible for $x >$ 0.3 in $Ba_2(In_{1-x}M_x^{3+})_2O_5$ H_2O, further increase in the thermal stability of the hydrate compound by the substitution of In cannot be expected. Tetragonal $Ba_2Sc_2O_5$ and cubic $BaScO_{3-\delta}$ also form hydrates $Ba_2Sc_2O_5 \cdot 0.60H_2O$ and $BaScO_{3-\delta} \cdot 0.37H_2O$, but the dehydration temperatures are even lower than that for $Ba_2In_2O_5$ [647,648].

3.1. Substitution of Al_2O_3 by Cr_2O_3

Cr_2O_3 and Al_2O_3 are sesquioxides having the same corundum crystal structure, which is approximately hexagonal, involving close-packed oxide ions with $Cr^{3+}$and Al^{3+}ions occupying two thirds of available octahedral sites [649,650]. The solid solution ($(Al_{1-x}Cr_x)_2O_3$) obeying **Vegard's law**[2] [651] is formed over the entire range of compositions[3] in Al_2O_3 – Cr_2O_3 sys-

2 Empirical rule that enables to calculate the crystal lattice parameter of solid solution or alloy from the concentration of constituents.

tem. With regard to the positive influence of chromia (Cr_2O_3) on the sintration as well as physical and refractory properties of Al_2O_3 ceramics [652-656], the effect of substitution of alumina by chromia in the strontium aluminate cement was investigated.Cr-doped structures have been widely studied as ceramic pigments [649] and catalyzers [657-659].

Cr_2O_3 is not stable and changes to CrO_3 in an atmosphere in which the oxygen partial pressure is relatively high such as air. CrO_3 exhibits high vapor pressure, so it is difficult to obtain a dense sintered body of Cr_2O_3 in an air atmosphere. In order to obtain a dense sintered body of Cr_2O_3, the sintering should be performed in an atmosphere with low oxygen partial pressure in which Cr_2O_3 remains stable [653].

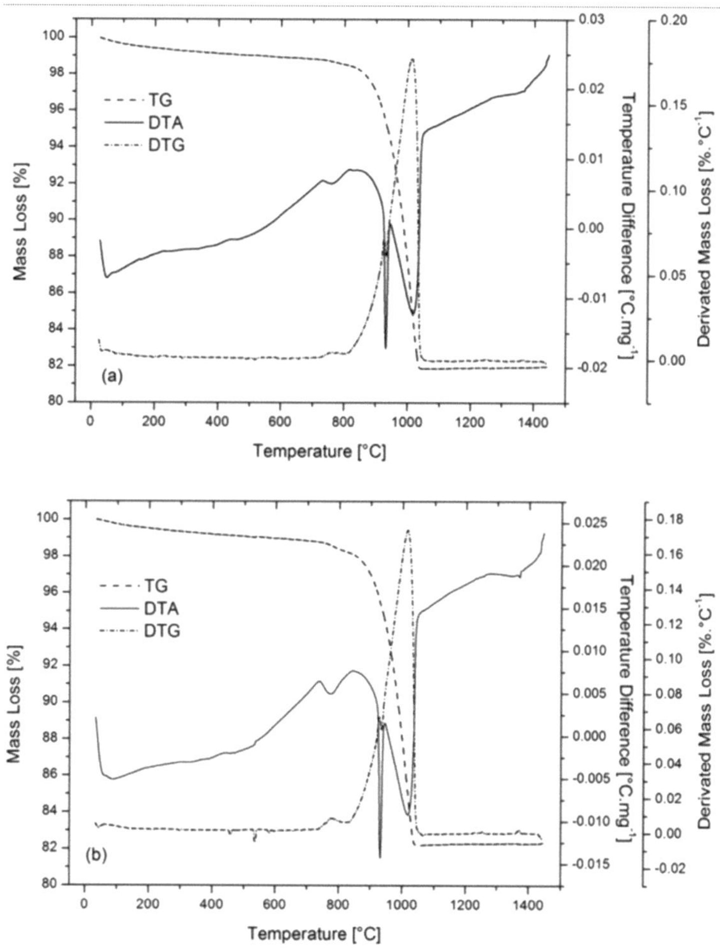

Figure 19. Thermal analysis of clinker prepared with 5 % (a) and 10 % Cr_2O_3 (b).

3 An immiscibility gap is present below 950°C, where two crystalline phases exist (alumina-rich and chromia-rich solid solutions), but the actual decomposition of homogeneous solid solution into two phases proceeds very slowly [651].

The toxicity and the transport behaviour of Cr depend strongly on its valence. The most stable oxidation states in the environment are hexavalent and trivalent. Cr(VI) exists primarily in the form of $HCrO_4^-$(bichromate) and CrO_4^{2-}(chromate), which are strong oxidants and can cause kidney tubule necrosis and, by inhalation, lung cancer. Cr(VI) compounds are typically soluble in groundwater, and thus mobile and bioaccessible. Where the intermediates to low redox potentials exist, Cr(VI) can be reduced to Cr(III) which usually forms insoluble oxides and oxyhydroxides being less bioaccessible [660].

Thermal analysis of raw materials with the substitution of 5 and 10 % Al_2O_3 by Cr_2O_3 (Fig. 19) shows increased intensity and peak temperature (from 735 to 785 °C) of endothermic process, which is probably related to the formation of chromia spinel phase. To compare with raw material without chromia (Fig.23 in Chapter 4), there is sharp transformation effect of α-$SrCO_3$ to β-$SrCO_3$ (please compare with Fig.12 in Chapter 4). It is possible to use the thermolysis of mixed alums in order to prepare mesoporous chromia-alumina [661] as pure and high reactive precursor for the synthesis. That may lead to the assumption that the addition of Cr_2O_3 into the raw meal suppresses the decomposition of strontium carbonate and leads to less reactive clinker containing the sintering additive (similarly to the effect described in Chapter 6.1.4).

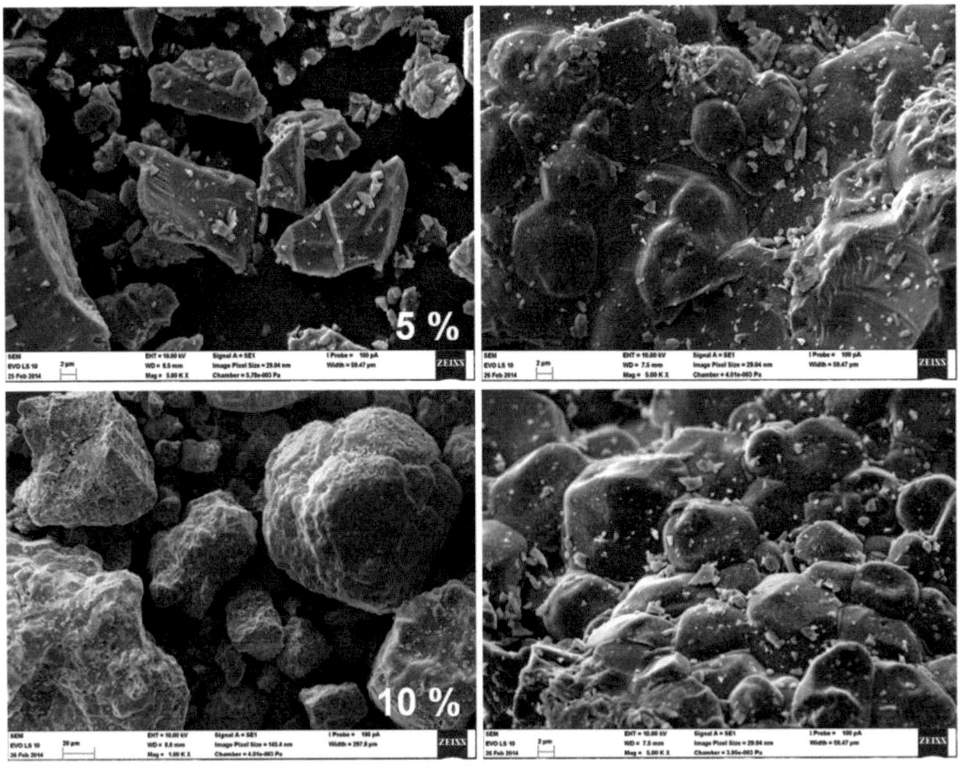

Figure 20. Scanning electron microscopy of clinker prepared with 5 % (a) and 10 % Cr_2O_3 (b).

The addition of Cr_2O_3 in to the raw meal leads to the formation of tetragonal strontium oxide chromate ($3SrO \cdot 3Al_2O_3 \cdot SrCrO_4$) [662] and strontium hexaaluminate causing the detriment of strontium aluminate as the main clinker phase (Fig.21). This endothermic process should be described by the following reaction:

$$\begin{aligned} &8\,SrCO_3 + 6\,Al_2O_3 + Cr_2O_3 + 3/2\,O_2(g) \rightarrow \\ &2\,(3SrO \times 3\,Al_2O_3 \times SrCrO_4) + 8\,CO_2(g) \end{aligned} \qquad (3)$$

The lack of SrO supports the formation of strontium hexaaluminate.

That behaviour also enables to explain the missing effect related to the formation of strontium aluminate, which is formed at higher temperature probably via the diffusion of Al^{3+}ions through Cr_2O_3-Al_2O_3 layer of formed solid solution. The opposite model, in which the spinel layer would be formed on Cr_2O_3 particles cannot explain the suppression of the formation of strontium aluminate revealed from the TG-DTA experiment.

$3SrO \cdot 3Al_2O_3 \cdot SrCrO_4$ is a pigment with hexavalent chromium and provides prepared clinker with yellow-greenish color. Since the transformation temperatures and the temperatures of endothermic effect are the same for both samples, increasing content of Cr_2O_3 seems to have no influence on the thermal decomposition of strontium carbonate (after the formation of layer on the surface of $SrCO_3$ grains the increasing abundance of Cr_2O_3 doesn't have any effect).

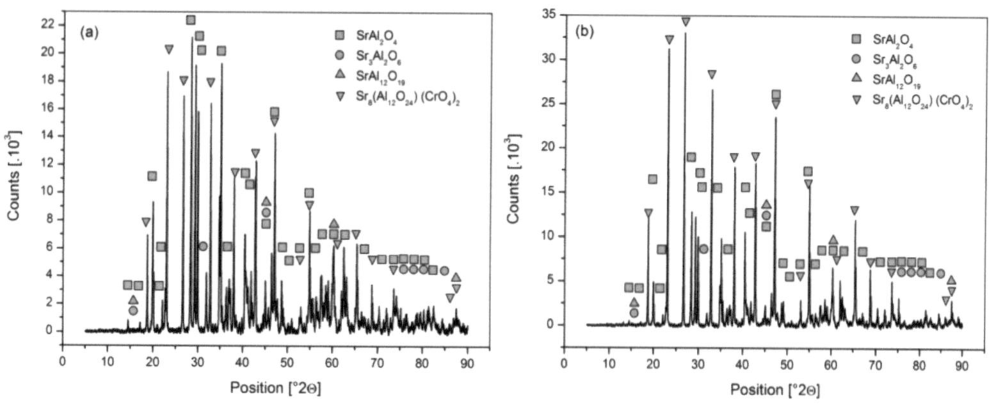

Figure 21. XRD analysis of strontium aluminate clinker doped by chromia.

The raw meal was treated to the temperature of 1500 °C under static air and ground after cooling. Scanning electron microscopy of ground clinker shows the formation of globular aggregates of rounded particles with the surface showing intersections of spherical shapes, which are better developed in the sample prepared with 10 % of Cr_2O_3 (Fig.20). Ground particles show brittle fracture through the glassy-like phase of clinker particles.

The diffraction lines of strontium oxide chromate ground clinker prepared with 10 % of chromia reaches even higher intensity than strontium aluminate. The amount of formed

SrA_6 phase seems to be decreasing with increasing content of Cr_2O_3 in the raw meal. Considerable amount of formed phases without hydraulic activity leads to the conclusion that the raw meal containing chromia should be prepared in slightly reducing atmosphere.

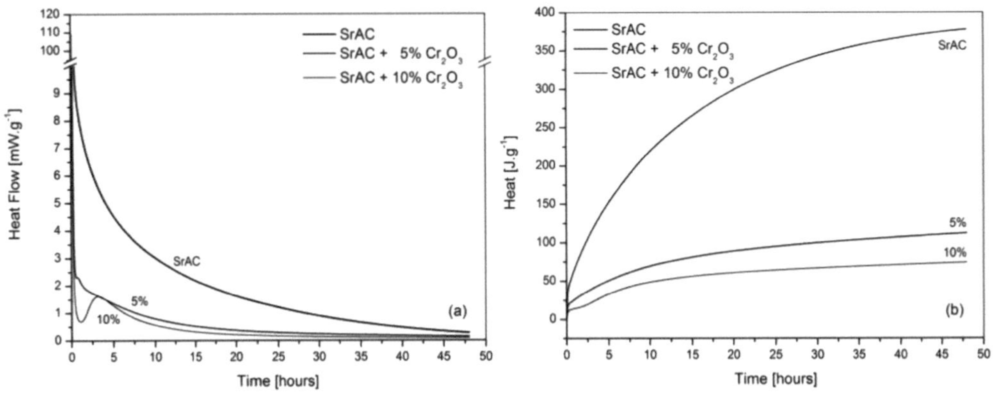

Figure 22. Heat flow (a) and heat of hydration (b) for the samples doped with chromia.

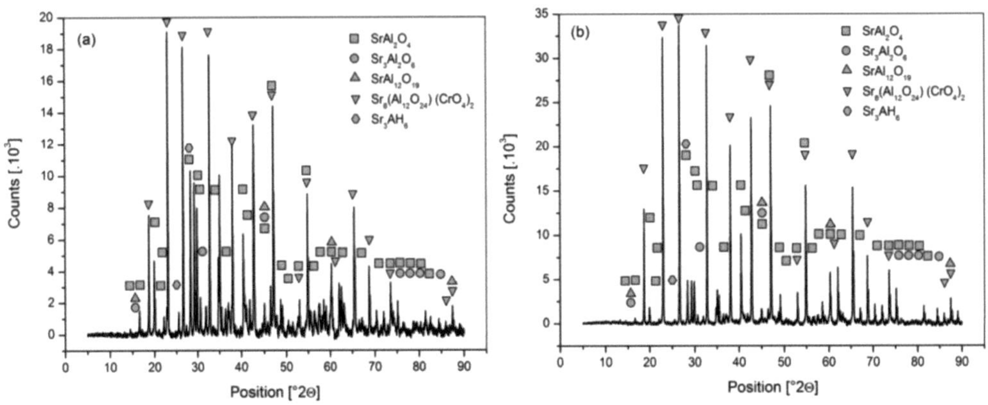

Figure 23. XRD analysis of strontium aluminate clinker after hydration.

The influence of the substitution of Al_2O_3 by Cr_2O_3 on the properties of strontium aluminate cement was investigated by isothermal calorimetric measurement of hydration at 25 °C (Fig.22).

The results show that the introduction of Cr_2O_3 into the structure of SrA significantly reduces the hydraulic activity and leads to the appearance of induction period. That behaviour is probably caused by the formation of chromium spinel (please see Chapter 10.12).

This behaviour results from decreasing content of strontium aluminate in the clinker with increased amount of added chromia. That means that newly formed phase of strontium alu-

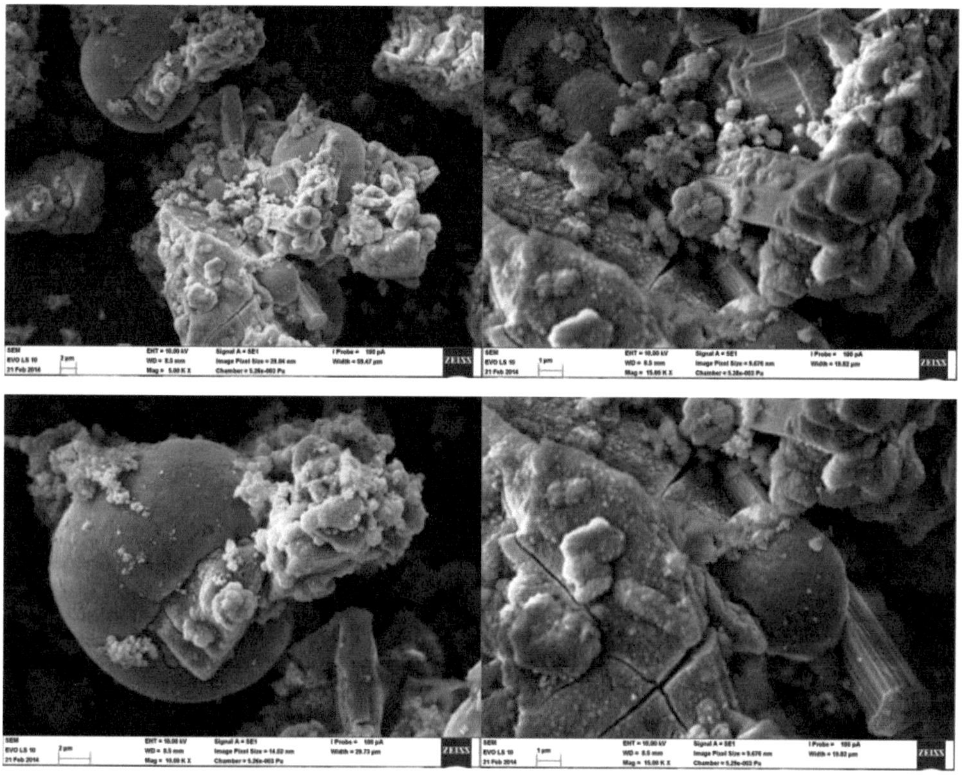

Figure 24. SEM images of hydrated cement stone for the sample prepared with 5% of Cr_2O_3.

minium oxide chromate has no hydraulic activity. X-ray diffraction analysis of hydrated cement stone supports the conclusion that strontium oxide chromate does not show significant hydraulic activity, if any. The main product of hydration remains tri-strontium aluminate hexahydrate as for the clinker prepared without chromia and the intensity of diffraction lines of strontium oxide chromate is not changed.

Electron microscopy shows that the morphology of hydration product is different from strontium aluminate cement without the addition of Cr_2O_3 into the raw meal (Fig.6 in Chapter 5). Moreover, the external appearance of hydrates changes with the amount of added chromia. Long columnar crystals, typical for the sample prepared with 5 % of Cr_2O_3 (Fig.24) are not present in the hydrated cement stone prepared from ground clinker with 10 % Cr_2O_3 (Fig.25). With increasing content of Cr_2O_3 the size of spheres was also significantly reduced.

3.2. Substitution of Al_2O_3 by Fe_2O_3, B_2O_3 and Y_2O_3

Strontium aluminate cements with the substitution of 5 % of Al_2O_3 by Fe_2O_3, B_2O_3, Y_2O_3, V_2O_5, ZrO_2, MnO_2 and ZnO were prepared in order to evaluate the influence of these compounds on the behaviour during hydration. The samples were pelletized and thermally

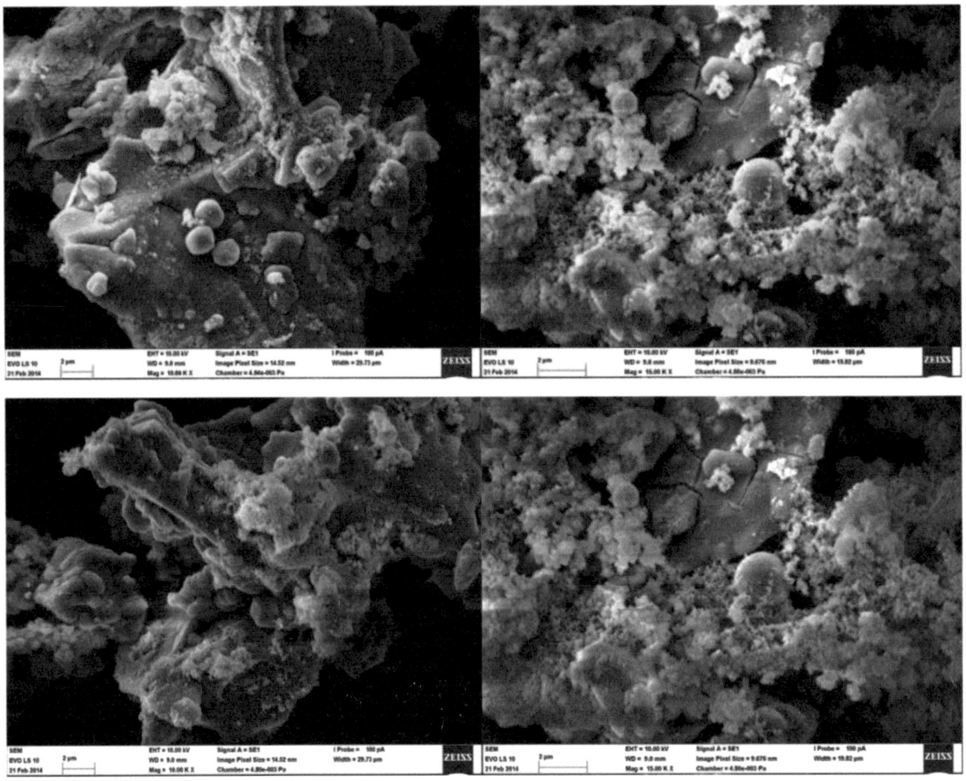

Figure 25. SEM images of hydrated cement stone for the sample prepared with 10 % of Cr_2O_3.

treated to the temperature of 1500 °C (1 hour). The thermal analysis of raw materials upon thermal treatment is shown in Fig.26.

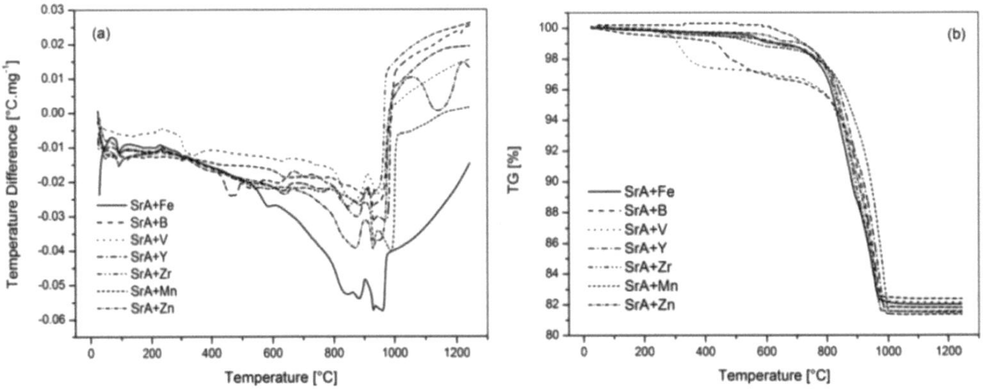

Figure 26. Thermal analysis of raw meals heated at the rate of 10 °C min^{-1}: DTA (a) and TG (b).

All investigated samples show almost the same mass loss of 8 %. The major part of mass loss is caused by the thermal decomposition of strontium carbonate that takes place within the temperature range from 700 to 1 000 °C.

Dopant element	Mechanism		R^2	E_a	A
	$g(\alpha) = kt$	Abr.		[kJ·mol⁻¹]	[s⁻¹]
Fe	$-\ln(1-\alpha)^3$	A_3	0.9925	583.06	$3.04 \cdot 10^{23}$
B	$1-(2\alpha/3)-(1-\alpha)^{2/3}$	D_4	0.9996	379.56	$3.79 \cdot 10^{12}$
V	$-\ln(1-\alpha)^3$	A_3	0.9911	664.46	$7.07 \cdot 10^{26}$
Y		A_3	0.9941	613.93	$1.54 \cdot 10^{24}$
Zr		A_3	0.9976	611.70	$5.28 \cdot 10^{24}$
Mn	$1-(1-\alpha)^{2/3}$	$F_{1/3}$	0.9977	169.61	$8.13 \cdot 10^{3}$
Zn	$-\ln(1-\alpha)^3$	A_3	0.9922	613.93	$3.74 \cdot 10^{24}$

Table 1. Influence of dopants on the thermal decomposition of strontium carbonate.

There is a significant difference compared to the sample prepared with the addition of chromia (Chapter 7.3.1) since the intensive exothermic peak related to the synthesis of strontium aluminate arises for all samples. On the contrary, this peak is smaller for the sample prepared with B_2O_3.

The model fitting method (Chapter 1.6.1) applied to the TG results provides the kinetic data listed in Table 1. The kinetics of thermal decomposition of strontium carbonate is changed by the presence of dopants (please consult with the results in Chapter 4.2). It can be seen that Fe, V, Y, Zr and Mn have similar effect on the mechanism (random nucleation) and kinetics of thermal decomposition of $SrCO_3$. The addition of B_2O_3 changes the mechanism to the non-steady state 3D diffusion and MnO_2 leads to the process driven by the rate of chemical reaction of one-third order.

The admixtures lead to the decreasing intensity of strontium aluminate diffractions (Fig. 27(a)) in the clinker after the thermal treatment. Based on this effect, they can be ordered as follows:

$$B_2O_3 > ZnO > MnO_2 > ZrO_2 > Fe_2O_3 > Y_2O_3 > V_2O_5.$$

The shift of the most intensive diffraction lines of $SrAl_2O_4$([-211], [220] and [211], Fig.27) for the samples prepared with ZnO and Fe_2O_3 indicate the substitution of $Zn^{2+} \leftrightarrow Sr^{2+}$ and $Fe^{3+} \leftrightarrow Al^{3+}$, respectively. While, the clinker prepared with ZnO shows well developed $SrAH_6$ crystals in hydrated cement stone (Fig.29), the clinker doped with Fe_2O_3 exhibits the highest heat released upon hydration (Fig.28).

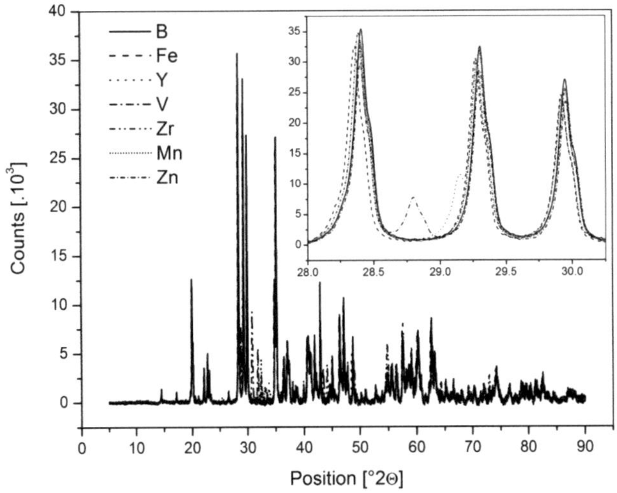

Figure 27. X-ray analysis of strontium aluminate clinker prepared with dopants before hydration.

Observed influence of B_2O_3 on the synthesis and hydration (Fig.28) of strontium aluminate clinker is caused by the ability to incorporate BO_4 unit into the framework of $SrAl_2O_4$ instead of AlO_4 tetrahedra [663]. $(BO)_4$ tetrahedra are of more ionic nature than $(AlO)_4$ due to smaller size ($r_{B^{3+}} < r_{Al^{3+}}$) and higher electronegativity (X_B=2 and X_{Al}=1.6, Pauling´s scale) of boron [377,664]. B_2O_3 is also known as a glass former, which has low melting point around 460 ∘C. It is regarded as an excellent flux to facilitate the material diffusion (pleasse see Table 1). Thus, B_2O_3 is usually added in the preparations of $SrAl_2O_4$ in order to reach lower forming temperature [663].

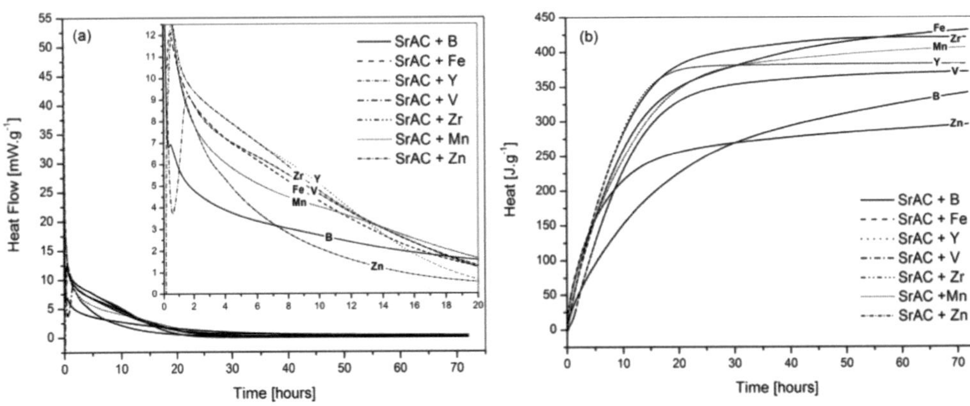

Figure 28. Hydration of clinker prepared with dopants.

The hydration of prepared samples was investigated by isothermal calorimetry at the temperature of 25 for the first 72 h of hydration (Fig.28). Since there is no induction period and large amount of heat is released immediately after mixing with water, the course of hydration of clinkers doped with ZnO and MnO_2 is similar to that of pure strontium aluminate. The heat flow then asymptotically decreases to the baseline. According to the decreasing heat of wetting, the samples can be ordered as follows:

$$ZnO >> MnO_2 > B_2O_3 > Fe_2O_3 \approx Y_2O_3 > ZrO_2 >> V_2O_5$$

Other clinkers show the induction periods the lengths of which decrease with following dopants applied:

$$V_2O_5 >> ZrO_2 > Fe_2O_3 > Y_2O_3 >> B_2O_3.$$

According to the time of maximum of main hydration peak and its height, the samples should be ordered as follows:

- Time: $V_2O_5 >> ZrO_2 \approx Y_2O_3 \approx B_2O_3 > Fe_2O_3$.
- Intensity: $Y_2O_3 \approx ZrO_2 \approx Fe_2O_3 > V_2O_3 > B_2O_3$.

Electron microscopy of hydrated cement stone (Fig.29) shows the formation of globular (grape-like) aggregates. But also the formation long columns of layered aggregates (*a*, *b* and *c*), long needle-like crystals (*b*) and well developed $SrAH_6$ crystal (*g*, *h*) can be observed. The stacked columnar structure of gibbsite aggregate indicates the option to use doped strontium aluminate cement for the preparation of iron-and aluminium based double layered hydroxides (LDH [665-672]).

Figure 30(a) describes typical Raman spectrum of pure monoclinic $SrAl_2O_4$, where the main peak at 467 cm^{-1} belongs to bending of O-Al-O in $[AlO_4]$ tetrahedra [795,1009]. The partial substitutions of Al_2O_3 and $SrCO_3$, as the reactants in solid state reaction with different metal oxides lead to theformation of new phases as is indicated by the shifts of the main peak of $[AlO_4]$ tetrahedral (Fig.30(b)).

The significant shifting was observed for the structure doped with MnO_2 and should be described as the distortion of monoclinic lattice structure. The experimentally determined influence of various metal oxides in lattice shift is ordered as follows:

$$MnO_2 > ZnO > B_2O_3 > Y_2O_3 > ZrO_2 > pure\ SrA\ phase > V_2O_5 > CaO.$$

Increasing content of boron in prepared clinker leads to the formation of large hexagonal plates during hydration (Fig.31). According to EDX (Energy Dispersion Spectroscopy), the composition of these crystals corresponds to Sr_2AH_x. The formation of well developed crystals of Sr-analogue of $C_2AH_8$was observed only for the samples doped by high amount of boron.

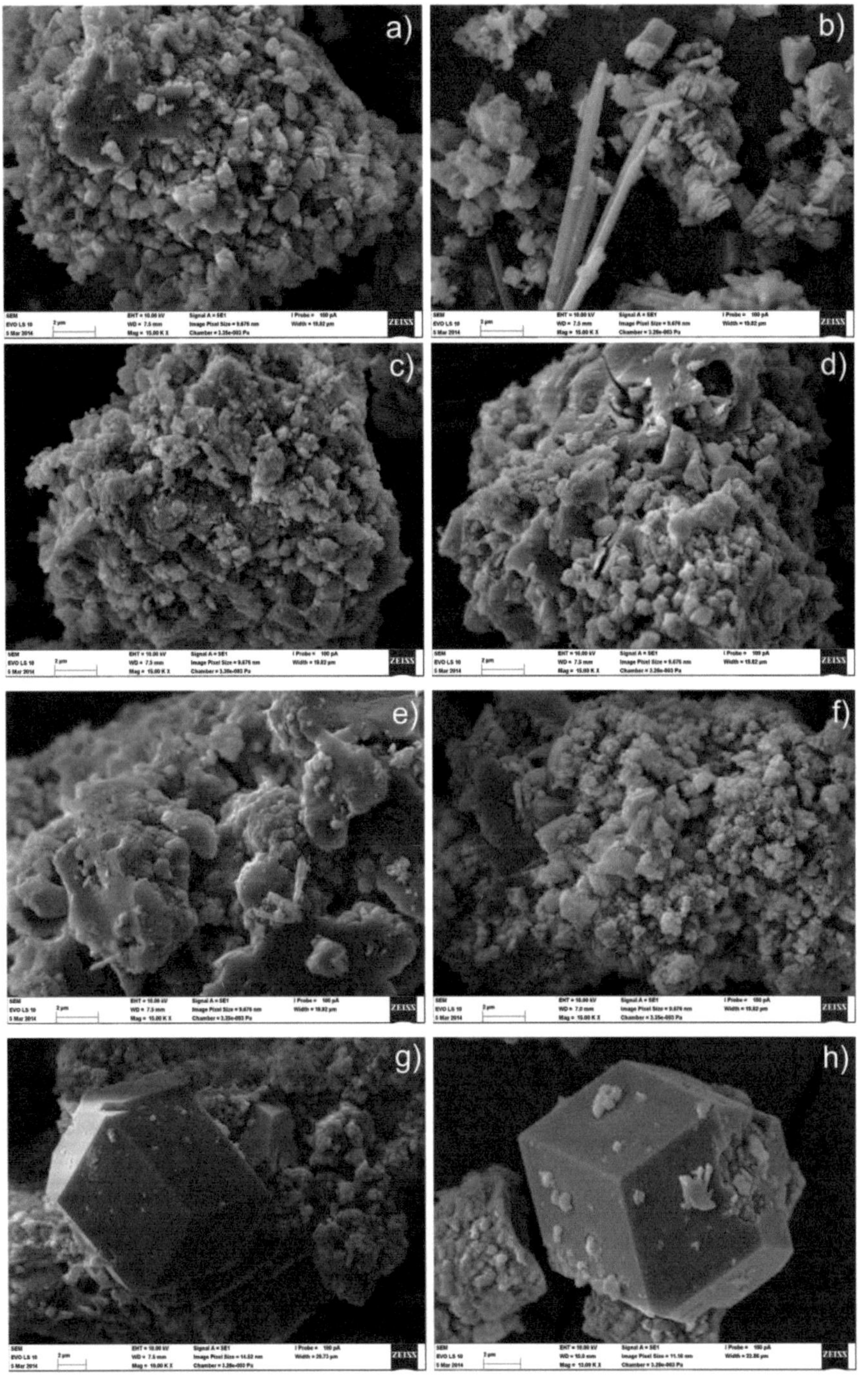

Figure 29. Electron microscopy of hydrated cement stone of clinker prepared with 5 % of B_2O_3(a), Fe_2O_3 (b) Y_2O_3(c), V_2O_3 (d), ZrO_2 (e), MnO (f) and ZnO (g) with well developed crystal of Sr_3AH_6 (h).

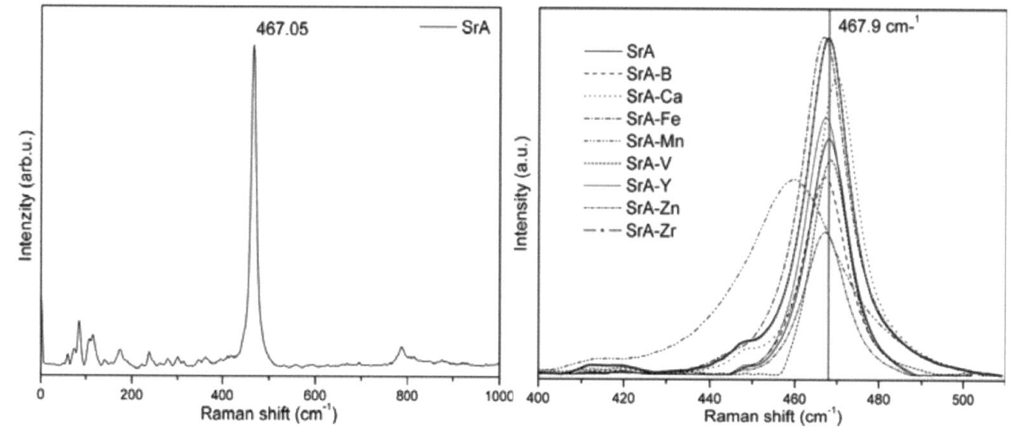

Figure 30. Raman spectra of $SrAl_2O_4$ phase (a) and doped Sr Al O_4 structures (b) (where=Ca, Zn, Mn and=B, Fe, V, Y, Zr in substitution of 5 wt.%).

Figure 31. Large hexagonal plates of $Sr_2AH_x$of hydrated cement stone with 10 % of boron.

3.3. Substitution of SrO by CaO

The substitution of Sr^{2+}in the structure of strontium aluminate by Ca^{2+} (1, 5 and 10 %) introduced as $CaCO_3$ into the raw meal leads to the substitutions in the Sr_3A during thermal processing of clinker. The content of calcium in formed solid solution of $(Sr_{3-x}Ca_x)A$ (where=0.34 (1%), 1.02 (5%) and 1.12 (10%)) increases with the extent of substitution, whereas the content of the $(Sr_{3-x}Ca_x)A$ phase decreases (Fig.32(a)). In other words, the introduction of $CaCO_3$ into the raw meal suppresses the formation of tristrontium aluminate phase during the thermal processing of clinker.

The main diffraction lines of strontium aluminate (-211, 220 and 211) are shifted to the higher position on the °2Θ scale with increasing amount of Sr^{2+}cations substituted by Ca^{2+} (Fig. 32(b)). Therefore the introduction of smaller ions (Table 1 in Chapter 8) of higher electrone-

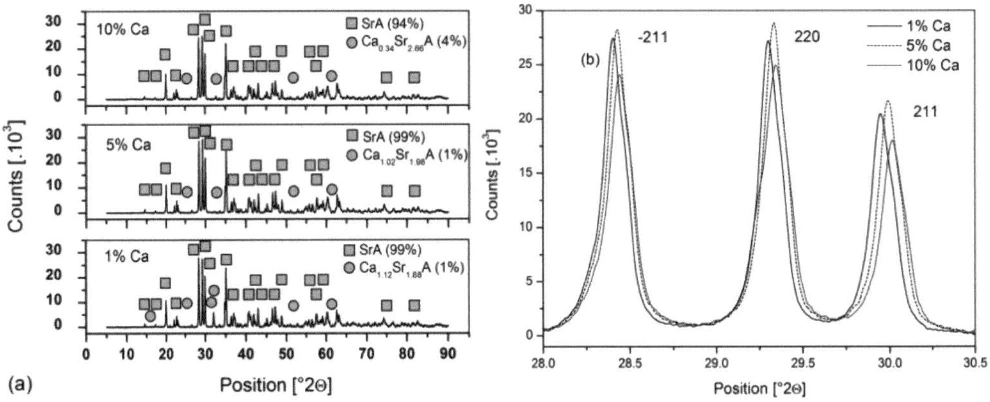

Figure 32. XRD analysis of strontium aluminate clinker doped by calcium.

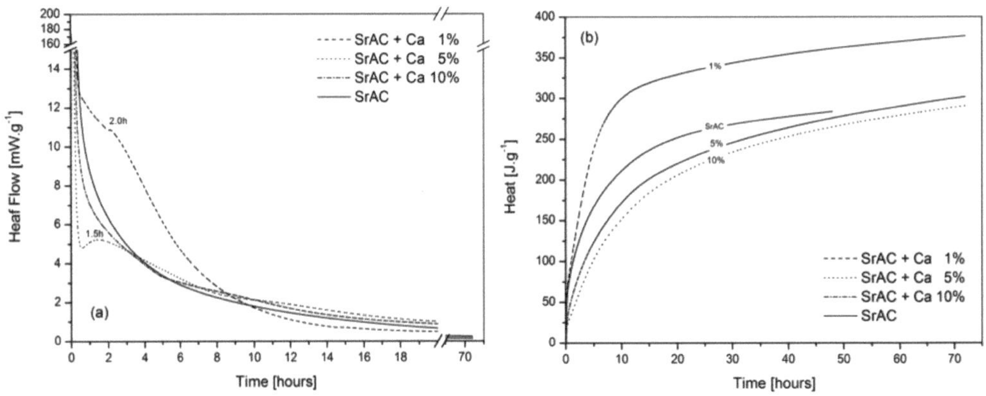

Figure 33. Heat flow (a) and heat of hydration (b) for the samples doped with calcium.

gativity leads to the decrease of volume of basic cell unit of strontium aluminate. That leads to lower reactivity with water (Fig.33, please refer to the discussion in Chapter 5.6 related to the size of opened channels in the structure of SrA and their influence on the reactivity with water).

The hydration of prepared samples was investigated by isothermal calorimetry at the temperature of 25°C for the first 72 h of hydration (Fig.33). The decrease of hydration heat (b) for the sample prepared with 5 and 10 % of calcium can be observed as the consequence of higher thermodynamic stability (please see the discussion in Chapter 1.2.4) as well as decreasing amount of formed $(Sr_{3-x}Ca_x)A$ solid solution.

The course of hydration of the sample prepared with 1% of calcium supports the results of X-ray diffraction analysis (Fig.32) indicating that the substitution in the Sr_3A structure proceeds prior to the substitution in the structure of SrA. That can be also explained via the mechanism of formation of SrA phase (please reffer to Fig.24 and the corresponding discus-

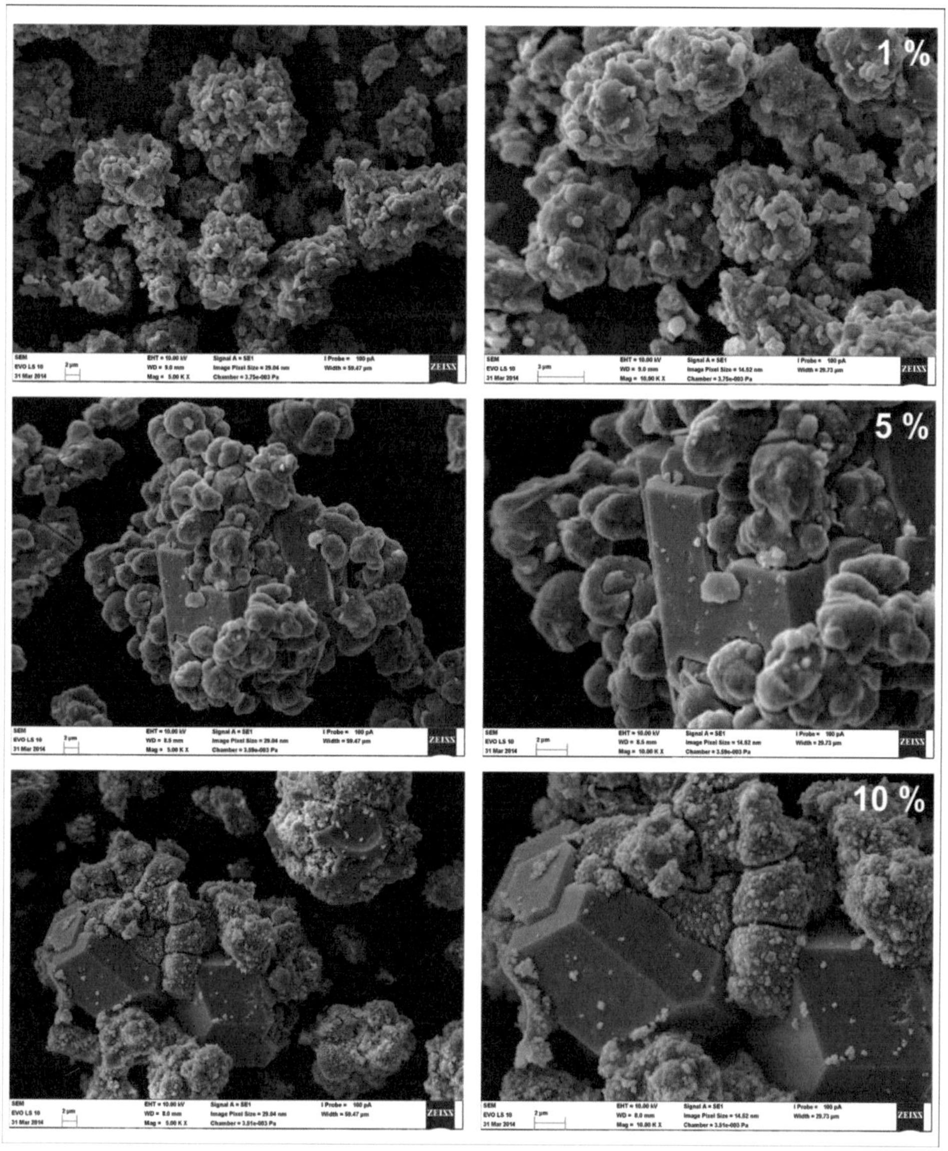

Figure 34. Electron microscopy of hydrated cement stone of clinker prepared with 1, 5 and 10% of Sr^{2+}substituted by Ca^{2+}.

sion in Chapter 4). The highest content of formed solid solution of $(Sr_{2.66}Ca_{0.34})A$[4] (Fig.32a) which has lower reactivity with water than SrA provides the explanation for the heat flow released from the sample upon hydration (please compare the heat released from SrA and Sr_3A, Fig.12 in Chapter 5).

4 Calcium has approximately two-fold dissolution heat than strontium.

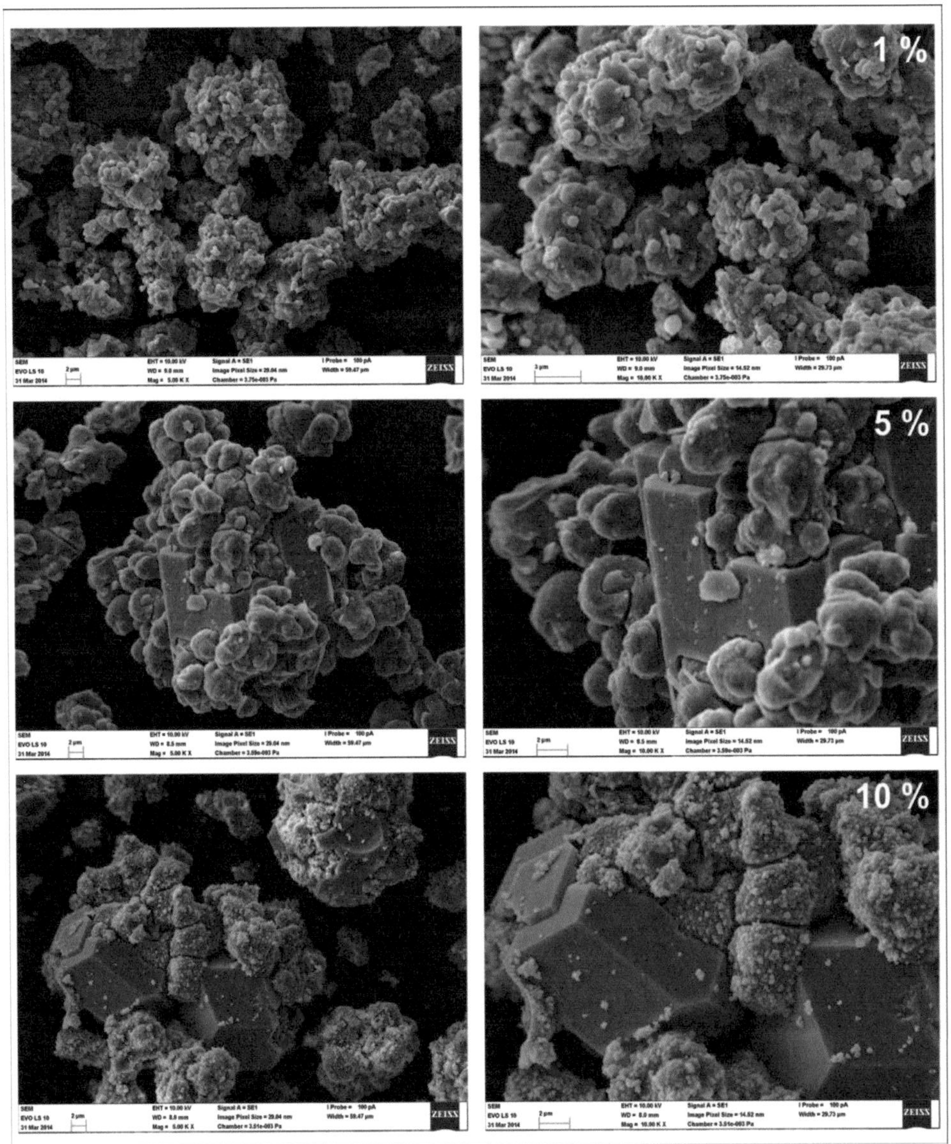

Figure 35. WDX analysis of hydrogarnet crystal formed during the hydration of strontium aluminate clinker prepared with 10 % of Ca^{2+}.

The maximum of main hydration effect, which appearsat 1.5 h for the sample prepared with 10% of calcium indicates, that the hydration properties are more similar to those of CAC as the content of Ca^{2+}ions in formed solid solution increases.

The hydrogarnet phase $(Sr_{3-x}Ca_x)AH_6$ and AH_3 are recognized as the main products of hydration of prepared clinker. Electron microscopy (Fig.34) shows the formation of well developed crystals of hydrogarnet phase in the samples prepared with 5 and 10% of Sr^{2+}

substituted by Ca^{2+}. Combining these results with previously described influence of Ca^{2+} ions on the formation of Sr_3A phase, the crucial role of lack of tristrontium aluminate and lower reactivity of Ca-substituted SrA in the nucleation and growth of large hydrogarnet crystals at the beginning of hydration process can be supposed.

Fig.35shows the composition of hydration product of strontium aluminate clinker prepared with 10% of Ca^{2+}. The (Sr+Ca):Al atomic ratio ~3:2 corresponds to the formation of Ca-substituted tristrontium aluminate hexahydrate with the composition given by the formula $(Sr_{2.7}Ca_{0.3})AH_6$. Therefore, Ca^{2+} ions substitute Sr^{2+} in the hydrogarnet phase in the ratio corresponding to the amount of cations substituted in the original clinker (10 %).

Chapter 8

Expansive Strontium Aluminate Cements

1. Expansive cements

Cement hydration is a complex set of exothermic chemical reactions. Each chemical compound may react with water in different ways, each reaction has its own volume stoichiometry, and the reaction (or hydration) products may undergo the transition to different forms. There are several mechanisms by which a cement paste changes its volume (shrinks or swells) during the hydration process under sealed conditions, or without the moisture transfer from paste to the environment. The major mechanisms of total shrinkage include [12,278,673-676]:

- Hydration shrinkage;
- Self-desiccation shrinkage;
- Thermal contraction or expansion;
- Crystallization swelling;
- Shrinkage during the phase transition or expansion.

Autogenous shrinkage is a phenomenon in which cementitious materials shrink at constant temperature without any changes in weight [677]. It is well-known that the volume changes in a cement paste or concrete mentioned above may occur simultaneously. It is the overall shrinkage behaviour that controls the concrete performance [278]. Furthermore, the influence of this phenomenon on total shrinkage depends on the size of specimen [414]. Most chemical shrinkage measurements reported are often performed at short duration and suggested water to cement ratio of 0.40 as required by ASTM C 1608 [678].

Autogenous shrinkage in concrete with low water-cement ratio is the major factor that generates larger shrinkage and early cracking [673,677,679]. A small increase in the chemical shrinkage at "infinite time" was found with increasing water-cement ratio. The influence of temperature was found to be twofold: Increasing temperature caused increasing rate of the development of chemical shrinkage and the decrease of the chemical shrinkage at "infinite time" [680]. The mineral admixtures in concrete affect the concrete shrinkage in different ways. Fly ash and limestone powder can reduce the autogenous shrinkage while silica fume may lead to increased shrinkage [679,681].

At very early age and during the setting, two phenomena are directly responsible for the shrinkage of material in sealed conditions [648]:

1. Capillary depression in the porous space;
2. Le Chatelier's contraction caused by the density difference between hydrates on one hand and cement and water on the other hand.

Drying shrinkage is relatively slow process. The driving force is the loss of water from the pores due to lower humidity prevailing outside the concrete. The diffusion of water through the material depends mainly on its porous structure, on the size and shape of the specimen, on the duration of drying and on prevailing atmospheric conditions [414,682].

The shrinkage can be compensated by several compounds acting through different mechanisms, but the principle remains similar-the incidence of expansion pressure on the setting paste. It can be achieved via the formation of new crystalline phases, predominantly ($3CaO \cdot Al_2O_3 \cdot 3CaSO_4 \cdot 32H_2O$, $C_6AS_3H_{32}$, AFt), swelling gel of portlandite ($Ca(OH)_2$,CH [12,688]) or brucite ($Mg(OH)_2$,MH [12,683-687,690]), or by the osmotic pressure generated by semi-permeable membrane of hydration products on the surface of anhydrous grains under suitable conditions.

The expansion admixture has to produce the expansive stress after the start of setting, at the stage when the paste attains certain rigidity to be able to transfer the expansion stress as a plastic material. If ettringite occurs homogeneously and immediately (within hours) after the start of setting, it does not cause any significant localized disruptive action (early ettringite formation, EEF). In the case of the expansion, which appears after the solidification (Delayed Ettringite Formation, DEF) it reduces the mechanical performance, or in extreme case, it leads to the material failure. DEF occurs when the constituents of concrete provide initial source of sulphates as a result of undesirable heating of the concrete to a temperatures higher than 65-70 °C. It has been found that the expansion time is inversely proportional to the fineness of expansive component and directly proportional to the amount of added sulfate [12,688-691,738,739].

The mechanisms mentioned above are based on the formation of ettringite where the expansion occurs due to the crystallization pressure as well as the swelling pressure which is caused by the selective absorption of ions combined with large number of water molecules on tiny ettringite crystals. The growing mechanism, the size and the shape of ettringite crystals are important factors, once the shrinkage compensation or the expansion effect shall be reached. The needle-like form of ettringite crystals growing on the surface of C_3A grains by the topochemical process after the solidification of cement slurry can convert the resulting strain into the desired volume expansion. In the case that ettringite crystals precipitate from the solution in the pores between grains of clinker phases, the expansion or shrinkage compensation effect does not occur [12].

Three types of expansive cements are defined by ASTM C 845 (Standard Specification for Expansive Hydraulic Cement) – K, M and S type [692]. The main aluminate phase that is responsible for the formation of ettringitevaries with the cement type. The Klein's compound

($3CaO{\cdot}3Al_2O_3{\cdot}CaSO_4$, $C_4A_3\bar{S}$, so-called "kleinite") is present in K-type cement [12,693-697], monocalcium aluminate ($CaO{\cdot}Al_2O_3$, CA) in M-type cement [12,693,697] and tricalcium aluminate ($3CaO{\cdot}Al_2O_3$, C_3A) in S-type cement [12,693,697].

Gypsum ($CaSO_4{\cdot}2H_2O$, $C\bar{S}H_2$) [698],α or β hemihydrates ($CaSO_4{\cdot}0.5H_2O$, $C\bar{S}H_{0.5}$) [699] and α or β anhydrites ($CaSO_4$, $C\bar{S}$) [699,370] are used as the sources of SO_4^{2-}ions. The decreasing dissolution rate of anhydrite results inthe lack of sulphate ions necessary for the formation andstability of ettringite, which is sensible to such factors as temperature, pH, $Al(OH)_4^-$ to $SO_4^{2-}$and $Ca^{2+}$to SO_4^{2-}ratio [700-703].

2. Ettringite and analog of ettringite

Ettringite ($3CaO{\cdot}Al_2O_3{\cdot}3CaSO_4{\cdot}32H_2O$, AFt as calcium aluminate or ferrite tri-substituted hydrates of general formula $3CaO{\cdot}(Al,Fe)_2O_3{\cdot}3CaX_2{\cdot}nH_2O$) is formed in hydrated cement at early hydration stage. The AFt designation was first suggested by Smolczyk in 1961 in order to distinguish the high sulfate hydrate phases from the low sulfate phases, which are collectively termed AFm ($3CaO{\cdot}Al_2O_3{\cdot}CaSO_4{\cdot}12H_2O$). From its natural occurrences, ettringite has been known since 1874 [637,704-707].

Ettringite, like other AFt phases, forms hexagonal prismatic or acicular crystals. The crystals have two distinct structural components:

Columns: $\{Ca_6[Al(OH)_6]_2 \times 24\,H_2O\}^{6+}$

Channels: $\{(SO_4^{2-})_3 \times 2\,H_2O\}^{6-}$

The columns consist of $AI(OH)_6$ octahedra alternating with triangular groups of edge-sharing CaO_8 polyhedra. The column alignment is along the c-axis of the trigonal-hexagonal unit cell. Eight oxygensatoms in the polyhedra come from the Ca coordination with four OH ions shared with the $AI(OH)_6$ octahedra and from the coordination with four H_2O molecules. The hydrogen atoms from the molecules of water form the cylindrical surface of the columns. The channels contain four sites per formula unit of the column structure which contains six calcium atoms. Three of these sites are occupied by $SO_4^{2-}$and one by two H_2O molecules [706-709].

Both calcium sulfoaluminate hydrates (AFt and AFm) have the ability to combine with a number of cations and anions, which makes them important with regard to the waste immobilization in cementitious matrices [710-712]. Ions available for the substitution in the ettringite structure are Ca^{2+}, Al^{3+}, $SO_4^{2-}$and OH^-. They are listed in Table 1.

Reported analogs of ettringite are listed in Table 2.

Site	Ion size [Å]	Site	Ion size [Å]	Site	Ion size [Å]	Site	Ion size [Å]
Ca^{2+}	1.12	Al^{3+}		SO_4^{2-}		OH^-	
Sr^{2+}	1.26	Cr^{3+}		$B(OH)_4^-$		O^{2-}	
Ba^{2+}	1.42	Si^{4+}		CO_3^{2-}			
Pb^{2+}		Fe^{3+}		Cl^-			0.62 – 1.38
Cd^{2+}	0.90 – 1.42	Mn^{3+}	0.42 – 0.76	OH^-			
Co^{2+}		Ni^{3+}		CrO_4^{2-}			
Ni^{2+}		Co^{3+}		AsO_4^{3-}	0.62 – 3.71		
Zn^{2+}		Ti^{3+}		SeO_4^{2-}			
				VO_4^{3-}			
				BrO_3^-			
				NO_3^-			
				MnO_4^-			
				CrO_3^-			
				SO_3^{2-}			
				IO_3^-			

Table 1. Ion substitutions in ettringite minerals according to [706].

Ion	Name/ formula	Reference
$B(OH)_4^-$	High-boro-ettringite ($3CaO{\cdot}Al_2O_3{\cdot}Ca(B(OH)_4)_2{\cdot}Ca(OH)_2{\cdot}36H_2O$), B-AFt)	[737,713,714]
	Low-boro-ettringite ($3CaO{\cdot} Al_2O_3{\cdot}Ca(B(OH)_4)_2{\cdot}2Ca(OH)_2{\cdot}36H_2O$)	
HBO_3^{2-}	Boro-monophase ($4CaO{\cdot}Al_2O_3{\cdot} \frac{1}{2} B_2O_3{\cdot}12H_2O$, B-AFm)	[737]
CrO_4^{2-}	Chromate ettringite ($3CaO{\cdot}Al_2O_3{\cdot}CaCrO_4{\cdot}nH_2O$)	[736,715]
	Chromate monophase ($3CaO{\cdot}Al_2O_3{\cdot}CaCrO_4{\cdot}15H_2O$)	[736]
SeO_4	Selenate ettringite($Ca_6[Al(OH)_6]_2(SeO_4)_3{\cdot}31.5H_2O$)	[712,716,717]
	Selenatemonophase ($3CaO{\cdot}Al_2O_3{\cdot}CaSeO_4{\cdot}xH_2O$)	[712,718]
CO_3^{2-}	Carbonate ettringite ($Ca_6Al_2(CO_3)_3(OH)_{12}{\cdot}26H_2O$)	[719]
Al^{3+}	Solid solution of Al- and Fe-ettringite ($Ca_6[Al_{1-x}Fe_x(OH)_6]_2(SO_4)_3{\cdot}26H_2O$)	[708]

Table 2. Reported analogs of ettringite.

According to the stability field diagramof the system $CaO–Al_2O_3–SO_3–H_2O$ (Fig.1(a)) established by Hampson and Bailey [720], high sulphate concentrations stabilize gypsum and only when its concentration is low, then ettringite becomes the stable phase [735]. The thermodynamic investigation of the $CaO-Al_2O_3-CaSO_4-H_2O$ (Fig.1(b)) system at 25 °C [721], 50 °C and 85 °C was performed by Damidot and Glasser [722]. The effect of Na_2O and K_2O on the equilibrium state is disused in works [721] and [723], respectively.

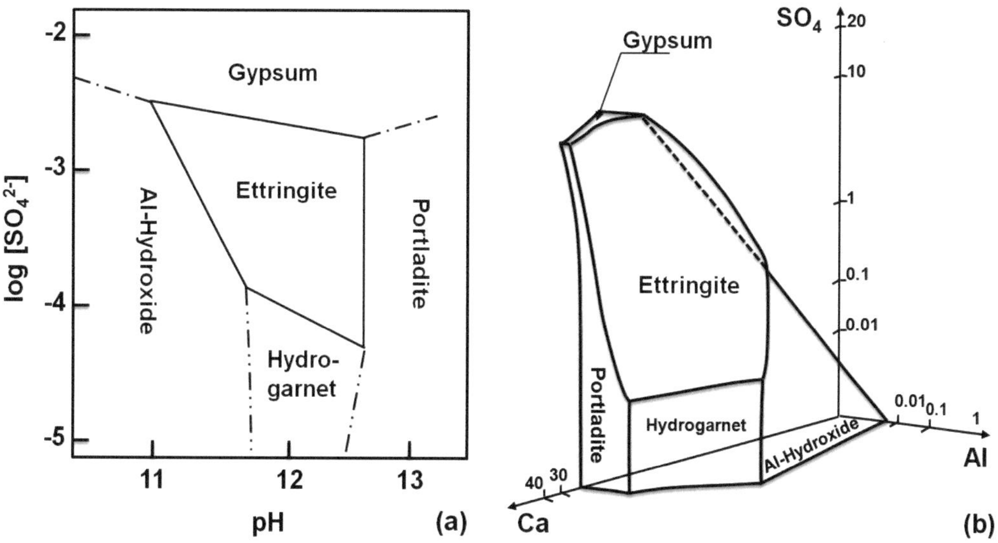

Figure 1. Ettringite stability in alkaline environments [720] (a) and three-dimensional representation of $CaO-Al_2O_3-CaSO_4-H_2O$ system [721] (b).

The following solid phases are possible in this system [721]:

- Ettringite (triphase or AFt in cement nomenclature): $3CaO \cdot Al_2O_3 \cdot 3CaSO_4 \cdot 32H_2O$;
- Monosulfate (monophase or AFm): $3CaO \cdot Al_2O_3 \cdot CaSO_4 \cdot 12H_2O$;
- Gypsum: $CaSO_4 \cdot 2H_2O$;
- Hydrogarnet: $3CaO \cdot Al_2O_3 \cdot 6H_2O$ (other calcium alumina hydrates are also possible but they are metastable with respect to hydrogarnet);
- Portlandite: $Ca(OH)_2$;
- Gibbsite: $Al(OH)_3$.

The reaction of cement minerals with sulphate and carbonate ions can lead to the formation of thaumasite ($Ca_6[Si(OH)_6]_2(CO_3)_2(SO_4)_2$ $24H_2O$ or $CaSiO_3$ $CaCO_3$ $CaSO_4$ $15H_2O$) at temperatures below 15 °C. This reaction can destroy the C-S-H matrix and result in the disintegration of concrete (thaumasite sulfate attack, TSA). Ettringite and thaumasite have very similar crystal structures. Thaumasite is hexagonal (a=11.54 Å, a=10.401 Å) with the columns of $[Ca_3Si(OH)_6$

$12H_2O]^{4+}$and sulfate and carbonate ions in an ordered arrangement in channels between the columns. Si is octahedrally coordinated by oxygen [724-730].

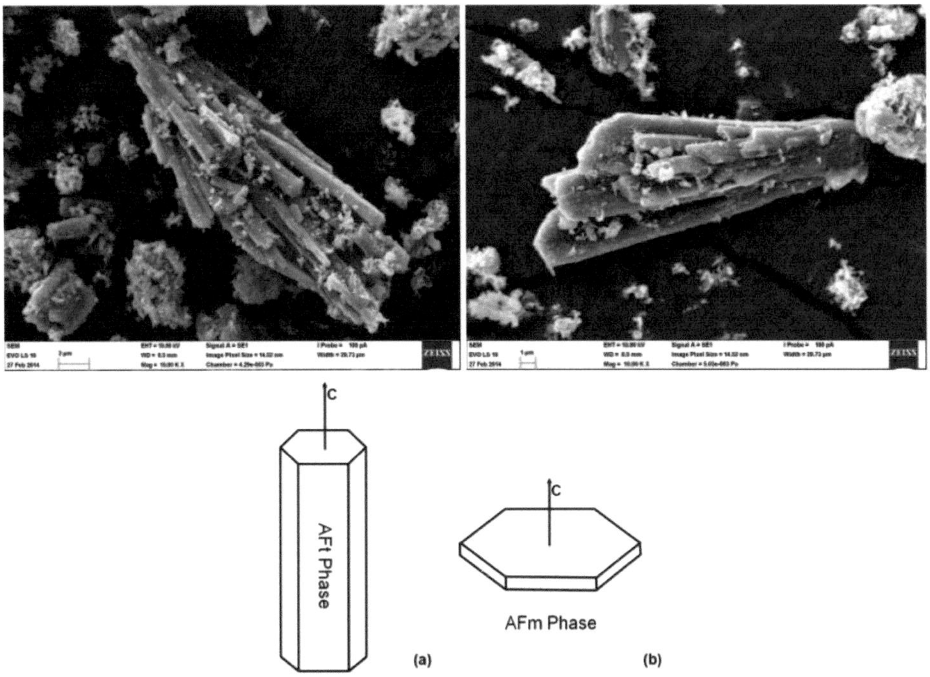

Figure 2. SEM image of ettringite crystals prepared by the saccharate method andcharacteristic crystal morphologies of AFt (a) and AFm (b) [737].

Most synthesis experiments reported in literature are based on the "saccharate" method reported originally by Carlson and Berman [731]. A CaO supersaturated, 10 % sucrose solution is mixed with Na_2O Al_2O_3 and Na_2SO_4 in the stoichiometric amounts. Sodium sulfate can be replaced by oxyanion salt to obtain substituted ettringite [713,732]. The same process without the addition of sucrose is referred to as the "paste reaction" method [713].

Ettringite can be also synthesized by mixing the calcium hydroxide slurry and the aluminium sulphate solution [733,734]:

$$\begin{aligned}&6\,Ca(OH)_2 + Al_2(SO_4)_3 \cdot 16H_2O + 10\,H_2O \rightarrow \\ &Ca_6[Al(OH)_6]_2(SO_4)_3 \times 26H_2O\end{aligned} \tag{1}$$

At high sulphate concentrations, gypsuminitially controls the maximum concentration of sulphate in solution by the reaction [375]:

$$Ca^{2+} + SO_4^{2-} + 2\,H_2O \longleftrightarrow CaSO_4 \times 2H_2O \tag{2}$$

At elevated pH (~12) the dissolution of gypsum is promoted while ettringite remains stable. Gypsumis used afterwards as a source of sulphate in the synthesis of ettringite according to the reaction given by:

$$\begin{aligned} &3\,Ca^{2+} + 2\,Al(OH)_4^- + 3\,CaSO_4 \times 2H_2O + 4\,OH^- + 20\,H_2O \longleftrightarrow \\ &3CaO \times Al_2O_3 \times 3CaSO_4 \times 32H_2O \end{aligned} \tag{3}$$

At low sulphate concentrations the ettringite synthesis proceeds following the reaction given by the equation:

$$\begin{aligned} &3\,Ca^{2+} + 2\,Al(OH)_4^- + 3\,SO_4^{2-} + 4\,OH^- + 26\,H_2O \longleftrightarrow \\ &3CaO \times Al_2O_3 \times 3CaSO_4 \times 32H_2O \end{aligned} \tag{4}$$

The reaction of tricalcium aluminate ($3CaO \cdot Al_2O_3$) with gypsum ($CaSO_4 \cdot 2H_2O$) is known to be rapid and complete at room temperature (within 30 to 60 min) [705].

Ettringite appears as a hydrate phase of Portland cement in the first stage of hydration process. In stage one, the aluminate phase reacts with water in the presence of calcium sulfate to form ettringite [706]:

$$\begin{aligned} &3CaO \times Al_2O_3 + 3\,CaSO_4 \times 2H_2O + 26\,H_2O \rightarrow \\ &3CaO \times Al_2O_3 \times 3CaSO_4 \times 32H_2O \end{aligned} \tag{5}$$

Ettringite can also be formed through the reaction of ferrite hydrate phase. Ettringite is formed within minutes after the start of cement hydration. Its concentration increases to a maximum within about 1 day and then it begins to decrease and may even approach zero. As ettringite disappears, mono-sulfate (AFm) hydrate phase takes its place due to lower sulfate concentration in the pore solutions. The ettringite produced in this first hydration stage tends to be in the form of stubby, prismatic crystals close to the aluminate surface as well as away from these surfaces. Depending on the cement composition and availability of CO_3^{2-} ettringite either persists or is converted into monosulfate [7,12,706,736,737]:

$$\begin{aligned} &3CaO \times Al_2O_3 \times 3CaSO_4 \times 32H_2O + 2\left(3\,CaO \times Al_2O_3\right) + 4\,H_2O \longleftrightarrow \\ &3\left(3CaO \times Al_2O_3 \times CaSO_4 \times 12H_2O\right) \end{aligned} \tag{6}$$

Structurally, AFm phases are similar to the mineral hydrocalumite, the crystals of which are morphologically recognizable as thin hexagonal plates (Fig.2). Basal spacing in AFm crystal structure depends on the type of anion, such as CO_3^{2-}, OH^-, Cl^- and SO_4^{2-}, within the interlayer region.

There are two different types of DEF-related damages depending on the sulfate source [738-740]:

1. External (ESA) sulfate attack, which occurs when environmental sulfate (from water or soil) penetrates into a concrete structure. It can be avoided or reduced by using impermeable concrete.

2. Internal (ISA) sulfate attack, which occurs also in sulfate free-environment by latesulfate ions release from either cement or gypsum contaminated in aggregates. The ISA-induced damage caused by the late sulfate release from the cement can be related either to the thermal decomposition of "primary" ettringite or to sulfate release from C-S-H.

Several forms of ettringite phase which may be formed duringthe hydration of Portland cement are listed in Table 3.

Ettringite	Time of formation	Conditions of formation	Al_2O_3 source	Texture	Expansive behaviour
Early primary	First hour of hydration	Normal conditions of hydration	C_3A	Microcrystalline	no
Late primary	First days or weeks of hydration	Excessive amount of gypsum is cement	C_3A	Microcrystalline	yes
External	Any time	Migration of SO_3 from outside	Mono-sulphate	Microcrystalline	yes
Delayed	Months after thermal curing	After curing at temperatures >70 °C	Mono-sulphate	Microcrystalline	yes
Secondary	Months after thermal curing	Recrystallization of ettringite in formed cracks upon wetting and drying	Ettringite	Microcrystalline	No (?)

Table 3. Forms of ettringite [741].

The formation of ettringite, which induces swelling, may also cause failures in stabilization of soil, where within the mechanical treatment CaO, $Ca(OH)_2$ or cement are used. Ettringite shows very high water retention potential. Therefore, as ettringite grows in stabilized soil, its presence enhances the water retention of soil, causing that the sample exerts more force in order to take in the water–sulfate solution from lower soil strata [742]. Although lime treatment can be effective in stabilizing gypseous soils and in enhancing their mechanical properties, the amount of added lime must be adjusted according to the mineralogical composition and the gypsum content of soil [743].

3. Expansive strontium aluminate cements

The additions of sulfate bearing phases such as anhydrite to strontium aluminate cement can be performed in order to prepare expansive cements based on strontium aluminate.The effect of gypsum onto the hydration of SrAC was investigated by isothermal calorimetric experiment using the blend of strontium aluminate cement with 2% and 5% of $CaSO_4 \cdot 2H_2O$ (Fig.3).

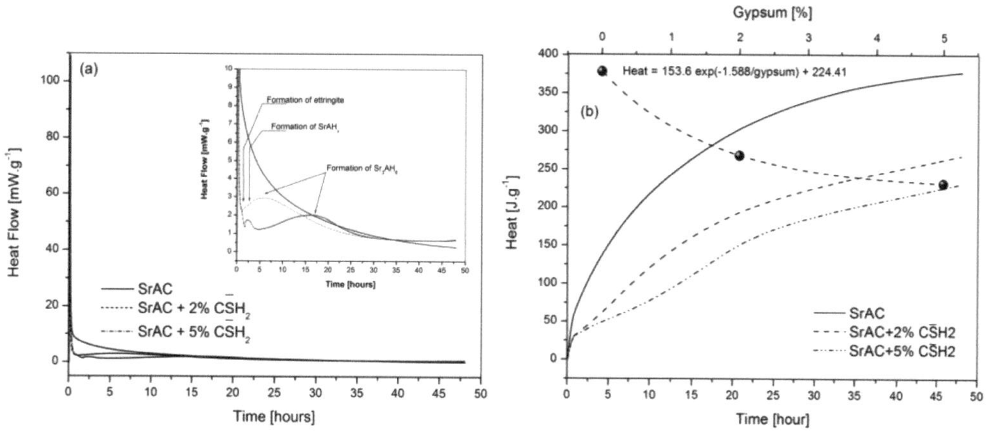

Figure 3. Effect of gypsum on the hydration of strontium aluminate cement.

In comparison with pure SrAC, the changes of heat flow indicate that gypsum has a significant effect on the course of hydration of strontium aluminate cement. The calorimetric data indicate the changes in the mechanism of hydration process and those caused by the formation of ettringite (Figs.5, 7 and 8). Fast precipitation of ettringite leads to the deficiency of $[Al(OH)_4]^-$ ions and Sr^{2+} ions which substitute for Ca^{2+} ions in the structure of $\{Ca_6[Al(OH)_6]_2 \cdot 24H_2O\}^{6+}$ columns. That explains the formation of $SrAH_x$ (Fig.5(b)) and the absence of gibbsite. It is obvious that the intensity of effect related to the formation of $SrAH_x$ increases with the amount of gypsum added to SrAC. The addition of gypsum changes the main hydration products at least during initial stages of hydration.

The diffusion barrier formed on the surface of strontium aluminate phase leads to a broad hydration effect where the time and temperature related to the maximum of released heat decreases with increasing content of gypsum. Increasing amount of added gypsum leads increased formation of $SrAH_x$ and ettringite while the amounts of formed Sr_3AH_6 and gibbsite decrease. Fig.3(b) shows that the heat released during the first 50 hours of hydration of strontium aluminate cement decreases exponentially with increasing content of gypsum. This behavior can provide the explanation for the hydration of the blend of strontium aluminate cement with SrAC (please see the discussion related to Fig.7 in Chapter 7).

The hydration of the samples after the calorimetric assessment was stopped by repeated washing of ground samples in acetone in order to remove water followed by the evaporation of solvent and drying upon silica-gel in desiccator. The storage of sample in the desiccator also restricts the course of carbonation [409,744,745].

The composition of hydrated cement stone was further investigated by thermal analysis, X-ray diffraction analysis, infrared spectroscopy and scanning electron microscopy in order to explain the effect of gypsum on the course of hydration and to evaluate the changes in the equilibrium composition of hydration products of SrAC.

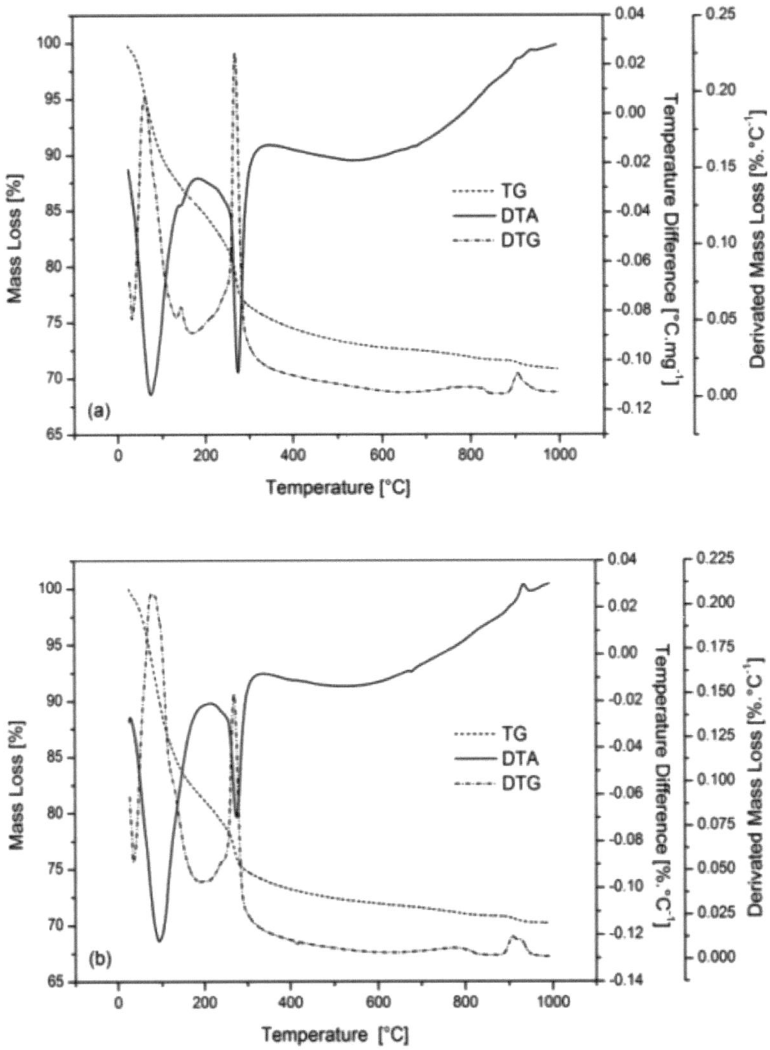

Figure 4. DTAof hardened cement stone prepared from the mixture containing 2 % (a) and 5 % (b) of gypsum.

The TG-DTA results are shown in Fig.4. They reveal higher weight change in the first step of thermogravimetric curve for the sample with 5 % of gypsum. The intensity of the effect related to the dehydration of gel phase and of main product of hydration (Sr_3AH_3) decreases (Chapter 5.1). That indicates that $[Al(OH)_4]^-$ ion formed during the hydration is used to form ettringite that slows down the hydration of strontium aluminate cement.

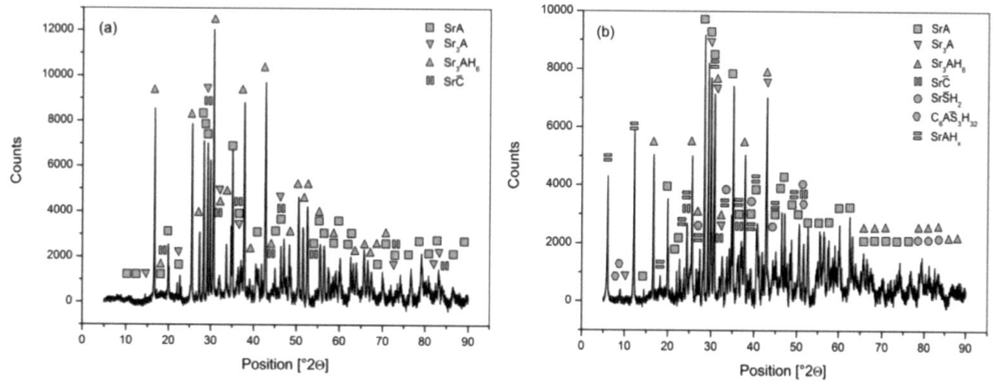

Figure 5. X-ray analysis of cement stone after 50 h of hydration: sample with 2 % (a) and 5 % (b) of gypsum.

The results of X-ray diffraction analysis of cement stone after 50 hours of hydration at the temperature of 25°C is shown in Fig.5. Higher intensity of diffraction lines of unreacted strontium aluminate in sample (b) compared to (a) indicates the retardation effect of gypsum on the hydration of strontium aluminate cement. Moreover, both samples still showed the diffraction features of tri-strontium aluminate after 50 h of hydration process, while this compound was not recognized in pure sample of strontium aluminate cement without gypsum (Fig.6). Both samples are slightly carbonated.

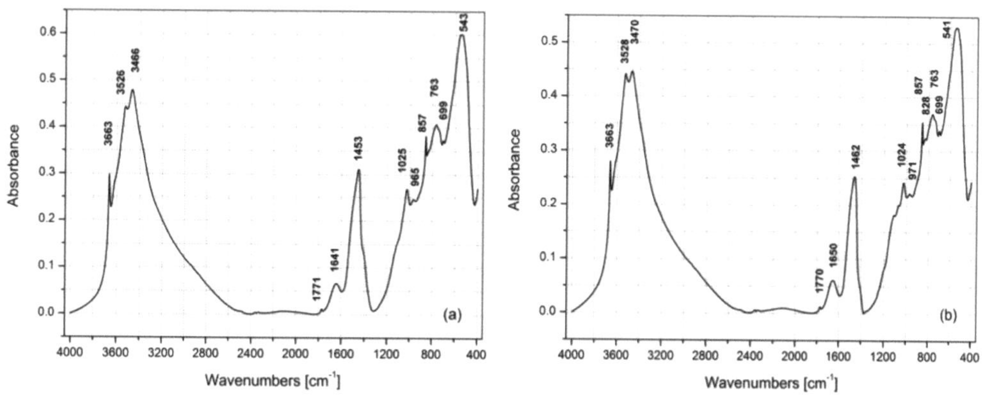

Figure 6. Infrared spectrum of hydrated cement stone of sample containing 2 % (a) and 5 % (b) of gypsum.

Infrared spectrum (Fig.6) shows the features typical for hydrated strontium aluminate cement (tri-strontium aluminate hexahydrate and strontium carbonate) which are described in

Chapter 5.1. Fundamental stretching modes of SO_4^{2-} ions are located on the left shoulder of multiple band below 1300 cm^{-1} (1250 – 950 and 700 – 580 cm^{-1}).

The bands of CO_3^{2-}anion with C_s symmetry occur in the spectra of both samples due to the carbonation to $SrCO_3$: the band of antisymmetric stretching ν_3 (1453 cm^{-1}), symmetric stretching ν_1 (1071cm^{-1})[1], out of plane bending ν_2 (857 cm^{-1}), in plane ν_4 (699 cm^{-1}) and the combination band ν_1+ν_4 at 1771 cm^{-1} [170,746-748]. The bands of hydration products are described in Chapter 5.1 (please see the discussion to Fig.4).

Figure 7. Strontium aluminate cement with 2 wt.% of gypsum.

1 Vibration mode is IR active due to the aragonite structure of $SrCO_3$ (Please see Chapter 2).

X-ray diffraction analysis does not provide the evidence about the formation of ettringite in the sample prepared from the mixture of strontium aluminate cement with 2 % of gypsum. On the other hand, SEM images (Fig.7) show rare and small clumps of needle-like crystals which occur between the plates of the main product of hydration, therefore a very small amount of strontium sulfate or ettringite is probably formed.

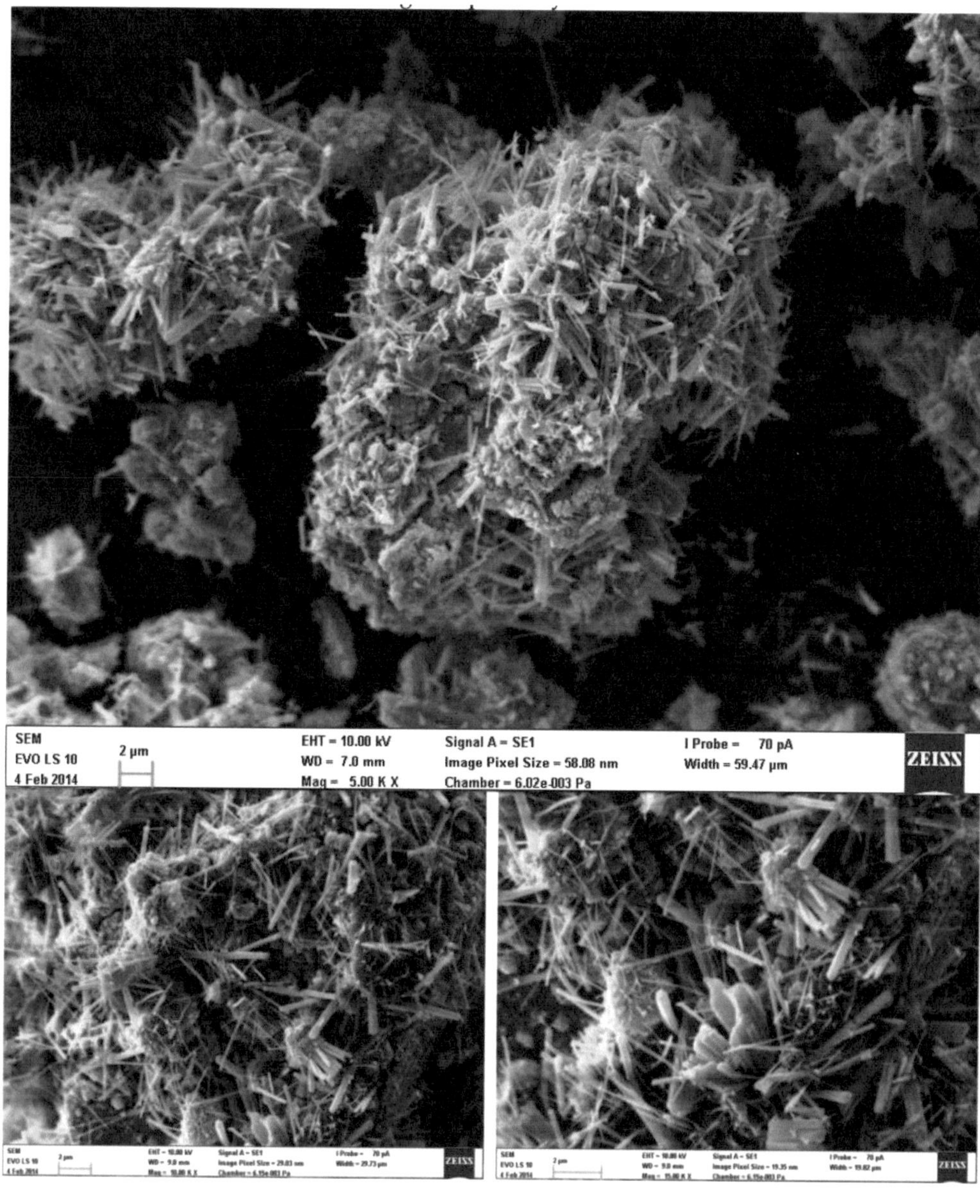

Figure 8. Strontium aluminate cement with 5 wt.% of gypsum.

The sample prepared with 5 % of gypsum shows the diffraction lines of both, strontium sulfate and ettringite. Higher content of sulfate bearing phase, compared to cement with 2 % of gypsum, is also well visible from scanning electron microscopy images, where large amount of needle-like crystals grow on the surface of clusters consisting of partially hydrated grains of strontium aluminate cement.

The comparison of Fig.7 and Fig.8 as well as the results of calorimetry (Fig.3) indicate the change in hydration mechanism from the crystallization to the diffusion. There is a way to prepare expansive strontium aluminate cement, but the control of setting time of the paste is necessary in order to reach required rigidity, which enables to handle the expansion stress. The positive effect was observed for calcined clay (Chapter 7.2).

Chapter 9

Advanced Techniques for Preparation of Strontium Aluminate Phases

1. Techniques using powder precursors

The solid state synthesis described in Chapter 4.3 is the traditional way for the synthesis of strontium aluminate phase. Small amounts of fluxing agents (Fig.1), such as B_2O_3 (H_3BO_3) or LiF can be used [749,810].

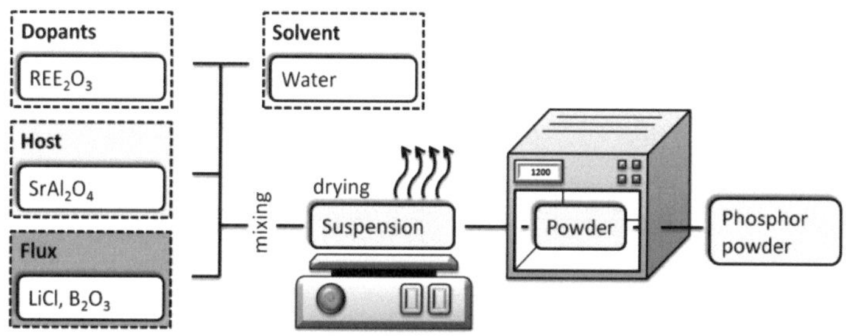

Figure 1. Flow chart of the preparation of REE doped $SrAl_2O_4$ phosphor powders [749].

There are also advanced techniques of the synthesis of strontium aluminate phases including the sol-gel process, the precipitation from solution, the pyrolysis and combustion techniques, the microwave synthesis, the mechano-chemical route, etc. It must be pointed, that the techniques mentioned in this chapter are the sophisticated synthesis routes which are tailored mainly for the processing of nano-scale powders, doped with rare earth elements (REE ions such as Ce^{4+}, Eu^{2+}, Dy^{3+}...), in order to prepare the precursors for strontium aluminate based phosphors for long-persistent luminescence materials, electronic and structural applications (Chapter 10.1). Therefore, the suitability of the techniques mentioned above for the synthesis of strontium aluminate clinker is limited by their complexity, time-consuming preparation process, labor productivity and by the cost of applied chemicals.

The Sol-gel process is a method based on the turn of the *"free dispersion system"* with particles of colloidal size (**sol**) into *"bound dispersion system"* (**gel**) with the particles cross-linked into the three-dimensional network. In principle there are two ways of the formation of gel:

1. Linking of growing particles into chains, connection of chains into branched chains and then into domains of three-dimensional networks (**microgel**). As the regions of formed

gel phase grow and are connected to each other, the continuous network of gel (**lyogel**) is formed. The volume of formed gel is approximately the same as that of the original sol. The process of gradual displacement of liquid phase from gel is termed as **syneresis**.

The particles in sol may lose the stability (**DLVO theory**[1] [750]) due to ageing of sol (spontaneously) or by the change of pH or temperature, addition of electrolytes and mechanic action (mixing, ultrasound) and the formation of flocs takes place. The process of formation of these clusters is termed as **flocculation** (**agglomeration**, **aggregation** and more generally as **coagulation**). The gel (**coagel**) is formed by the connections between aggregates after sedimentation

The capillary forces generate the strain stress in gel during the evaporation of solvent and the solid framework falls into pieces of **xerogel**[2]. Subsequent mechanical (grinding) and thermal treatment (calcination) lead to the powder precursor. According to the applied chemical the alkoxide-, semialkoxide-, Pechini-[751] and modified Pechini-route were recognized.

The sol–gel process is an efficient technique for the syntheses of phosphors due to good mixing of initial materials and relatively low reaction temperature resulting in more homogeneous products than those obtained by solid-state reaction synthesis method. The technique uses alkoxides and salts listed in Table 1. There are numerous modifications in applied solvent, pH, chelating agent as well as in the temperature and time of treatment, but the most general scheme of the synthesis includes these steps:

a. **Dissolving of initial materials:** the kind of solvent or liquid depends on initial materials. Oxides are usually dissolved in HNO_3 of various concentrations. Water or water slightly acidified by the addition of several HNO_3 drops (in order to avoid the hydrolysis) is used for the dissolution of nitrates. Alkoxides are mixed with organic solvent (methanol, ethanol, propanol, ethylene glycol…) which enables to control the process of hydrolysis via defined amount of water (water to alkoxide ratio). Other important parameters are the catalysis, the temperature, the time and intensity of stirring.

b. **Mixing of constituents:** chelating agents and pH adjusting chemicals are often used in order to avoid the undesirable precipitation and to improve the system homogeneity. For the Pechini and modified Pechini method [751,756] the ratio between chelating agent and metal cations is one of the most important parameters. Usually the value higher than one is used.

c. **Concentrating of the solution and the formation of gel by the evaporation of solvent under stirring or by condensation reaction:** the prepared system is stirred for several hours. The required temperature (mostly in the range from 60 to 80 °C) and pH are kept constant during this time. The condensation reaction between chelating agent (most usually citric acid) and glycol requires an increase in temperature (100 – 150 °C), which is necessary for Pechini in order to obtain clear transparent color of viscous polymer [67,752,753]:

1 Theory is named after Derjaguin B, Landau L, Verwey E and Overbeek J.

2 There is an option to avoid these changes by drying gel under critical conditions in order to prepare aerogel.

Chemical		Compound name	Function	References
Salts	$Sr(CH_3(CO_2))_2$	Strontium acetate monohydrate	Starting chemical	[758,759]
	$Sr(NO_3)_2{\cdot}4H_2O$	Strontium nitrate tetrahydrate		[755,756,760,765]
	$SrCO_3$	Strontium carbonate		[757]
	$Al(NO_3)_3{\cdot}9H_2O$	Aluminium nitrate heptahydrate		[755,756,758]
Alkoxides	$Al(i\text{-}OC_3H_7)_3$	Aluminium isopropoxide		[755,760]
Water and organic liquids	H_2O	Water	Solvent	[759,760]
	HNO_3 (aq) [3]	Nitric acid		[759,765]
	C_2H_5OH	Ethanol		[765]
	$C_3H_8O_2$	2-methoxyethanol		[760]
Additives [1]	$C_2H_5OCH_2CH_2OH$	Ethylene glycol monoethyl ether		[759]
	$C_3H_8O_3$	Glycerol, propan-1,2,3-triol		[759]
	$C_{2n}H_{4n+2}O_{n+1}$	PEG, Polyethylene glycol		[755-758]
	NH_4HCO_3	Ammonium bicarbonate	pH adjusting	[755]
	NH_4OH	Ammonium hydroxide		[757]
	CH_3COOH	Acetic acid	Chelating agent	[758]
	$C_6H_8O_7$	Citric acid		[756,757]
	$C_6H_{10}O_3$	Ethyl acetoacetate [4]		[760]
	$(C_6H_9NO)_n$	PVP, Polyvinylpyrrolidone		[765]
	C_3H_5NO	Acrylamide		[761]
	$C_7H_{10}N_2O_2$	Methylene bisacrylamide		[761]
	$(NH_4)_2S_2O_8$	Ammonium persulfate	Radical initiator	[761]
	$C_6H_{16}N_2$	N,N,N0,N0-tetramethyl-ethylenediamin, TMEDA		[761]
Dopants [2]	$Dy(NO_3)_3{\cdot}xH_2O$	Dysprosium (III) nitrate hydrate	Modification of lumi-nescent properties	[755,756]
	$Eu(NO_3)_3{\cdot}xH_2O$	Europium(III) nitrate hydrate		[755,760,756]
	$Ce(NO_3)_3{\cdot}6H_2O$	Cerium(III) nitrate hexahydrate		[758]

[1] These chemicals form the polymeric network (Pechini and modified Pechnini methods), avoiding undesirable precipitation via the formation of chelates with metal ions (complexing agent), initiators or chemicals maintaining the pH in the system.

[2] Dopants enhance nitrates of REE element, the product is termed as $ArAl_2O_4$:REE (REE=Rare Earth Elements).

[3] Dissolution of oxides.

[4] In order to avoid the hydrolysis of alkoxide cations before mixing.

Table 1. Initial materials for the preparation of strontium aluminates via advanced techniques.

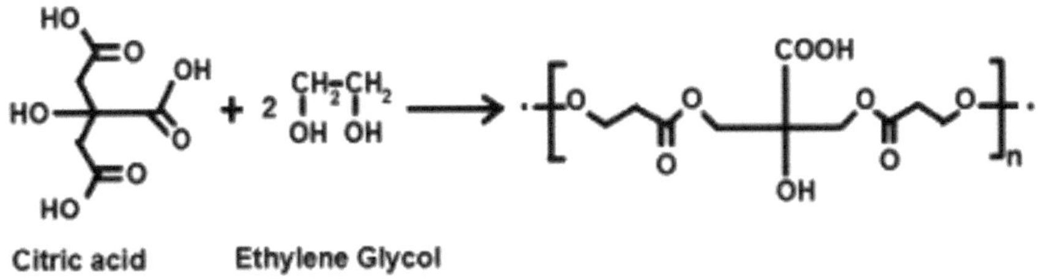

(1)

Modified Pechini methods uses EDTA [754] as chelating agent or does not apply glycol.

d. Drying of gel to xerogel: the temperatures in the temperature range from 100 to 200 °C are often used. The time of drying varies from several hours to a few days.

e. Thermal treatment: the pyrolysis of organic compounds and the calcination are two main processes which take place during this step. The temperatures higher than 400 °C applied for several hours are usually necessary in order to complete the removal of organics, but the course of process is highly affected by the nature of applied chemicals, by heating rate and by kiln atmosphere (static, inert, oxidative or reductive). The temperatures about 600 °C are usually applied. The removal of carbon formed under the reductive conditions (or site reduction conditions) often requires much higher temperatures and affects the oxidation state of metal cations via the p_{CO}/p_{CO2} ratio. Low partial pressure of oxygen also supports the formation of various organic compounds during the pyrolysis. At sufficient temperature, the solid-state reaction takes place during the process of calcination. Much higher temperature may cause the material to reach an undesirable degree of sintering which complicates the preparation of fine powder of the precursor. The calcination temperature of 900 °C is usually used for the preparation of strontium-aluminate precursor.

f. Mechanical treatment of xerogel into the fine powder of precursor: the calcined product is ground into fine powder of the precursor. This operation does not require any extensive force as porous and very brittle product is prepared. Some times the hand grinding in porcelain or agate dish is performed.

The prepared powder of precursor can be shaped by pressing, cold-isotactic pressing [762] or hot-isotactic pressing (HIP) [371]. Another option is the preparation of paste or suspension which can be deposited onto the substrate (e.g. by dip-or spin-counting [763,764], electrophoretic deposition, slip-or solution casting [761,764]…) or casted into the mound. The subsequent thermal treatment including the synthesis of required phases and the sintering can be performed in the furnace or by special sintering techniques. The method is schematically shown in Fig.2.

The possibility to prepare pure and fine product of good homogeneity under low processing temperature is the main advantage of the sol-gel method. The synthesis of $SrAl_2O_4$:REE fibers via the combination of electro-spinning with the sol-gel process were described by Cheng at al [765]. Ceramics can not be electrospun directly to fibers from the solution. Recently, the

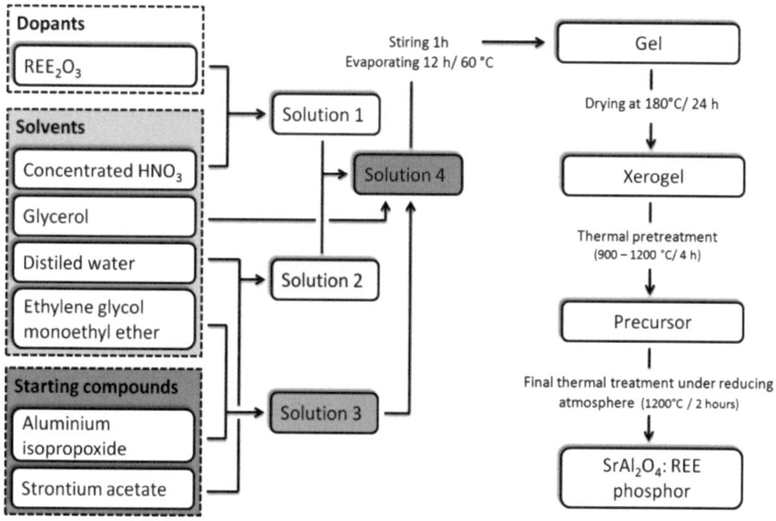

Figure 2. General scheme of the sol-gel process.

electro-spinning combined with the sol-gel process is being widely employed to fabricate the ceramics fibers.

The **precipitation method** or the chemical precipitation method exhibits some advantages, such as low processing temperature, high homogeneity and the purity of products. By this method, the nano-sized particles which are uniformly distributed, could be prepared [766]. In this method, a solution of the precursor reactants is mixed with dopants in an acid solution. Once the desired compound precipitates, the sample is centrifuged and washed repeatedly. The precipitate is treated at high temperature, then cooled and ground to the precursor of desired grain size [767]. The solvothermal co-precipitation synthesis uses an autoclave in order to prepare the precipitate under hydrothermal conditions [768,769].

The **combustion synthesis** techniques are classified according to the physical nature of the initial reaction medium [776]:

- Solution-combustion synthesis (SCS) where the initial reaction medium is aqueous solution.
- Conventional SHS method where the reactants are in solid state (condensed phase combustion).
- Flame synthesis, i.e. gas phase combustion.

The **solution-combustion method** is based on the preparation of mixture of nitrates of metal cations ($Sr(NO_3)_2{\cdot}4H_2O$, $Al(NO_3)_3{\cdot}9H_2O$) and dopants such as $REE(NO_3)_3$ with organic fuel (urea, $CO(NH_2)_2$). The temperature ranging from 60 to 80 °C is mostly used for this operation. The continuous stirring, lasting from several minutes to one hour, ensures the homogenization of mixture. The evaporation and recrystallization of the product provide the solution, gel, foam or fine powder for subsequent thermal treatment. The combustion is fast process (a few minutes) which takes place after introducing the sample into the furnace heated to the temperatures from 400 to 600 °C. The temperature during the combustion may reach 1100 °C. The voluminous foamy ash after the combustion is then ground and used as a precursor for the flame-pyrolysis of oxide powder for the synthesis of strontium aluminates [770-773].

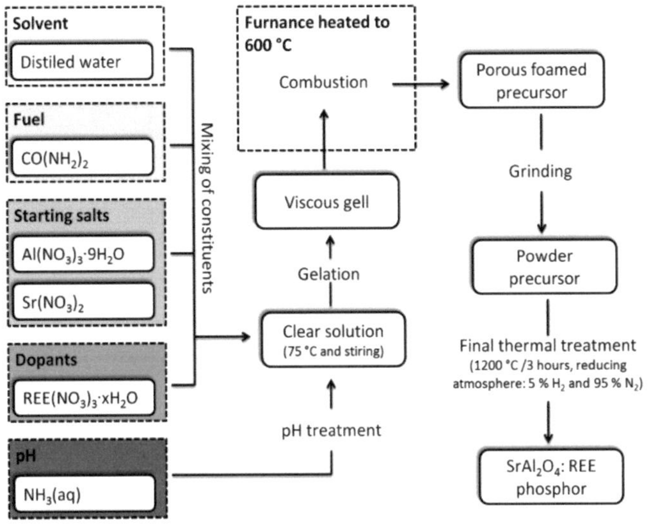

Figure 3. Scheme of the combustion method.

Final thermal treatment in a weak reductive atmosphere leads to the $SrAl_2O_4$:REE phosphor [772]. It is obvious that the preparation is much faster if compared to the sol-gel process, but there is only a limited chance to control the conditions of thermal treatment. Fast formation of large amount of gases and the swelling of material may cause the losses of precursor. The method is schematically shown in Fig.3.

The **Detonation method** of the synthesis of $SrAl_2O_4$ was described by Li at al [774]. The detonation method is an efficient technique for the preparation of nanometer powders because of good mixing of the initial materials. At the same time, the size of prepared particles can be controlled to a certain extent by adjusting the detonation parameters; the production cost is lower compared to sol-gel process and the process can be finished in a short interval of time. During the detonation synthesis process, the mixtures composed of raw materials can undergo complex physical and chemical reactions and have series of changes, such as the conglomeration, the crystallization and the phase-translation at high temperature.

When exploding and decomposing, nitrates can release great quantum of oxygen. So they can be used as oxidizers or accessorial oxidizers in dynamite. The main equations of the reaction of detonation can be expressed as follows:

$$Sr(NO_3)_2 + 2\,Al(NO_3)_3 \cdot 9H_2O \rightarrow SrAl_2O_4 + 4\,N_2 + 18\,H_2O + 10\,O_2 \quad (2)$$

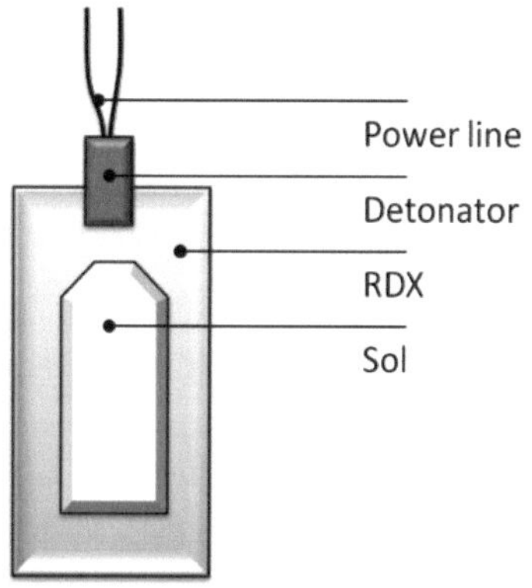

Figure 4. Schematic presentation of detonation method for the synthesis of strontium aluminate [774].

$$2\,CO(NH_2)_2 + 3\,O_2 \rightarrow 2\,CO_2 + 2\,N_2 + 4\,H_2O \qquad (3)$$

$$2\,C_3H_6N_6O_3 + 3\,O_2 \rightarrow 6\,H_2O + 6\,N_2 + 6\,CO_2 \qquad (4)$$

It is believed that the detonation products do not have enough time to grow into large and perfect crystallites, when the mixture is subjected to shock waves. The detonation products mainly consist of particles with very small sizes containing high density of defects and dislocations [774,775].

The **self-propagating high-temperature synthesis** (SHS) is one of the combustion methods [776-779] which uses strong exothermic reactions for the preparation of oxide ceramics [780], non-oxide ceramics [778,781-785] and cermets [786-789]. After mixing the powder of initial solid reactants and shaping (pressing, isotactic pressing) it the exothermic self-catalysed reaction is initiated. Once initiated, the combustion front becomes self-sustained, traveling through the sample due to the sufficient liberation of reaction heat and the final product is formed progressively with no additional heat required. The method can be applied whenever a sufficiently exothermic chemical reaction is available.

For example, strong affinity of aluminium to oxygen causes, that the formation of aluminium oxide is a strong exothermic reaction. The reaction is well known and used for the alumino-thermic reactions[3], which are of huge practical demand for the production of ferroalloys and welding of rail tracks. A tempting option arises to obtain the heat for the synthesis of strontium-aluminate via self-propagating high-temperature synthesis. $SrAl_2O_4$ was prepared by SHS method by Sathaporn at al [790] according to the general scheme in Fig.5.

Milling is known as one of important unit operations and is widely used in various processing of materials such as minerals, food, medicine, chemicals and building materials. As an

3 Discovered by Nikolay Beketov.

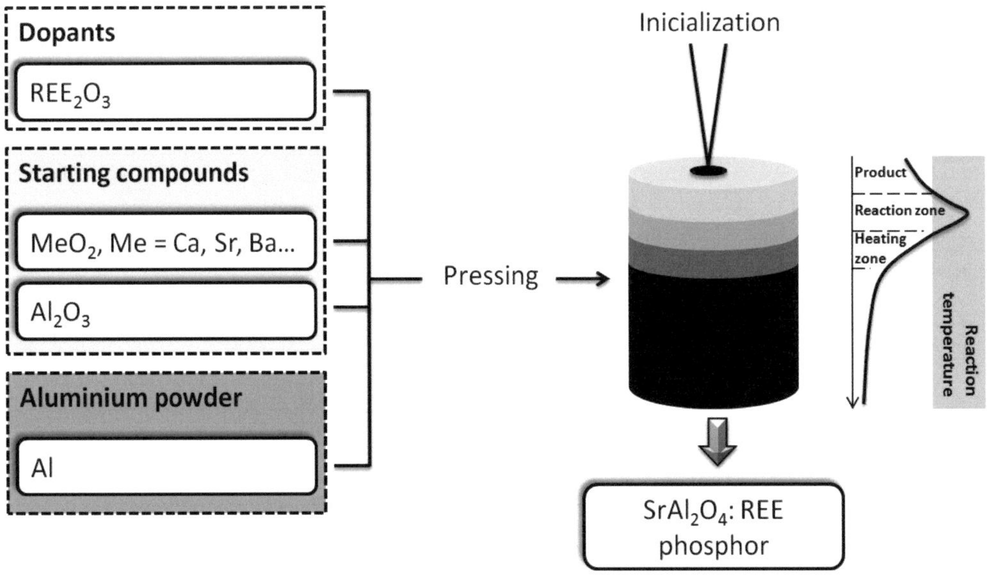

Figure 5. The flow diagram for the SHS synthesis of strontium aluminate.

extended branch in milling operations, the field of mechanochemistry has attracted much attention in recent years and the research papers on this topic have been increasing. One of unique phenomena in mechanochemistry is the solid state reaction among two or more multi-components without heating, to produce a constituent compound. A potential application for this solid state reaction is not only the material synthesis but also the separation and recovery of chemical species and components from minerals and waste materials treated by the reaction through another chemical and/or physical operation [791].

The boundary of two or multi-components of solid material may be activated by the following reasons: When the sample powders are trapped and crushed between two balls colliding inside a ball mill pot, they undergo plastic deformation, and are repeatedly flattened, cold-welded, fractured and rewelded. The force of the impact acts on the powder particles, leading to the breakage of crystallographic bonds and new surface is produced. The new created surfaces enable the particles to weld together easily and this leads to an increase in the rate of dissolution of solid material. Mechanical deformation produces new surface via the formation of fragments and the increase of surface energy of the material. Other profound changes affecting the surface as well as the chemical, physico-chemical and structural properties may also take place. This is proved by the presence of variety of crystal defects such as increased number of grain boundaries, dislocations, vacancies and interstitial atoms, stacking faults, and deformed and ruptured chemical bonds. The presence of such defect structure enhances the diffusivity of solute elements. Consequently, grinding a mixture of two or more solids substances results in the micro-homogenization of initial components, and sometimes, it induces the formation and synthesis of new fine powders [791-793].

The preparation of $SrAl_2O_4$ nanoparticles by CO_2 laser vaporization (LAVA) technology and laser melting method is described in works [794,795]. The LAVA technique is suitable for the preparation of a variety of ceramic nano-powders from coarse initial powders (usually oxides) although no specially designed precursors are required. The great advantage of the laser synthesis is the possibility to one-step, fast synthesis of these materials in air at the atmospheric pressure.

2. Preparation of crystals

The bulk single crystals of strontium aluminate can be prepared by the floating zone (FZ) or laser floating zone (LZF) method [796,797] or by the Czochralsky technique [26,798]

3. Preparation of thin layers

Thin film phosphors have several advantages over powders, such as higher lateral resolution from smaller grains, better thermal stability, reduced out gassing and better adhesion to solid substrates [799]. The following methods can be used for the preparation of thin films:

i. **The Pulsed laser ablation (PLD):** is a preparatory technique with several attractive features, including the stoichiometric transfer of target material, the generation of quality plume of energetic species, the hyper thermal reaction between ablated cations and molecular oxygen in the ablation plasma and the compatibility with background pressures ranging from UHV to 100 Pa. The plasma produced during the pulsed laser ablation is very energetic and its mobility can be easily controlled by changing the processing parameters [799-804]. The process involves various phenomena such as the target heating, the material removal and the plasma shielding. The absorption of laser radiation heats the target and the vaporization of target material takes place. Plasma shielding causes the drop in the laser intensity reaching the target surface [805].

ii. **Pulsed ion beam evaporation (IBE):** the method uses the high-power pulsed ion beam which is focused on the target. Formed high density ablation plasma enables to prepare thin film of material [806].

iii. **Electron beam bombardment (EBB**) [807].

iv. **RF magnetron sputtering** [808,809].

Chapter 10

Other Technical Applications of Strontium Containing Materials

The properties and application fields of other strontium containing materials are shortly described in this chapter. Although these material and substances shouldn't be directly related to strontium aluminate cements or strontium aluminates, this survey may be illustrative for the importance and mutual relationships between individual strontium bearing materials.

1. Phosphors

Some amorphous calcium aluminates are photosensitive and thus are potential candidates for optical information storage devices [128,790,810-813], hence the research on the strontium aluminates is recently very intensive. Strontium based aluminate phosphors (Table 1) are well known for their high quantum efficiency, long-lived afterglow, good chemical stability, and other excellent luminescent features, which make them appropriate candidate to replace the traditional II–VI based phosphors. These materials are successfully used in various applications like luminous paints for highways, airports, buildings, ceramic products, textile industry, dial plates of glow watches, warning signs, escape routes, etc. [814]. The following properties are often mentioned with regard to these compounds:

- **Phosphorescence:** when the substances slowly re-emit absorbed electromagnetic radiation (usually UV) in the form of visible light. The same effect, but absorbed energy is re-emitted immediately, is termed as **fluorescence**.
- **Thermoluminescence:** is a process at which the substance releases high-energy radiation in the form of visible light upon heating, which enables electrons to return to their positions.
- **Mechanoluminescence:** is a process at which the mechanical action causes emitting of light.

In order to investigate their properties, various strontium aluminate phases occurring in the $SrO - Al_2O_3$ system were synthesized using the solid-state synthesis as well as nontraditional synthesis routes described in Chapter 9.

According to the **Blasse's theory** [827] on the energy transfer mechanism in oxide phosphors, the critical energy transfer distance (R_c) can be calculated from the concentration quenching data using the following equation:

$$R_c = 2\left(\frac{3V}{2\pi x_c N}\right)^{1/3} \quad (1)$$

Where x_c is the critical concentration, N is the number of host cations in the unit cell of volume V. The luminescence of $Sr_{3-n}Ca_nAl_2O_6:Eu^{2+}$ phosphor that is completely solid solution of $Sr_3Al_2O_6$ and $Ca_3Al_2O_6$ can be shifted from 618 (orange) to 655 nm (red) with the increase of Ca/Sr ratio [378].

Phase	Dopant/ activator				
	Eu^{2+}/Eu^{3+}	**Dy^{3+}**	**Sm^{3+}**	**Ce^{4+}**	**Gd^{3+}**
$Sr_4Al_2O_7$	Red/orange [815, 816]	Blue and yellow [816]	---	---	---
$Sr_3Al_2O_6$	Red/ orange [814, 817-819]			Blue [820]	
$SrAl_2O_4$	Green [821, 822]	Green [823, 824]1)	Orange [749]2)	Green [763]3)	Blue [821]4)
$SrAl_4O_7$	[825]5)				
$SrAl_6O_{19}$	Violet [826]6)	---	---	---	---

1) Luminescence at 560 nm (green) and 595 (orange).

2) Enhanced by B_2O_3. Three main emission peaks at 562, 596 and 643 nm. The most intensive in orange.

3) The second excitation peak in ultraviolet 381 nm.

4) $SrAl_2O_4:Eu^{2+}$, Gd^{3+}phosphor.

5) Green doped by Cu.

6) Intensive emission at 397 nm and two weak bands at 594 and 615 nm (orange).

Table 1. Luminescence of some phases of $SrO-Al_2O_3$ system activated by commonly used ions.

REE activated strontium silicates, aluminosilicates, borates, zirconates ($AZrO_3$, A=Ca, Sr, Ba) and hafnates (SHO) such as strontium metasilicate ($SrSiO_3$:REE) [828-831], strontium orthosilicate (Sr_2SiO_4: REE) [832-838], glasses and glass-ceramics [839-847], $SrZrO_3$ [848-851] and $SrHfO_3$ [851-853] also show the photoluminescent properties.

2. Strontium containing bioactive cements

Although strontium is not considered as an essential element and does not have any known biological role, it is present in all living organisms. Strontium resembles the calcium element in its properties; like calcium, it is taken up and preferentially located in bones. Here, strontium can have both beneficial and deleterious effects in humans depending on the amount taken up [854]. Strontium ranelate ($C_{12}H_6N_2O_8SSr_2$), a pharmacologic agent used to treat individuals with osteoporosis, is indicated to be used in Europe and Australia but not in Canada or in the United States. Similar efficiency to delivery strontium to bones of animals was found for strontium citrate [855].

Injectable acrylic bone cements are widely used in orthopedic surgery to fix artificial prostheses. Conventional polymethyl methacrylate (PMMA) bone cement has been successfully used in arthroplasties of hip, knee and other joints for the fixation of polymer or metallic prosthetic implants to living bone; however, it still has some potential problems and risks, such as poor adhesion of bone cement to bone surface and high exothermic reaction during the polymerization [856, 857].

The size and morphology of individual particles, as well as the kind and extent of isomorphous substitutions greatly affect many physico-chemical properties of apatites. The great variety of possible cationic and anionic substitutions is justified by high stability and flexibility of the apatite structure. The unit cell of stoichiometric crystalline hydroxyapatite hosts 10 cations arranged in two non-equivalent positions: four at the M(1) site aligned in the column, each surrounded by nine oxygen atoms, and six at the M(2) site arranged at the apexes of "staggered" equilateral triangles, each surrounded by seven oxygen atoms. Among the bivalent cations which can replace calcium in Ca-HA, strontium has attracted a remarkable interest for its possible biological role [858-860].

Apatites are a family of inorganic crystalline compounds of general formula $M_{10}(XO_4)_6Y_2$, where M is usually a bivalent cations, such as Ca^{2+}, Sr^{2+}, Ba^{2+}, Cd^{2+}, Pb^{2+}, but monovalent and trivalent cations, such as Na^+, K^+ and Al^{3+}, can be hosted as well. XO_4 is usually PO_4^{3-}, VO_4^{3-} or AsO_4^{3-}, but possible substitutions also include SiO_4^{4-}, CO_3^{2-}, and SO_4^{2-}. Y denotes monovalent anions[1] OH^-, F^-, Cl^- or Br^- [860, 861].

Strontium stimulates the bone formation and has been recently used to treat osteoporotic spinal fractures and other related osteoporosis [862,863]. The bioactive bone cements based on strontium hydroxyapatite ($(Sr, Ca)_5(PO_4)_3(F, OH)$, Sr-HA or Sr-HAP, space group $P\ 6_3/m$) powder and Bis-GMA based resin were extensively studied as promising biomaterials [864-868]. The precipitation techniques [859] or hydrothermal syntheses [869-871] are usualy applied for the preparation of Sr-HA. The hydroxyapatite cements are considered as stronger than brushite cements [872] discussed below.

Calcium phosphate cements (CPC) are suitable materials for local delivery systems in osseous tissue since they can simultaneously promote the bone regeneration and prevent it from infectious diseases by releasing the therapeutic agents. Recent advances in CPC technology result in the enhancement of manipulation, applications and osteoconductive properties of these cements. These improvements allow the CPCs to be tested as carriers for local delivery of drugs and biologically active substances such as growth factors. Low-temperature setting of CPCs allows the incorporation of heat-labile medicaments and substances into the cement matrix during its preparation. Secondary CPCs which set (Eq.2, [881]) by the entanglement of brushite crystals ($CaHPO_4 \cdot 2H_2O$) draw the attention as drug-delivery systems mainly due to

1 Fluoroapatites obtained by the simultaneous substitution of Ca and PO_4^{3-} by REE and SiO_4^{4-} are described in Chapter 10.3.

their biodegradability. Ionic substitution in ceramic biomaterials is a reliable approach to alter the properties like crystallinity, solubility and biological performance. The Sr-substituted calcium phosphate cement releases Sr^{2+} ions at the dose of 12–30 ppm with a zero-order kinetics [873-875].

$$Ca_3(PO_4)_2 + Ca(H_2PO_4)_2 + 8\,H_2O \rightarrow 4\,CaHPO_4 \times 2\,H_2O \quad (2)$$

Brushite is formed on the calcite surfaces in caves in nature, where ammonium hydrogen phosphate or other related chemical in bat guano is the source of phosphate ions [876]:

$$(NH_4)_2HPO_4 + Ca^{2+} + 2\,H_2O \rightarrow CaHPO_4 \times 2H_2O + 2\,NH_4^+ \quad (3)$$

It is supposed, that brushite act as the precursor at the formation of human bones and teeth [877]. Brushite is also used as food additive and a component of tooth paste. It can be synthesized in high purity by slow mixing of calcium chloride and disodium hydrogen phosphate [878].

However, brushite is metastable under physiological conditions [879] and the in vivo studies have shown that the brushite cement is highly resorbable [875]. The brushite crystal growth inhibition results in smaller-sized crystals, allowing them to pack closely together and improving brushite cement's mechanical properties. Furthermore, increasing of brushite cement density improves the cement mechanical properties by reducing the cement porosity that is inversely proportional to the cement strength. This can be achieved by increasing the conversion of β-TCP (calcium triphosphate, $Ca_3(PO_4)_2$) to brushite [872,874,879-881]. Although TCP exhibits the bioactivity [882] its instability above the pH of 4.2 generally leads to the interaction with water on its surfaces in body fluid [883,884]:

$$4\,Ca_3(PO_4)_2 + 2H_2O \rightarrow Ca_{10}(PO_4)_6(OH)_2 + 2Ca^{2+} + 2\,HPO_4^{2-} \quad (4)$$

Brushite crystals show great variety of morphologies depending on the chemical conditions. Prismatic crystals precipitate from highly concentrated solutions, whereas tabular crystals are formed from more diluted solutions. Irregular crystals, limited by rounded edges or incomplete faces are common at all concentrations. The formation of twins according to [010] is also common [876].

Anhydrous form of brushite (monetite, $CaHPO_4$), is also useful biomaterial for the bone regeneration. Monetite can be prepared via two ways [876,885]:

1. Thermal dehydration of already set brushite cements that can be expressed by equations:

$$CaHPO_4 \times 2H_2O \xrightarrow{110-215°C} CaHPO_4 + 2\,H_2O \quad (5)$$

$$2\,CaHPO_4 \xrightarrow{450^{\circ}C} Ca_2(P_2O_7) + H_2O \qquad (6)$$

2. Modifying the precipitation conditions of brushite cements.

Strontium ions can be successfully incorporated into brushite crystals, but they have a negative effect on brushite cement cohesion. Both strontium and pyrophosphate ions have a synergetic retardant effect on the brushite cement setting reaction. As the ionic radius of strontium is higher than that of calcium, the brushite lattice parameters increase with the increase of strontium content [828].

3. Britholites

Britholites are phosphosilicate apatites, formed by the simultaneous substitutions of rare earth elements (REE=Ln^{3+}, La^{3+}, Nd^{3+}, etc.) and $SiO_4^{4-}$for Ca^{2+} and PO_3^{4-}. Britholites are considered to be particularly promising to be used as the matrices for the confinement of nuclear waste or for the preparation of luminescent pigments. Several processes were developed for the preparation of britholites. Usually, they are obtained via the solid state reaction at high temperature [886-889]. As the mechanochemical synthesis leads to the carbonated britholites, the heat treatment up to 1100 °C is needed for the preparation of pure britholites [890,891].

The synthesis of strontium fluorobritholites can be described by the following equation [890]:

$$\begin{aligned} &SrF_2 + 3\,SrCO_3 + (6 - x)/2\,Sr_2P_2O_7 + x/2\,La_2O_3 + x\,SiO_2 \\ &\rightarrow Sr_{10-x}La_x(PO_4)_{6-x}(SiO_4)_x\,F_2 + 3\,CO_2 \end{aligned} \qquad (7)$$

4. Ceramics and fiber-reinforced Sr-celsian matrix composites

Barium feldspar ($BaAl_2Si_2O_8$, celsian, BAS_2) and strontium feldspar ($SrAl_2Si_2O_8$, Sr-celsian, strontian, strontium anorthite, $SrAS_2$) exhibit a very similar polymorphism. For this reason, the terms monocelsian (celsian) and hexacelsian are also often used in literature to denote the monoclinic and hexagonal species of $SrAl_2Si_2O_8$ and $BaAl_2Si_2O_8$, respectively. Paracelsian is another known monoclinic polymorph of celsian [893,892].

Celsian and Sr-celsian (Fig.1) are the materials of great technological interest due to their excellent thermal and electrical properties. The monocelsian form exhibits high melting point (1650°C) and low coefficient of thermal expansion ($2.5 \cdot 10^{-6}$ K^{-1}). The hexacelsian form, the high-temperature metastable phase of strontium aluminosilicate, shows high coefficient of thermal expansion ($8 \cdot 10^{-6}$ K^{-1}) and reversible phase transformations within 600–800 °C causing the volume expansion of approximately 3% [893-895,1012].

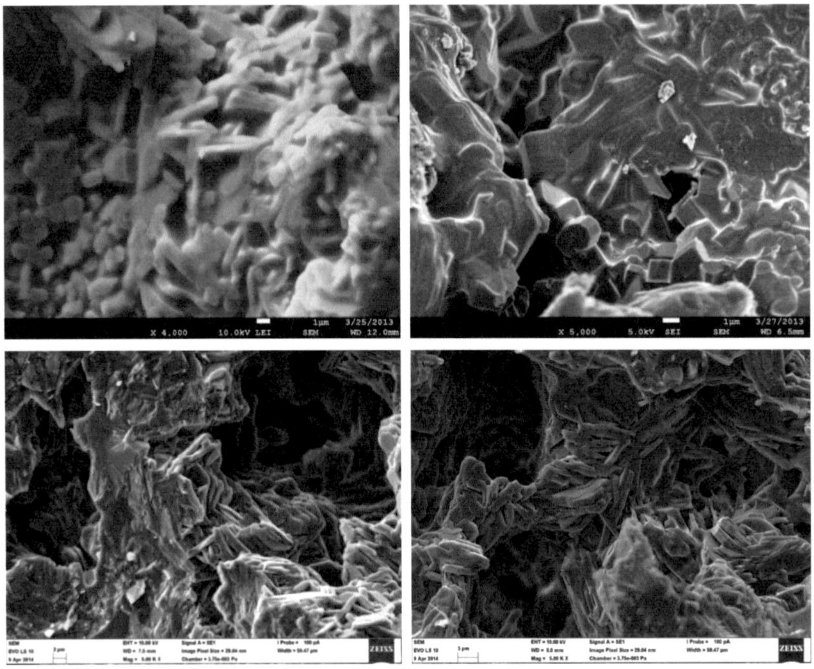

Figure 1. Synthetic anorthite [1011] and strontian prepared by the thermal treatment to 1300 and 1400 °C, respectively.

The properties of all three important feldspars, namely anorthite, strontian and celsian, are listed in Table 2.

Feldsapar	Formula, space g. 1)	Composition [% wt.]2)			Density	Temperature [°C]	
		MeO	Al_2O_3	SiO_2	[$g \cdot cm^{-3}$]	$T_{M \to H}$ 3)	Melting
Anorthite	$CaAl_2Si_2O_8$ ($P\bar{1}$, $I\bar{1}$)	20.2	36.6	43.2	2.77	--- 4)	~1550
Strontian	$SrAl_2Si_2O_8$ (I2/c)	31.8	31.3	36.9	3.08	758	~1650
Celsian	$BaAl_2Si_2O_8$ (I2/c)	40.8	27.2	32.0	3.39	1590/ 1650	~1760

1) Space group of monoclinic phase.

2) Calculated according to the feldspar stoichiometry.

3) Transformation temperature of monocelsian to hexacelsian.

4) Triclinic anorthite has metastable hexagonal and orthorhombic polymorphs [1013].

Table 2. Properties of $MeAl_2SiO_8$ feldspars.

In strontium aluminosilicate the hexacelsian phase may also exist as metastable. The existence of ferroelastic phase transformation to monocelsian at about 573 K followed by large expansion and the formation of microcraks is the main disadvantage for the application of hexacelsian. The kinetics and mechanism of transformation from hexacelsian to monocelsian, with and without doping with mineralizers, were reported, as well as the effects of mineralizers such as Li_2O, LiF, MoO_3 and TiO_2.

The addition of SrO can enhance the transformation of hexacelsian to celsian. Moreover, the addition of SrO can effectively reduce the thermal expansion coefficient of the (Ba, Sr)O–Al_2O_3–$2SiO_2$ ceramics [895,896,1022]. The processing and the properties of fiber-or whisker-reinforced celsian matrix composites were reported in the following papers [897-899].

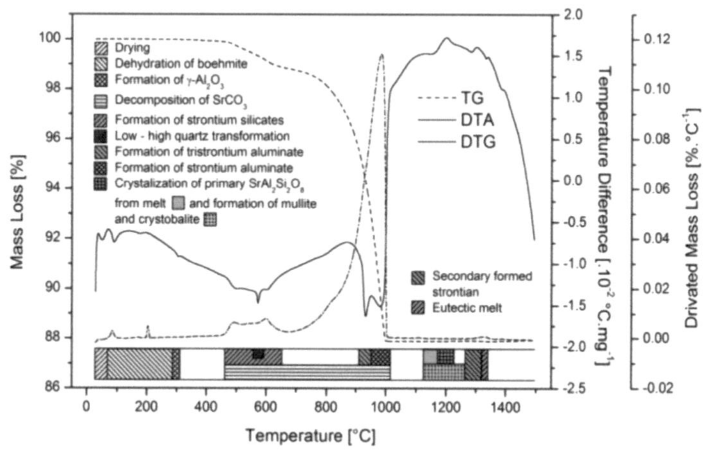

Figure 2. Thermal analysis of formation of strontian, the assessment using the heating rate of 10 °C min-1.

Fig.2 shows the results of thermal analysis of mixture of strontium carbonate ($SrCO_3$), reactive alumina (the mixture of boehmite and X-Al_2O_3) and quartz (low quartz) in the ratio corresponding to the stoichiometric composition of strontian. Drying as well as the dehydration of boehmite and the formation of γ-Al_2O_3 take place up to the temperature of 450°C. Partial thermal decomposition of strontium carbonate within the temperature interval from 450 to 650 °C is associated with the formation of strontium silicates. The formation of strontium aluminates within the temperature range from 900 to 1020°C shows the same features as for SrAC raw meal (please see Fig.23 in Chapter 4).

The results indicate that strontian can be formed as primary and secondary strontian. Primary strontian crystalizes from non-equilibrium melt at the temperature of 1200 °C and the process is accompanied by the formation of mullite ($3Al_2O_3 \cdot 2SiO_2$) and the transformation of high quartz to cristobalite (c-SiO_2). The establishment of the equilibrium composition with increasing temperature leads to the formation of secondary strontian via the solid-state synthesis, where the reaction between mullite, silicate phases and strontium aluminate takes place. At the temperature of 1384 °C, the eutectic melt appears.

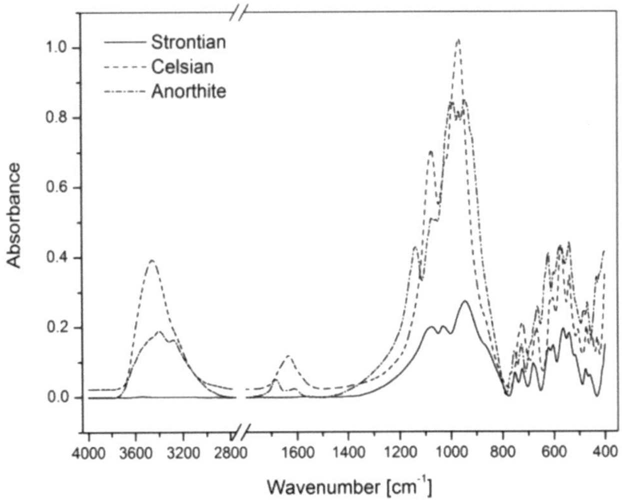

Figure 3. Infrared spectrum of strontian, celsian and anorthite.

Fig.3 shows the infrared spectrum of strontian after the thermal treatment, which is compared to the spectrums of celsian and anorthite. The spectrums of all samples show similar features with three main groups of bands [1025-1029]:

1. The spectral region from 1400 to 800 cm^{-1} is related to the symmetric and antisymmetric stretching of Si-O-Si and Si-O-Al bonds in SiO_4 and AlO_4 tetrahedra.
2. The spectral region from 800 to 650 cm^{-1} shows the features of harmonic vibration of Si-Si(Al) bonds in corner sharing tetrahedra.
3. The spectral region from 650 to 400 cm^{-1} is related to the vibrations of tetrahedral ring, O-Si(Al)-O bending and coupling between the deformation and stretching modes.

All feldspars are tectosilicates and share similar "mirrored crankshaft-chain" frameworks of polymerized Al/Si tetrahedra. Therefore, the number and intensity of bands in the spectrum of feldspars increase with increasing degree of Al/Si ordering [1028]. The lattice modes and the bands related to M-O motions are located in the FAR infrared region [1029].

5. Glasses and glass ceramics based of strontian

The glass-ceramic materials, which are employed in the applications where good wear and erosion resistance are required, often at high temperature and in chemically hostile environments, are of increasing interest. Their optical properties are described in Chapter 10.1. For these reasons the knowledge of their chemical composition and mechanical properties is important, in order to determine in particular how various materials respond to the surface contacts. This kind of materials can be made either by traditional glass-forming technique, followed by the nucleation/crystallization heat treatment, or by the sintering of glass powders, followed by the crystallization at higher temperatures. The latter technique is used especially where unusual product shapes are required, which cannot be obtained by common ways of glass shaping. Powdered glass, with the grain size less than 40-70 μm, with suitable binder is formed by pressing, casting in plaster molds or extruding [900-902].

Homogeneous and transparent glasses can be prepared within the $SrAl_2Si_2O_8$ (strontian)-$Sr_2Al_2SiO_7$ (helenite)-$SrSiO_3$ (strontium metasilicate) region of $SrO\text{-}Al_2O_3\text{-}SiO_2$ system [1014]. The thermodynamic model derived for the binary $SrO - SiO_2$ melts by R.-Serrano at al. [1023] is based on the consideration of the depolymerization reaction of SiO_2:

$$M^{2+}O^{2-} + \equiv Si\text{-}O\text{-}Si \equiv \rightarrow \equiv Si\text{-}O \quad (M^{2+}) \quad O\text{-}Si \equiv \tag{8}$$

or

$$O^{2-} + O^{0} \rightarrow 2\,O^{-} \tag{9}$$

where M=Ca, Mg, Sr and Ba, O^{2-} is free oxygen ion, O^0 is bridging oxygen bonded to two silicon atoms and O^- is oxygen bonded to one silicon atom.

Glass-ceramics based on strontian can be applied in aircraft industry for the preparation of radio-technical devices protected by radio-transparent barriers with the resistance to high temperatures due to their temperature stability, low thermal expansion coefficient, reduced electrical conductivity and dielectric losses. The selective growth of aluminium titanate (Al_2TiO_5, tialite[2]) and monoclinic $SrAl_2Si_2O_8$ is enabled by keeping the ratio of $SrO : Al_2O_3$ lower than one. That prevents the $SrO\text{-}TiO_2\text{-}Al_2O_3\text{-}SiO_2$ system from the crystallization of rutile and hexagonal polymorph of $SrAl_2Si_2O_8$ [1030].

$SrO\text{-}Al_2O_3\text{-}La_2O_3\text{-}SiO_2\text{-}B_2O_3$, $La_2O_3\text{-}Al_2O_3\text{-}B_2O_3\text{-}SiO_2$ and other strontium bearing glasses are concluded promising sealing materials for SOFC (Solid Oxide Fuel Cell) applications [1033-1035]. The preparation and the properties of mica glass-ceramics in the system $SrO\cdot 4MgO\cdot xAl_2O_3\cdot 6SiO_2\cdot 2MgF_2$ (x=1, 1.5 and 2) were described by Mllik at al. [1036]. The investigation of the crystallization and properties of the non-ferroelectric piezoelectric glass ceramics in the $PbO\text{–}SrO\text{–}TiO_2\text{–}SiO_2$ system demonstrates that the crystallization temperature of glass decreases with increasing mole fraction of PbO [1037].

6. Strontium containing bioactive glasses and glass ceramics

Bioactive glasses and glass–ceramics are widely studied due to their particular property of direct bonding to human bones through the formation of biologically active apatite layer at the bone/implant interface [903].

---

2 Tialite ceramics and composites are intensively studied in the field of engineering ceramics for their low thermal expansion coefficient and high thermal shock resistance. Tialite is usually prepared from its constituting oxides mixed in the equimolar ratio and treated to the temperatures above 1300 °C. The product is used as a precursor for further fabrication process (shaping and sintering) [1031,1032].

The substitution of calcium by strontium results in a small but significant expansion of the glass network, which is associated with larger size of Sr^{2+} cation compared to Ca^{2+}. The expansion of the glass network results in the weakening of the glass network and in associated decrease in the glass transition temperature and dilatometric softening point and in the increase in the thermal expansion coefficients [904]. The expansion of the network structure of glass increases the ion dissolution rates and has significantly pronounced effect on bone cells in vitro and in vivo. Therefore, bioactive response should be greater due to higher dissolution rate and biological effect of Sr on bone-forming cells. The slight supersaturation of physiological fluids with respect to Sr, Ca and phosphate causes the hydroxyapatite layer to occur more quickly [905]:

$$5\,Sr^{2+} + 5\,Ca^{2+} + 6\,PO_4^{3-} + 2\,OH^- \longleftrightarrow Sr_5Ca_5(PO_4)_6(OH)_2 \qquad (10)$$

The behaviour during the crystallization of glass–ceramic materials based on Na_2O–CaO–P_2O_5–SiO_2 glass system modified by SrO/CaO and K_2O/Na_2O shows that potassium ions could be accommodated in the wollastonite structure while strontium ions might be incorporated in the sodium calcium silicate phase. The bioactivity studies on the prepared glass–ceramics samples show that the materials are capable of bonding with human bone due to the formation of apatite layer after the immersion in SBF solution. K_2O reduces the formation of apatite layer in the comparison with the potassium-free variety [903].

7. Strontium containing phosphorescent glasses and glass-ceramics

Glasses embedded with rare-earth (RE) ions are very attractive materials for the development of many optoelectronic devices, e.g. in fiber amplifiers, lasers, fluorescent devices and white light emitting diodes (LED) [843,906,907].

Long lasting phosphorescence (LLP) due to Eu^{2+} and Ce^{3+} ions was observed in Eu^{2+}, Ce^{3+} co-doped strontium borate glasses. The phosphorescence of Eu^{2+} decreases with the addition of Ce^{3+} because of the competition to obtain the trapped electron, and the phosphorescence of Ce^{3+} in the sample decays more quickly than that of Eu^{2+}, which is suggested as the reason for the emission energy of Ce^{3+} to be higher or the distance between Ce^{3+} and electron traps of the glasses to be longer [908].

8. Strontium metal-organic framework

The research in the field of metal–organic frameworks (MOFs) has grown exponentially in the last decade because of the numerous potential applications such as in gas adsorption and storage, drug delivery, catalysis, fabrication of molecular sieve membranes and luminescent properties [909-911].

The strontium metal–organic framework of $[Sr_2(BTEC)(H_2O)_4]\bullet 2H_2O$ was synthetized by Lo et al [911]. Sr^{2+} ion was nine-coordinated in the shape of a tricapped trigonal prism, SrO_9, and it bound six oxygen atoms from the carboxylate groups belonging to four BTEC ligands (H_4BTEC=benzene-1, 2, 4, 5-tetracarboxylic acid) and three oxygen atoms from the coordinated water.

9. Strontium titanate and other perovskites

Strontium titanates can be used as dielectric materials in high performance metal–insulator–metal (MIM) capacitors for analogue applications. The grain boundary barrier layer capacitors [912,913], oxygen gas sensors [914,915], substrates [916,917], catalysts [918] and solid oxide fuel cells (SOFC) anodes [919,920] are some of the most typical applications of these materials. Strontium titanate is also one of promising candidates for host materials in low-voltage electron-excitation displays because of its chemical and compositional stability and its optical band gap (3.3 eV).

Four phases of $SrTiO_3$ (ST, cubic perovskite with space group *Pm3m*), $Sr_4Ti_3O_{10}$ (tetragonal structure with space group *I4/mmm*), $Sr_3Ti_2O_7$ (tetragonal, double perovskite structure with space group *I4/mmm*) and Sr_2TiO_4 (quasi-two-dimensional K_2NiF_4 structure of Ruddlesden–Popper type, space group *I4/mmm*) were described in binary system of $SrO-TiO_2$. These phases can be prepared from the mixtures of $SrCO_3$ and TiO_2 in the ratio corresponding to the stoichiometry of required phase [921]. For the formation of those four compounds from their component oxides TiO_2 with rutile structure and SrO, the standard Gibbs free energy changes are given by [922]:

$$\Delta G^{o}_{(ox)}(SrTiO_3) \pm 89 / (J \cdot mol^{-1}) = -121878 + 3.881\,(T / K) \tag{11}$$

$$\Delta G^{o}_{(ox)}(Sr_4Ti_3O_{10}) \pm 284 / (J \cdot mol^{-1}) = -409197 + 14.749\,(T / K) \tag{12}$$

$$\Delta G^{o}_{(ox)}(Sr_3Ti_2O_7) \pm 190 / (J \cdot mol^{-1}) = -285827 + 10.022\,(T / K) \tag{13}$$

$$\Delta G^{o}_{(ox)}(Sr_2TiO_4) \pm 110 / (J \cdot mol^{-1}) = -159835 + 3.770\,(T / K) \tag{14}$$

Among the solid-state syntheses mentioned above, using the convectional [912,921,922] or the microwave sintering [923], the preparation of strontium titanate or its solid solutions via the sol gel method [918,924,925,926], the coprecipitation [926], the Pechini [920,927,928] or modified Pechini method [929], the solvothermal synthesis [926], the spray pyrolysis [930], and the solid mechanosynthesis [914,915,931] were reported. The mechanosynthesis of strontium aluminate is feasible only under highly energetic milling and failed for less energetic milling due to the thermodynamic restriction of the process [931].

Titanate-based perovskite-type oxides can be described by the general formula of $MTiO_3$, where M=Pb, Ba, Sr, Cd, Fe, Zn. Strontium titanate is paraelectric at room temperature and does not show the transition to ferroelectric phase when cooled to liquid helium temperature while barium titanate is ferroelectric at room temperature and shows the transition temperature of 120 °C. Depending on the Ba/Sr ratio, the properties of the solid solution (transition temperature, optical and electrical properties) of barium titanate and strontium titanate ($Ba_xSr_{1-x}TiO_3$, BST) differ over a broad range [924,932,933]. The addition of B_2O_3 to BST powder leads to the decrease of sintering temperature from 1350 to 900 °C. The temperature of maximum permittivity and the Curie temperature are not affected [934]. Plumbic titanate (PT) has high dielectric constant, but it is unsuitable for the applications mentioned above due to higher (490 °C) transition temperature (from ferroelectric tetragonal to paraelectric cubic phase). Depending on increasing Pb/Sr ratio in the solid solution ($Pb_xSr_{1-x}TiO_3$, PST), the Curie temperature decreases linearly with increasing content of Sr^{2+}[935].

Due to their catalytic activity, ferroelectric properties, piezoelectric behaviour chemical and thermal stability and electrooptic propeties, perovskyte-type oxides (ABO_3) such as strontium niobate [936-938], cerate ($SrCeO_3$) [939-943], chromate ($SrCrO_3$) [944,945], molybdate ($SrMoO_3$) [944,946,947], $SrMnO_3$ [948,949], $SrFeO_3$ [950], $SrCoO_3$ [951,952], $SrRuO_3$ [948], etc. are intesively studied. These compounds belonging to the homologous series $A_nB_nO_{3n+2}$ are a special group of perovskite-related layered materials [938].

10. Strontium ferrites

Ferrite materials may be classified into three different classes; spinel ferrites, garnet ferrites and hexagonal ferrites. The magnetic spinel has the general formula of MFe_2O_4, where M is the divalent metal ion, usually Ni, Co, Mn, or Zn. The garnet ferrites have general composition given by the formula $A_3B_3X_2O_{12}$, where A ions occupy distorted cube sites, while B and X are placed in octahedral and tetrahedral sites, respectively. In rare-earth iron garnets (REIG) $Ln_3Fe_5O_{12}$ and yttrium iron garnet $Y_3Fe_5O_{12}$ (YIG), Fe^{3+}ions show octahedral or tetrahedral environment, while rare-earth ions are in the centers of deformed cubes [953-955].

Hexagonal ferrites, $MFe_{12}O_{19}$ (M=Ba, Sr, Pb) with the magnetoplumbite structure have been important permanent magnet materials in microwave, small motor, and, more recently magnetic recording applications since their discovery in 1950s. It is well known that the coercive force of hexaferrites depends mainly on the crystallite size; high coercivity can be reached when the size of crystallites is ~100 nm. In order to improve the fundamental magnetic properties of hexaferrites, many studies have also been carried out concerning the cationic substitution. Some experiments used light rare-earth ions such as La ($La_xSr_{1-x}Fe_{12}O_{19}$), Pr and other metal cations in the substitution for Sr (Ba) and Fe, respectively, taking into account the ionic radius of the elements [956,957].

Hexagonal ferrites, so-called hard ferrites, possess relatively high saturated and remnant magnetizations and exhibit high magnetocrystalline anisotropy. High heating effect can thus be expected. Iron ions occur at five different sites: the octahedral sites, crystallographically known as 2a, 12k, and $4f_2$, and the tetrahedral sites $4f_1$ and 2b. In the magnetically ordered state in $BaFe_{12}O_{19}$ or $SrFe_{12}O_{19}$, the 12k, 2a, and 2b sites have their spins aligned in parallel to each other and to the crystallographic c-axis, whereas those of $4f_2$ and $4f_1$ point the opposite directions [958].

Their structures of a polytypic nature can be conventionally often described in the terms of an intergrowth of the blocks S, R and T. The S block is a (1 1 1) slice of the common spinel structure, while R and T are the blocks of hexagonal symmetry which represent weak links in the magnetic exchange interactions. Among them let us mention the M phase as example, arising from the stacking of ...SRS*R*... type and the Y phase formed by stacking of ...$S_1T_1S_2T_2S_3T_3$... type. Hexagonal ferrites are complex magnetic systems and offer several kinds of non-equivalent sites, octahedral, bipyramidal and tetrahedral, which can be occupied by diverse magnetic or non-magnetic cations. Their magnetic properties thus can be readily modified by changing the stacking of individual blocks and the composition in a desirable way [959].

Ferite materials can be prepared by various techniques including the solid-state synthesis [854,856,858], the sol-gel method [960,961], the coprecipitation method [962,963] and the combustion synthesis [964].

For the microwave applications, the dielectric properties such as dielectric constant and dielectric loss are very important as the dielectric constant affects the thickness of microwave absorbing layer and the dielectric loss factor (*tan* δ) of a material determines the dissipation of electrical energy. This dissipation may be due to electrical conduction, dielectric relaxation, dielectric resonance and loss from non-linear processes. High performance devices require low dielectric loss because the lower it is the higher the efficiency and the lower the noise. The porosity of material also increases the dielectric loss and that may be due to the absorption of humidity by the pores of that material [963,965,966].

11. High-temperature ceramic superconductors

In 1986 Bednorz and Müller discovered the high temperature superconductivity by predicting that the JT (Jahn–Teller) distortion plays an important role in lanthanum copper oxides (LSCO) and thus the electron–lattice interactions are very strong in cuprates [967].

It was shown by the first-principle variational calculation of the spin-density-functional approximation, that when Sr^{2+} ions were substituted for La^{3+} ions in LSCO, the optimized distance between apical O and Cu in LSCO decreased with Sr concentration. As a result the elongated CuO_6 octahedrons by the JT interactions shrank by doping holes. This shrinking effect against the Jahn–Teller distortion is termed as "**anti-Jahn–Teller effect**" [968].

Strontium is used for the synthesis of ceramic superconductors for various systems. Some examples of these compounds are listed in Table 3. It should be noted, that all examples are ceramic conductors, but only the system with critical temperature (T_c) higher than 30 K can be considered as high-temperature superconductors. Although $\boldsymbol{T_c}$ is the most known parameter, there are two other parameters: $\boldsymbol{B_c}$ (critical magnetic field) and $\boldsymbol{I_c}$ (critical current density),

Superconducting system	Superconductor (Abbreviation)		T_C [K]	Reference
Bi–Sr–Ca–Cu–O (BSCCO)	$Bi_2Sr_2CuO_{6+\delta}$ (Bi 2201)		< 20	[969, 970, 971]
	$Bi_2Sr_2CaCu_2O_{8+\delta}$ (Bi 2212)		85 (110)[1)]	[969-972]
	$Bi_2Sr_2Ca_2Cu_3O_{10+\delta}$ (Bi 2223)		110	[969-971, 973]
Tl–Sr–Ca–Cr–Cu–O	$TlSr_2CuO_5$(Tl 1201)		70	[974]
	$(Tl_{1-x}Cr_x)Sr_2CaCu_2O_{7-\delta}$	Tl 1212	103	[974, 975]
	$(Tl_{0.5}Pb_{0.5})Sr_2Ca_2Cu_3O_9$ [3)]	Tl 1223	130	[975, 976, 977]
	$Tl_2Sr_2CaCu_2O_{8-\delta}$	Tl 2212	110	[974, 975, 978]
	$(Tl_{1-x}Hg_x)_2Sr_2CaCu_2O_{8-\delta}$		<45	[974, 975]
		Tl 2223	125	[975]
Sr-doped La-Cu-O [2)]	$La_{2-x}Sr_xCuO_{4-\delta}$	LSCO	~12	[968, 979-981]

[1)] In the Pb-doped systems ([Bi, Pb]:Sr:Ca:Cu).

[2)] High temperature superconducting cuprates (HTSC).

[3)] The highest critical temperature in the system.

Table 3. Examples of strontium bearing ceramic superconductors.

which both together provide the phase diagram for high temperature superconductors of the first and second type.

12. Heisenberg antiferromagnets

Two-dimensional triangular Heisenberg antiferromagnets (2DTHAF) have been the subject of intense research activity for a long time, both from the theoretical and experimental viewpoints [982,983].

From the experimental point of view, an important class of quasi-2DTHAF involves S=3/2 spins associated with Cr^{3+}, in two families of layered chromites:

a. Ordered rock-salt structures ($ACrO_2$ with A=Li, Na);

b. So-called delafos-site compounds ($MCrO_4$ with M=Cu, Ag, Pd).

Recent works demonstrate that the α-ACr_2O_4 compounds (with A=Ca, Sr, Ba) should be included in the same class of materials.

AM_2O_4 compounds usually crystallize in the $CaFe_2O_4$ type, but using high preparation temperature (higher than 2000°C), SrO reacts with Cr_2O_3 to form the compound with a new type of crystal structure. $SrCr_2O_4$ is a metastable substance stabilized by entropic effects, which crystallizes in the form of extremely thin two-dimensional sheets. The attempts to increase the

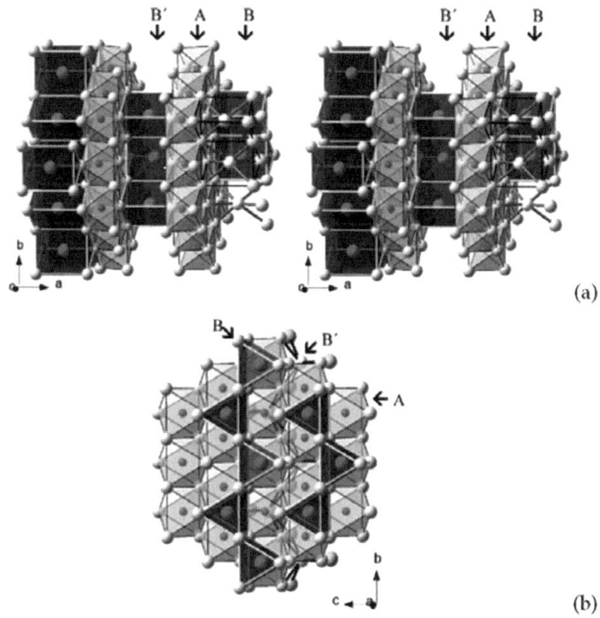

Figure 4. Crystal structure of $SrCr_2O_4$: (a) connection layers of CrO_6-octahedra (layer A), chains of SrO_6 trigonal prisms (layer B and B′) and (b) connections of SrO_6 prism chains with CrO_6 octahedra of layer (A) using triangular polyhedral faces. The end of one SrO_6 prism chain is situated in layer (B′) above both sides of layer (A) [984].

thickness of the crystals by annealing at lower temperatures (900-1000 °C) led to the distortion of the metastable crystal structure in all cases [984].

The $SrCr_2O_4$ type exhibits the layers (A) consisting of close-packed planes of CrO_6-octahedra. These layers (A) are connected by SrO_6 trigonal prisms (Fig.4(a), dark polyhedra, bright edges). SrO_6-prisms themselves form $_\infty[SrO_4]$-chains crossing the crystal along [010]. The connections between $_\infty[SrO_4]$-chains and the CrO_6-octahedra are illustrated in Fig.4(b) [984].

13. Strontium carbide, nitride and silicide

Strontium carbide (SrC_2) was synthesized for the first time by Moissan [985-987] from $Sr(OH)_2$ or $SrCO_3$ with sugar coal in an electrical arc furnace at the end of 1800s. Binary carbides can be described by the general formula AC_2, where A=Mg, Ca, Sr, Ba, Rb. At ambient conditions SrC_2 crystallizes in the tetragonal CaC_2(I) structure (I4/mmm). The C_2^2 dumbbells are aligned along the tetragonal *c* axis in these structures. At the temperature about 643 K a phase transition to a cubic high temperature modification occurs, which is analogous to Fm $\bar{3}$ m structure of CaC_2(IV) [988-990]. At low temperatures a partial transformation of SrC_2 (I) to monoclinic low-temperature modification (C2/c, SrC_2 (II)) was observed. This transformation starts at room temperature and levels out at about 80 K [989].

Strontium nitride (Sr_2N) crystallizes in a layered structure of the anti-$CdCl_2$ type. In Sr_2N, close-packed Sr atomic layers sandwich nitrogen atoms to form the Sr_2N layers, and the gaps between the layers are considerably expanded (the Sr–Sr distance across the gap reaches 0.473 nm, while that in the Sr_2N layer is 0.352 nm), indicating the van der Waals character of the gaps (Fig.5). Recently, the series of new compounds such as Sr_4N_3, SrN, and SrN_2 have been reported as derivatives of Sr_2N [991, 992].

The metals from the second group (A) form nitrides with A–N bonding and the structures change dramatically with the descent in the group; the lighter metals (Be–Ca) form ionic, insulating or semiconducting, salt-like compounds whereas the heavier members of the (Ca–Ba) group form subnitrides with low-dimensional structures and metallic properties. The simple reaction chemistry of subnitrides is intriguing. Layered subnitrides A_2N (A=Ca, Sr, Ba) provide a structural basis for the inclusion of anions from simple spherical species such as halides (A_2NX, X=F, Cl, Br, I) to form more complex, anisotropic entities such as $N_2^{2-}$and CN_2^{2-} [993]. The optoelectronic applications, such as in the field of high performance light emitted diodes, are possible for strontium-germanium ternary nitrides (A_2MN_3, where A=Sr, Ba and M=V, Ta, Nb) α-and β –Sr_2GeN_2 [994] and Sr_2TaN_3 [995].

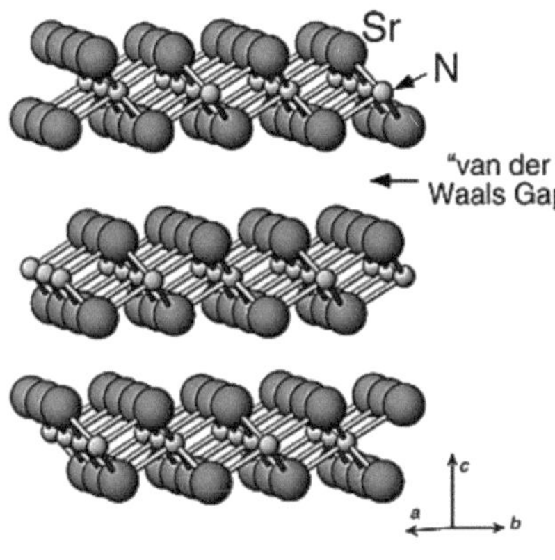

Figure 5. Crystal structure of Sr_2N [991].

In recent years an increasing attention has been drawn to the alkaline-earth metal (AEM, (Ca, Sr, Ba)Si_2) silicides, in particular with the ratios of 2:1, 1:1 and 1:2, due to their semiconducting and superconducting properties [996]. Pseudo-binary hexagonal **strontium silicides** $SrNi_xSi_{2-x}$ ($0.1< x <0.7$) were prepared by arc-melting. The phase shows the superconductivity with T_C of 2.6 K. Besides, for these silicides the non-conventional AlB_2-like phase may be stabilized by partial replacement of Si by *sp* metals (Al or Ga); in order to turn the lattice into pseudobinary silicides Sr(Al, Ge)$_x$$Si_{2-x}$. The superconducting transition at the temperature $\boldsymbol{T}_C$ of 5.1 K was found for this compound [997]. The ternary Sr-Al-Si system produced the clathrate type I structure by arc melting of the appropriate elemental combinations. The layers are interspersed with Si atoms. The $AAl_{2-x}Si_x$ (Al=Ca, Sr; x=1) phases crystallize as the AlB_2-type structures (P6/mmm) [998,999].

The formation of ternary hydrides (MTrTtH, where M=Ca, Sr, Ba, Tr=Al, Ga, In and Tt=Si, Ge, Sn) was also reported. SrAlSiH crystallizes in the layered structure (*P3m1*), where Al and Si are in ordered hexagonal sheet with hydrogen surronding a Al on one side only [1000]. SrAlGeH can be synthetized by hydrogenating the intermetallic AlB_2-type precursors (SrAlGe) or by the reaction of SrH_2 with a mixture of Al and Ge in the presence of pressurized hydrogen [1001].

14. The effect of strontium on the properties of alloys

Magnesium (Mg) alloys are known for their remarkable strength-to-weight ratio (density 1.74 g cm^{-3}) and are therefore being considered as excellent material candidates in aerospace and transportation industries, where the performance and fuel economy are of major importance. However, insufficient control of room temperature formability compared to currently used aluminum alloys for example, as well as plastic anisotropy has restricted the extended use of Mg alloys. In this context, several attempts have been made to introduce various alloying combinations, which linked with advanced processing techniques have the potential to minimize the formation of strong textures during hot deformation which are responsible for observed poor formability of Mg alloys. Relevant improvements of properties and corrosion resistance have been reported in the case of Rare Earth elements (REE) which, however, are relatively cost prohibitive [1038,1039].

Strontium and calcium are initially introduced to Mg alloys in order to replicate the effects of REE. Thermally resistant Sr-rich precipitates form the unique Mg–Al–Sr high temperature creep resistant alloys, which are successfully used in casting automotive engine blocks. Strontium was also used to develop wrought Mg alloys where thermally stable Sr-rich precipitates were the effective particle stimulated nucleation (PSN) sites during hot deformation and were capable of reducing the texture anisotropy by nucleating into randomly orientated grains. It was also noted that when the amount of Sr increased above certain values, such precipitates could also act as the crack nucleation sites and therefore reduce the ductility. Furthermore, the effect of Sr on the twinning of a Mg-Al alloy caused by the changes in solid solution composition was also investigated [1039-1040].

The effects of Sr in AZ31 are dependent on the concentration of Sr in the alloy and the type of second phase which precipitates. When added to AZ31, Sr forms Al–Sr, Mg–Al–Sr, Mg–Sr and Sr–Zn precipitates. In the Mg–Al–Zn system Sr prefers to bind to Al rather than to other elements. Because of this high affinity of Sr to Al, all the precipitates in the AZ + Sr system could be classified into two groups: precipitates with Al and precipitates without Al [1040].

The additions in the range of a few 100 ppm of Sr, Na, Ca, Ba or Eu modify the eutectic Si morphology of Al-Si alloys from coarse plate-like into fine fibrous and have a beneficial effect on both strength and ductility [1041]. Strontium decreases the volume percent, size and number of settled intermetallic compounds in A380 aluminium alloy. The distribution of alloying elements in Sr-modified melt is also more homogenous than in unmodified melt [1042].

15. Immobilization of strontium radionuclides

Strontium radioactive isotopes Sr-89 and Sr-90 emit high-energy beta radiation. They are extremely hazardous because they deposit in bones replacing calcium. Their radiation can damage bone marrow and blood-forming organs, inducing cancer [91]. The radionuclide solubility is one of the factors, along with the sorption and inventory, which can determine the 'source term' for potential migration to the far away field of a nuclear waste repository. Several processes can affect the concentration of radionuclide in water in pores of cement [1002,1003]:

1. Precipitation as the simple salt
2. Coprecipitation with other phases;
3. Incorporation in lattice of major cement hydration products (solid solution);
4. Sorption at hydrous surfaces (chemisorption, adsorption);
5. Complex and colloid formation in aqueous phase.

The absorption is the incorporation of a sorbate within a physical or molecular structure of a sorbent, i.e. the sorbate has the access to the bulk, not just to the surface, of sorbent. Physical adsorption can be divided into three types:

1. Surface complexation, which is a mixture of chemical and electrostatic interactions;
2. Ion exchange, which is an example of electrostatic adsorption due to columbic forces of attraction between charged sorbate species and the sorbent;
3. Other mechanisms, e.g. solid-solution formation, coprecipitation.

The binder most commonly used for the immobilization of toxic and radioactive waste is Portland cement. The most important reactions of cement with waste are those which lead to binding, i.e. chemical reaction and formation of thermodynamically stable or near stable product. The behaviour of Sr in cements might be expected to mirror that of Ca. However, it is soluble in NaOH or KOH to such extent that $Sr(OH)_2$ precipitation in cements is unlikely [12, 1002, 1004].

The major amount of Sr was incorporated into the calcium aluminium sulphate hydrates. Appreciable substitution of strontium in Aft and C_3AH_6 was demonstrated, and was thought by the authors to be likely in AFm-type phases. The structure of AFt has a range of structural sites which can be occupied by both, cations and anions. Therefore, cements high in Ca and Al, such as high alumina cements (HAC) or the mixtures of alumina cement with Portland cement, will have the best 'sorption' characteristics for Sr. The leaching experiments on Sr dopped pastes, in conjunction with the consideration of the effect of carbonation, indicated the coprecipitation of $SrCO_3$ to be a solubility-limiting process for the release. An ettringite analogue[3] with Sr replacing Ca was reported [12,1002,1004]. On the other hand, delayed formation of ettringite leads to cracking of the test specimens. This process is associated with increasing leachable surface and decreasing physical stability [12].

Cement based materials were used in the radioactive waste management to produce stable waste forms and to build engineering barriers in disposal facilities. The immobilization of low and intermediate level radioactive wastes in cementitious matrices is the most commonly used technique to produce inexpensive waste matrix that complies with the regulatory requirements. Immobilized waste forms can be safely isolated from the biosphere for a long time in a radioactivewaste repository. Conventional cementitious waste matrices consist of waste, cement, additives and water. The additives (blast furnace slag, fly ash, kaolin, zeolites and bentonite) are used to enhance the mechanical performance of the waste matrix and/or to reduce the leachability of problematic radionuclides e.g. ^{137}Cs, ^{60}Co, and ^{90}Sr [1005-1007].

The binding of radionuclides by a cement matrix is determined by their chemical nature and is not affected by their radiation characteristics. In most nuclear wastes the present radionuclides constitute only a small fraction of the total mass, and are intermixed with non-radioactive species. If some of them possess chemical toxicity, such products are classified as mixed wastes. The immobilization of heavy metals (As, Ba, Cd, Cr, Hg, Pb, Zn...) is generally attributed to the precipitation of insoluble hydroxides in the high pH environment [12].

Blended cement[4] with bentonite was proposed as a buffer and/or shaft seal in geological disposal for nuclear wastes due to its low permeability, chemical and physical stability, and compatibility with different host environments. The utilization of bentonite as liner and waste matrix additive was supported by the availability of smectite ion exchange and amphoteric surface complexation sites for the adsorption of radionuclides [1005, 1008]. The blends of Portland cement with zeolite are also promising ion exchange materials for the removal of radioisotopes from aqueous nuclear wastes because of their selectivity, radiation and thermal stability, and good compatibility with the cement matrix [1006].

The presence of blast furnace slag in Portland cement reduces the leachability of technetium and chromium. The addition of fly ash or silica fume into the cementitious system reduces the leachability of heavy metals, mainly by reducing the content of free calcium hydroxide in hardened paste [12].

16. The biological role of strontium

Natural strontium is a mixture of four stable isotopes: ^{84}Sr (0.56%), ^{86}Sr (9.86%), ^{87}Sr (7.02%), and ^{88}Sr (82.56%). Strontium comprises 0.02-0.03% of the earth's crust from where strontium dissolved in water derives. Its concentration in soil and drinking water varies between 0.001 and 39 mg dm^{-3}. Normal daily intake contains 2-4 mg Sr per day, most of it is derived from vegetables and cereals. Thus, the amount of Sr in food of Western countries is negligible compared to Ca. The intake of Sr depends on the Sr contents in nutrients, and for plants the concentration of Sr correlates with the Sr content in soil, which is much lower than that of Ca [1015].

---

3 Structure of ettringite and ettringite analogue are described in Chapter 8.2.

4 Blended cements were defined in the introduction of Chapter 7.

Sr has never been shown to be an essential element, that is, causing death when absent, but Sr may promote growth in some plants. Since strontium given as strontium ranelate (please see Chapter 10.2) augments bone Ca in experimental animals and reduces the fracture rate in osteoporotic patients, it could be hypothesized that one feature of osteoporosis may be a certain degree of Sr deficiency, but the data on normal content of Sr in bone are scarce. One investigation of trace elements in iliac crest biopsies of an experimental osteoporotic rabbit model did not show reduced bone Sr levels. However, it was shown that among the trace metals present in human bone, Sr was the only one that was correlated with bone compression strength [855,1015,1016].

The toxic symptoms in human due to overdosing on Sr have not been reported. However, the intravenous administration of high doses of Sr induces hypocalcaemia due to increased renal excretion of Ca. The only stable Sr-containing chemical that is considered to be harmful to humans in small amounts is strontium chromate, the toxicity being caused by the chromium which is a genotoxic carcinogen [854,1015,1017].

Strontium in human biology and pathology has attracted less attention than the other two important divalent metals calcium and magnesium, and over the years it has been an object of academic rather than clinical interest. Although this is still true, that there is an increasing awareness of the biological role of Sr after the development of the strontium ranelate drug, which has recently been shown to reduce the incidence of fractures in osteoporotic patients. Radioactive Sr isotopes are dealt with only when they are used for physiological or diagnostic purposes [1015,1018]. Strontium isotope ratio analysis also becomes common in archaeological investigations [1019-1021].

Chapter 11

Notations and Abbreviations

The following abbreviations and symbols are used in this book:

Symbol	Description
$\bar{q}$	The heat flux vector
[p]	Isobaric Conditions
[T]	Isothermal Conditions
A	Frequency factor [s^{-1}]
$a, b, c, \alpha, \beta, \gamma$	Parameters of phase unit cell
$B_{i,j}$	The primary breakage distribution
$c^{o}{}_{pm}$	Standard isobaric molar thermal capacity [$J \cdot K^{-1} \cdot mol^{-1}$]
C_i	Reaction Species, Reactant or Product
D	Diameter of mill
E_a	Activation energy [$J \cdot mol^{-1}$]
ETC, Λ	Effective thermal conductivity
F	The Number of Phases in Given Thermodynamic System
f_i	The feed rate of size fraction [$t \cdot h^{-1}$]
G	Gibbs Energy
$g(\alpha)$	Kinetic function ($g(\alpha) = kt$)
h	The order of matrix of constitution coeficients
HAC	High Alumina Cement
HCV	High Caloric Value [$J \cdot mol^{-1}$]
K	Boltzmann Constant, $k = R/ N_A = 8.314/ 6.023 \cdot 10^{23} = 1.381 \cdot 10^{-23}$ $J \cdot K^{-1}$.
k	The constant of reaction rate
K	Equilibrium constant
k_B	Boltzmann constant
LCV	Lower Caloric Value [$J \cdot mol^{-1}$]

Symbol	Description
LHV	Lower Heating Value [$J\cdot mol^{-1}$]
M_A	Alumina module
M_H	Hydraulic Module of Clinker
M_H	Hydraulic module
N	Number of Moles [mol].
n	Kinetic factor (kinetic exponent)
n_+, n_-	The number of cations, anions
N_A	Avogadros number ($6.02214\cdot 10^{23}$ mol^{-1})
NCV	Net Caloric Value [$J\cdot mol^{-1}$]
P	Pressure [Pa]
P_c, F_c	The sieve size passing 80% of clinker after and before crushing
PC, OPC	Portland Cement, Ordinary Portland Cement
P_D	The partial pressure of water vapor [Pa]
PSD	Particle packing density
R	Universal Gas Constant, $R = p_{st}\cdot V_{st}/ T_{st} = 1.0325\cdot 10^5 \cdot 22.414\cdot 10^{-3}/ 273.15 = 8.314\ J\cdot mol^{-1}\cdot K^{-1}$.
R	Number of independent reaction in the system (according to the Gibbs stoichiometric law).
R_c	Critical Energy Transfer Distance in Blasse's Theory
S	Number of Species in Given Thermodynamic System
$S°_m$	Standard molar entropy [$J\cdot K^{-1}\cdot mol^{-1}$]
SD_{SrO}	Strontium saturation factor
SI	The shape index of peak
S_i	Specific rate of breakage
T	Temperature [K]
t	Time [s]
T_m	The temperature of peak [K]
T_{pw}	The temperature of wet point [°C]
v	Number of Degrees of Freedom (according to the Gibbs phase law).
V	Volume
w/c	The water to cement ratio
$w_{1/2}$	The half-width of peak [K]
W_A	Absolute humidity of air [$kg\cdot m^{-3}$]
W_c	The energy consumed for crushing the clinker [$kWh\cdot t^{-1}$]

Symbol	Description
W_m	The mill specific output motor power [$kWh \cdot t^{-1}$]
W_R	Relative humidity of air [%]
W_s	Specific air humidity [$kg_w \cdot kg_{air}^{-1}$]
x_c	Critical Concentrations.
x_j	Molar Ratio (dimensionless, or 100 x_j [%])
Z	Number of formula per unit cell of phase
z	Stoichiometric factor
X	Pauling´s electronegativity
$\Delta_c H^\circ$	Heat of Combustion [$J \cdot mol^{-1}$]
$\Delta_f H^\circ$	The standard enthalpy of formation [$J \cdot mol^{-1}$]
$\Delta G^{\#}$	Gibbs energy of activated complex [J]
$\Delta H^{\#}$	Enthalpy of activated complex [J]
$\Delta_r G^\circ$	The standard Gibbs energy of reaction [J]
$\Delta_r G^{\circ(bo)}$	The standard Gibbs energy of reaction recalculated to one mol of basic oxides [$J \cdot mol^{-1}$]
$\Delta_r H^\circ$	The standard enthalpy of reaction [J]
$\Delta_r S^\circ$	The standard entropy of reaction [$J \cdot K^{-1}$]
$\Delta S^{\#}$	Entropy of activated complex [$J \cdot K^{-1}$]
Θ	Heating rate [$°C \cdot min^{-1}$]
α	The fractional conversion or degree of conversion (normalized on range from 0 to 1 or from 0 to 100 %)
ε	Porosity
ϕ_{ij}	The structure composition factor.
λ	The coefficient of thermal conductivity
λ_e	The effective thermal conductivity of porous materials
μ_i	Chemical Potential
μ_i°	Standard Chemical Potential
v_i	Stoichiometric coefficient for species C_i
v_i	Stoichiometric coefficient
ρ	Density [$kg \cdot m^{-3}$]
τ	The fraction of condensation energy transferred to the reactant at interface

The following cement chemistry notation is used in this book:

Oxide/ compounds	Formula	Abbreviated symbol
Aluminium oxide	Al_2O_3	A
Calcium oxide	CaO	C
Carbon dioxide	CO_2	$\bar{C}$
Iron oxide	Fe_2O_3	F
Calcium fluoride	CaF_2	$\bar{F}$
Water	H_2O	H
Potassium oxide	K_2O	K
Magnesium oxide	MgO	M
Sodium oxide	Na_2O	N
Phosphorus oxide	P_2O_5	P
Silicon oxide	SiO_2	S
Sulfur oxide	SO_3	$\bar{S}$
Titanium oxide	TiO_2	T

Chapter 12

References

[1] Wells LS, Clarke WF, McMurdie HF. Study of the system $CaO-Al_2O_3-H_2O$ at temperatures of 21° and 90° C. Journal of Research of the National Bureau of Standards 1943; 30 367-409.

[2] Zhang R, Zheng Sh, Ma Sh, Zhang Y. Recovery of alumina and alkali in Bayer red mud by the formation of andradite-grossular hydrogarnet in hydrothermal process. Journal of Hazardous Materials 2011;189(3) 827-835.

[3] Massazza F. Pozzolanic cements. Cement and Concrete Composites 1993;15(4) 185-214.

[4] Carlson ET. A study fo some strontium aluminates and calcium-strontium aluminate solid solutions. Journal of Research of the National Bureau of Standards 1955;54 329-334.

[5] Flint EP, McMurdie HF, Wells LS. Hydrothermal and X-ray studies of the garnet-hydrogarnet series and the relationship of the series to hydration products of Portland cement. Journal of Research of the National Bureau of Standards 1941;26 13-33.

[6] Carlson ET. Hydrogarnet Formation in the System Lime-Alumina -Silica -Water. Journal of Research of the National Bureau of Standards 1956; 56: 327-335.

[7] Taylor HFW. Cement chemistry, 2nd edition. Thomas Telford Publishing: London; 1997. ISBN: 0-7277-2592-0.

[8] Prodjosantoso AK, Kennedy BJ, Hunter BA. Phase separation induced by hydration of the mixed Ca/Sr aluminates $Ca_{3-x}Sr_xAl_2O_6$: A crystallographic study. Cement and Concrete Research 2002; 32(4) 647-655.

[9] Brandenberger E. Crystal structure studies on the calcium aluminates. Schweizer Mineralogische und Petrographische Mitteilungen 1933;13 569-570.

[10] Maekawa G. Journal of the Society of Chemical Industry, Japan 1943;45 751.

[11] Douy A, Capron M. Crystallization of spray-dried amorphous precursors in the $SrO-Al_2O_3$ system: a DSC study. Journal of the European Ceramics Society 2003;23(12) 2075-2081.

[12] Odler I. Special inorganic cements. 1st ed. London: E & FN. Spon; 2000. ISBN: 0-419-22790-3.

[13] Illyoukha N, Timofeeva V. Development of hydraulic zirconia cements and their applications for production of refractories items. Communication from the Academic Ceramic Centre, The Technical University, Kharkov, Ukraine, Personal Communication, 1998.

[14] Chatterjee AK. Re-examining the prospects of aluminous cements based on alkali-earth and rare-earth oxides. Cement and Concrete Research 2009;39(11) 981-988.

[15] Rezaie MR, Rezaie HR, Naghizadeh R. The effect of SiO_2 additions on barium aluminate cement formation and properties. Ceramics International 2009;35(6) 2235-2240.

[16] Mohapatra M, Pattanaik DM, Anand S, Das RP. Effect of barium to aluminium ratio on phases leading to barium aluminates. Ceramics International 2007;33(4) 531-535.

[17] Brito HF, Felinto MCFC, Hölsä J, Laamanen T, Lastusaari M, Malkamäki M, Novák P, Rodrigues LCV, Stefani R. DFT and synchrotron radiation study of Eu^{2+} doped $BaAl_2O_4$. Optical Materials Express 2012;2(4) 420-431.

[18] Franco M. System SrO-Al_2O_3, pp. 118; in Phase diagram for ceramics, Part 1,Edited by Levine EM, Robbins CR, McMurdie HF. The American Ceramic Society; Ohio; 1959.

[19] Ropp RC. Encyclopedia of the Alkaline Earth Compounds. Amsterdam: Elsevier; 2013. ISBN: 978-0-444-59550-8.

[20] Obukuro Y, Matsushima Sh, Nakamura H, Arai M, Yamada H, Xu Ch.-N. Electronic structure of Eu^{2+}-doped $SrAl_2O_4$ using modified Becke-Johnson exchange potential. Solid State Communications 2014;186 46-49.

[21] Stanciu ML, Ciresan MG, Avram NM. Crystal Field Analysis of Cr^{3+} doped $SrAl_2O_4$ spinel. Acta physica polonica A 2009;116(4) 544-546.

[22] Liu B, Barbier J. Structures of the stuffed tridymite derivatives, $BaMSiO_4$ (M = Co, Zn, Mg). Journal of Solid State Chemistry 1993;102(1) 115-125.

[23] Madhavi S, Ferraris C, White T. Synthesis and characterization of three-dimensionally ordered macroporous ternary oxide. Journal of Solid State Chemistry 2006;179(3) 866-872.

[24] Fukuda K, Fukushima K. Crystal structure of hexagonal $SrAl_2O_4$ at 1073 K. Journal of Solid State Chemistry 2005;178 2709-2714.

[25] Machida K.-I, Adachi G.-Y, Shiokawa J, Shimada M, Koizumi M. Structure of strontium tetraaluminate β-$SrAl_4O_7$. Acta Crystallographica Section B 1982;38(3) 889-891.

[26] Lindop AJ, Goodwin DW. The refined structure of SrO·2Al_2O_3. Acta Crystallographica Section B 1972; 28 2625-2626.

[27] Harindranath K, Viswanath KA, Chandran CV, Bräuniger T, Madhu PK, Ajithkumar TG, Joy PA. Evidence for the co-existence of distorted tetrahedral and trigonal bipyr-

amidal aluminium sites in $SrAl_{12}O_{19}$ from 27Al NMR studies. Solid State Communications 2010;150(5-6) 262-266.

[28] Venkateshwaran B, Yao M, Guo R, Bhalla A, Balachandran U. Low temperature dielectric properties of magnetoplumbite family of materials. International Journal of Inorganic Materials 1999;1(3-4) 213-217.

[29] Xie L, Cormack AN. Defect solid state chemistry of magnetoplumbite structured ceramic oxides: I. $SrAl_{12}O_{19}$. Journal of Solid State Chemistry 1989;83(2) 282-291.

[30] Chawla S, Yadav A. Role of valence state of dopant (Eu^{2+}, Eu^{3+}) and growth environment in luminescence and morphology of $SrAl_{12}O_{19}$ nano- and microcrystals. Materials Chemistry and Physics 2010;122(2-3) 582-587.

[31] Lindop AJ, Matthews C, Goodwin DW. The refined structure of SrO $6Al_2O_3$. Acta Crystallographica Section B 1975; 31 2940-2941.

[32] Verdun HR, Wortman DE, Morrison CA, Bradshaw JL. Optical properties of Nd^{3+} in single crystal $SrAl_{12}O_{19}$. Optical Materials 1997;7(3) 117-128.

[33] Iyi N, Göbbels M. Crystal Structure of the New Magnetoplumbite-Related Compound in the System SrO–Al_2O_3–MgO. Journal of Solid State Chemistry 1996;122(1) 46-52.

[34] Li YJ, Ma YY, Ye S, Hu GP, Zhang QY. Site-related near-infrared luminescence in $MAl_{12}O_{19}$ (M = Ca, Sr, Ba):Fe^{3+} phosphors. Materials Research Bulletin 2014;51 1-5.

[35] Chen X, Zhang Y, Zhong X, Xu Z, Zhang J, Cheng Y, Zhao Y, Liu Y, Fan X, Wang Y, Ma H, Cao X. Thermal cycling behaviors of the plasma sprayed thermal barrier coatings of hexaluminates with magnetoplumbite structure. Journal of the European Ceramic Society 2010;30(7) 1649-1657.

[36] Obradors X, Labarta A, Isalgué A, Tejada J, Rodriguez J, Pernet M, Solid State Communications 1988;65 189–192.

[37] Ishiwata Sh, Terasaki I, Azuma M, Takano M. High pressure synthesis and structure of a new magnetoplumbite-type cobalt oxide $SrCo_{12}O_{19}$. Journal of Solid State Chemistry 2008;181(5) 1273-1278.

[38] Verstegen JMPJ, Stevels ALN. The relation between crystal structure and luminescence in β-alumina and magnetoplumbite phases. Journal of Luminescence 1974;9(5) 406-414.

[39] Wang Ch.-H, Guo D.-F, Li Z.-F, Wang X.-M, Lin J.-H, Zeng Z.-Z, Jing X.-P. Crystal structure of $Sr_6Y_2Al_4O_{15}$: XRD refinements and first-principle calculations. Journal of Solid State Chemistry 2012;192195-200.

[40] Kimura K, Ohgaki M, Tanaka K, Morikawa H, Marumo F. Study of the bipyramidal site in magnetoplumbite-like compounds, $SrM_{12}O_{19}$ (M = Al, Fe, Ga). Journal of Solid State Chemistry 1990;87(1) 186-194.

[41] Yamaguchi O, Narai A, Shimizu K. New compound in system SrO–Al_2O_3. Journal of the American Ceramic Society 1986;69(2) C36-C37.

[42] Kahlenberg V. Synthesis and crystal structure of $Sr_{10}Al_6O_{19}$: a derivative of the perovskite structure type in the system SrO–Al_2O_3. Materials Research Bulletin 2002;37(4) 715-726.

[43] Kahlenberg V. The Crystal Structures of the Strontium Gallates $Sr_{10}Ga_6O_{19}$ and $Sr_3Ga_2O_6$. Journal of Solid State Chemistry 2001;160(2) 421-429.

[44] Nevskii NN, Glasser LD, Iliukhin VV, Belov NV. Determination of the crystal structure of strontium aluminate by means of vector subsystems." Soviet physics, Crystallography (Engl. Transl.) 1979;24(1): 93-95.

[45] Nadezhina TN, Pobedimskaya EA, Belov NV. Crystal structure of strontium aluminate $Sr_4Al_4O_2[Al_{10}O_{23}]$. Soviet physics, Crystallography (Engl. Transl.) 1976;21(4) 471-473.

[46] Takeda T, Takahashi K, Uheda K, Takizawa H, Endo T. Crystal Structure and Luminescence Properties of $Sr_2Al_6O_{11}$:Eu^{2+}. Journal of the Japan Society of Powder and Powder Metallurgy 2002;49(12) 1128-1133.

[47] Gibbs JW. On the equilibrium of heterogeneous substances. Transactions of the Connecticut Academy of Arts and Sciences, Volume III, published by the academy 1874 to 1878.

[48] Kim G.-H, Sohn I. Effect of Al_2O_3 on the viscosity and structure of calcium silicate-based melts containing Na_2O and CaF_2. Journal of Non-Crystalline Solids 2012;358(12-13) 1530-1537.

[49] Park JH, Min DJ. Effect of fluorspar and alumina on the viscous flow of calcium silicate melts containing MgO. Journal of Non-Crystalline Solids 2004;337(2) 150-156.

[50] Wiedenroth A, Rüssel Ch. The effect of MgO on the thermodynamics of the Fe^{2+}/Fe^{3+}-redox equilibrium and the incorporation of iron in soda-magnesia-aluminosilicate melts. Journal of Non-Crystalline Solids 2003;320(1-3) 238-245.

[51] Benne D, Rüssel Ch, Menzel M, Becker KD. The effect of alumina on the Sn^{2+}/Sn^{4+} redox equilibrium and the incorporation of tin in $Na_2O/Al_2O_3/SiO_2$melts. Journal of Non-Crystalline Solids 2004;337(3) 232-240.

[52] Gerlach S, Claußen O, Rüssel Ch. A voltammetric study on the thermodynamics of the Fe^{3+}/Fe^{2+}-equilibrium in alkali–lime–alumosilicate melts. Journal of Non-Crystalline Solids 1999;248(1) 92-98.

[53] Massazza F, Cannas M. System CaO-Al_2O_3-SrO, subsolidus. Annali di Chimica (Rome, Italy) 1959;49(7–8) 1342–1350.

[54] Massazza F, Sirchia E. Equilibriums at the temperature of fusion in the ternary system SrO-Al_2O_3-CaO. Annali di Chimica (Rome, Italy) 1959;49 1352–1370.

[55] Tarnopol'skaya RA, Gul'ko NV. The CaO-SrO-Al_2O_3-ZrO_2 system and its importance for refractories technology. Refractories and Industrial Ceramics 1967;8(11-12) 760-763.

[56] Pitak YN, Proskurnya EM. On the subsolidus structure in the $CaAl_4O_7$ – $Ca_7Al_6ZrO_{18}$ – $CaAl_2O_4$ – $CaZrO_3$ – $SrZrO_3$ – $SrAl_2O_4$ region of the CaO – SrO – Al_2O_3 – ZrO_2 system. Refractories and Industrial Ceramics 2000;41(9-10) 360-363.

[57] Kellogg HH, in: Applications of Fundamental Thermodynamics to Metallurgical Processes, G.R. Fitterer (Ed.), Gordon and Breach, London; 1967, p. 357.

[58] Latimer WM. Methods of estimating the entropies of solid compounds. Journal of the American Chemical Society 1951;73(4) 1480-1482.

[59] Spencer PJ. Estimation of thermodynamic data for metallurgical applications. Thermochimica Acta 1998;314(1-2) 1-21.

[60] Aronson S. Estimation of the heat of formation of refractory mixed oxides. Journal of Nuclear Materials 1982;107 343-346.

[61] Voňka P, Leitner J. A method for the estimation of the enthalpy of formation of mixed oxides in Al_2O_3–Ln_2O_3 systems. Journal of Solid State Chemistry 2009;182(4) 744-748.

[62] Zhuang W, Liang J, Qiao Z, Shen J, Shi Y, Rao G. Estimation of the standard enthalpy of formation of double oxide. Journal of Alloys and Compounds 1998;267(1-2) 6-10.

[63] Moiseev G, Leitner J, Šesták J, Zhukovsky V. Empirical dependences of the standard enthalpy of formation for related inorganic compounds enhancing glass formers. Thermochimica Acta 1996;280-281 511-521.

[64] Šesták J, Moiseev GK, Tzagareishvili DS. Oxide-phase simulated thermodynamics and calculation of thermochemical properties of compounds auxiliary to Y-Ba-Cu-O high Tc superconductors. Japanese Journal of Applied Physics 1994;33 97-102.

[65] Le Van M, Bulletin de la Société Chimique de France 1972;2 579.

[66] Balonis M, Glasser FP. The density of cement phases. Cement and Concrete Research 2009;39(9) 733-739.

[67] Xu Y, Lu P, Huang G, Zeng Ch. Synthesis of $SrAl_4O_7$ via citric acid precursor. Materials Chemistry and Physics 2006;95(1) 62-66.

[68] Boyko ER, Wisnyi LG. The optical properties and structures of $CaO\cdot(Al_2O_3)_2$ and $SrO\cdot(Al_2O_3)_2$. Acta Crystallographica 1958;11 444-445.

[69] Moukhina E. Determination of kinetic mechanisms for reactions measured with thermoanalytical instruments. Journal of Thermal Analysis and Calorimetry 2012;109(3) 1203-1214.

[70] Galwey AK. Theory of solid-state thermal decomposition react Scientific stagnation or chemical catastrophe? An alternative approach appraised and advocatedions. Journal of Thermal Analysis and Calorimetry 2012;109 1625-1635.

[71] Koga N, Málek J, Šesták J, Tanaka H. Data treatment in non-isothermal kinetics and diagnostic limits of phenomenological models. Netsu Sokutei 1993;20(4) 210-223.

[72] Tanaka H. Thermal analysis and kinetics of solid state reactions. Thermochimica Acta 1995;267(1) 29-44.

[73] Šesták J. Thermal analysis, Part D: Thermophysical properties of solids, their measurements and theoretical thermal analysis. Amsterdam and New York: Elsevier Science Publishers; 1984. ISBN: 0444996532.

[74] Xu B, Smith P. Dehydration kinetics of boehmite in the temperature range 723–873 K. Thermochimica Acta 2012;531 46-53.

[75] Ptáček P, Kubátová D, Havlica J, Brandštetr J, Šoukal F, Opravil T. The non-isothermal kinetic analysis of the thermal decomposition of kaolinite by thermogravimetric analysis. Powder Technology 2010;204 222-227.

[76] Coats AW, Redfern JP. Kinetic Parameters from Thermogravimetric Data. Nature 1964;201 68-69.

[77] Flynn JH, Wall LA. General Treatment of the thermogravimetry of polymers. Journal of research of the National Bureau of Standards, Sect. A, 1966;70 487-523.

[78] Kissinger HE. Reaction Kinetics in Differential Thermal Analysis. Analytical Chemistry 1957;29(11) 1702-1706.

[79] Blaine RL, Kissinger HE. Homer Kissinger and the Kissinger equation. Thermochimica Acta 2012;540 1-6.

[80] Friedman HL. Kinetics of thermal degradation of char-forming plastics from thermogravimetry. Application to a phenolic plastic. Journal of Polymer Science Part C: Polymer Symposia 1964;6 183-185.

[81] Starink MJ. The determination of activation energy from linear heating rate experiments: a comparison of the accuracy of isoconversion methods. Thermochimica Acta 2003;404 163-176.

[82] Málek J, Criado JM, F.J Gotor, Šesták J. Some comments about a correct estimation of the kinetic exponent for non-isothermal solid-state processes using Augis and Bennett method. Thermochimica Acta 1998;322(1) 77-82.

[83] Augis JA, Bennett JD. Calculation of Avrami parameters for heterogeneous solid-state reactions using a modification of Kissinger method. Journal of Thermal Analysis 13 (1978) 283–292.

[84] Ray CS, Yang Q, Huang W-H, Day DE. Surface and internal crystallization in glasses as determined by differential thermal analysis. Journal of the American Ceramic Society 1996;79(12) 3155-3160.

[85] Málek J. The applicability of Johnson-Mehl-Avrami model in the thermal analysis of the crystallization kinetics of glasses. Thermochimica acta 1995;267 61-73.

[86] Straszko J, Olszak-Humienik M, Możejko J. Kinetics of thermal decomposition of $ZnSO_4 \cdot 7H_2O$. Thermochimica Acta 1997;292(1-2) 145-150.

[87] Vlaev L, Nedelchev N, Gyurova K, Zagorcheva M. A comparative study of non-isothermal kinetics of decomposition of calcium oxalate monohydrate. Journal of Analytical and Applied Pyrolysis 2008;81253-262.

[88] Pourmortazavi SM, Kohsari I, Teimouri MB, Hajimirsadeghi SS. Thermal behaviour kinetic study of dihydroglyoxime and dichloroglyoxime. Materials Letters 2007;61 4670-4673.

[89] L'vov BV, Polzik LK, Ugolkov VL. Decomposition kinetics of calcite: a new approach to the old problem. Thermochimica Acta 2002;390 5-19.

[90] L'vov BV. Thermal decomposition of solids and melts—new thermochemical approach to the mechanism, kinetics and methodology. Berlin: Springer; 2007.

[91] Patnaik P. Handbook of inorganic chemicals; McGraw-Hill Companies, Inc.; 2002. ISBN 0-07-049439-8.

[92] Garcia-Guinea J, Crespo-Feo E, Correcher V, Cremades A, Rubio J, Tormo L, Townsend PD. Luminescence of Strontianite ($SrCO_3$) from Strontian (Scotland, UK). Radiation Measurements 2009;44(4) 338-343.

[93] Helz GR, Holland HD. The solubility and geologic occurrence of strontianite. Geochimica et Cosmochimica Acta 1965;29(12) 1303-1315.

[94] Casey WH, Chal L, Navrotsky A, Rock PA. Thermochemistry of mixing strontianite [$SrCO_3$(s)] and aragonite [$CaCO_3$(s)] to form $Ca_xSr_{1-x}CO_3$(s) solid solutions. Geochimica et Cosmochimica Acta 1996;60(6) 933-940.

[95] Massoni N, Rosen J, Chartier M, Cozzika T. Study of barytocalcite as a conditioning matrix for carbon 14: Comparison of several synthesis routes. Journal of Nuclear Materials 2013;441(1–3) 152-158.

[96] Kulik DA, Vinograd VL, Paulsen N, Winkler B. (Ca,Sr)CO_3 aqueous-solid solution systems: From atomistic simulations to thermodynamic modelling. Physics and Chemistry of the Earth, Parts A/B/C 2010;35(6–8) 217-232.

[97] Alía JM, Díaz de Mera Y, Edwards HGM, Martín PG, Andres SL. FT-Raman and infrared spectroscopic study of aragonite-strontianite ($Ca_xSr_{1-x}CO_3$) solid solution. Spectrochimica Acta Part A: Molecular and Biomolecular Spectroscopy 1997;53(13) 2347-2362.

[98] Fubini B, Renzo FD, Stone FS. Strontianite-aragonite solid solutions $Sr_xCa_{1-x}CO_3$: Effect of composition on the orthorhombic-rhombohedral phase transition and the conversion to oxide solid solutions $Sr_xCa_{1-x}O$. Journal of Solid State Chemistry 1988;77(2) 281-292.

[99] Fubini B, Renzo FD, Stone FS. Investigation of the thermal stability of $CaCO_3$-$SrCO_3$ solid solutions. Thermochimica Acta 1987;122(1) 23-28.

[100] Ruiz-Hernandez SE, Grau-Crespo R, Ruiz-Salvador AR, De Leeuw NH. Thermochemistry of strontium incorporation in aragonite from atomistic simulations. Geochimica et Cosmochimica Acta 2010;74(4) 1320-1328.

[101] Rushdi AI, McManus J, Collier RW. Marine barite and celestite saturation in seawater. Marine Chemistry 2000;69 19-31.

[102] Utigard T, Toguri JM, See Through The Hall-Heroult Cell During Electrolysis. Proceedings of the International Symposium on Quality and Process Control in the Reduction and Casting of Aluminum and Other Light Metals. Canada, Winnipeg: August 23-26; 1987, p.37-49.

[103] Tabereaux AT, Peterson RD. Aluminum Production. Treatise on Process Metallurgy. In: Industrial Processes; ed: Seetharaman S, p.839-917; 2014. ISBN: 978-0-08-096988-6.

[104] Chapman V, Welch BJ, Skyllas-Kazacos M. Anodic behaviour of oxidised Ni–Fe alloys in cryolite–alumina melts. Electrochimica Acta 2011;56(3) 1227-1238.

[105] Haupin W. Aluminum. In: Encyclopedia of Physical Science and Technology (Third Edition), p.495-518; 2003.

[106] Amigó JM, Serrano FJ, Kojdecki MA, Bastida J, Esteve V, Reventós MM, Martí F. X-ray diffraction microstructure analysis of mullite, quartz and corundum in porcelain insulators. Journal of the European Ceramic Society 2005;25(9) 1479-1486.

[107] Leriche A, Moortgat G, Cambier F, Homerin P, Thevenot F, Orange G, Fantozzi G. Preparation and characterization of a dispersion toughened ceramic for thermomechanical uses (ZTA). Part I: Material preparation. Characterization of microstructure. Journal of the European Ceramic Society 1992;9(3) 169-176.

[108] Piconi C, Condo SG, Kosmač T. Alumina- and Zirconia-based Ceramics for Load-bearing Applications. In: Advanced Ceramics for Dentistry; ed. Shen J; p219-253; 2014. ISBN: 978-0-12-394619-5.

[109] Djangang CN, Kamseu E, Ndikontar MK, Nana GLL, Soro J, Melo UC, Elimbi A, Blanchart P, Njopwouo D. Sintering behaviour of porous ceramic kaolin–corundum composites: Phase evolution and densification. Materials Science and Engineering: A 2011;528(29–30) 8311-8318.

[110] Katsavou ID, Krokida MK, Ziomas IC. Determination of mechanical properties and thermal treatment behavior of alumina-based refractories. Ceramics International 2012;38(7) 5747-5756.

[111] Gogtas C, Lopez HF, Sobolev K. Role of cement content on the properties of self-flowing Al_2O_3 refractory castables. Journal of the European Ceramic Society 2014;34(5) 1365-1373.

[112] Marinescu ID, Rowe WB, Dimitrov B, Ohmori H. Abrasives and abrasive tools Tribology of Abrasive Machining Processes (Second Edition). Imprint: William Andrew; p.243-311; 2013. ISBN: 978-1-4377-3467-6.

[113] Mata-Osoro G, Moya JS, Pecharroman C. Transparent alumina by vacuum sintering. Journal of the European Ceramic Society 2012;32(11) 2925-2933.

[114] Sarkar R, Ghosh A, Das SK. Reaction sintered magnesia rich magnesium aluminate spinel: effect of alumina reactivity. Ceramics International 2003;29(4) 407-411.

[115] Klatt M, Bellmann D, Kampmann R, Wagner R, Wolf C, Hübner H. Analysis of creep pore formation in liquid-phase sintered alumina. Materials Science and Engineering: A 1997;234–236 932-935.

[116] Goswami AP, Roy S, Das GC. Effect of powder, chemistry and morphology on the dielectric properties of liquid-phase-sintered alumina. Ceramics International 2002;28(4) 439-445.

[117] Goswami AP, Roy S, Mitra MK, Das GC. Influence of powder, chemistry and intergranular phases on the wear resistance of liquid-phase-sintered Al_2O_3. Wear 2000;244(1–2) 1-14.

[118] Mohammad-Rahimi R, Rezaie HR, Nemati A. Sintering of Al_2O_3–SiC composite from sol–gel method with MgO, TiO_2 and Y_2O_3 addition. Ceramics International 2011;37(5) 1681-1688.

[119] Sktani ZDI, Azhar AZA, Ratnam MM, Ahmad ZA. The influence of in-situ formation of hibonite on the properties of zirconia toughened alumina (ZTA) composites. Ceramics International 2014;40(4) 6211-6217.

[120] Vasudevan R, Karthik T, Ganesan S, Jayavel R. Effect of microwave sintering on the structural and densification behavior of sol–gel derived zirconia toughened alumina (ZTA) nanocomposites. Ceramics International 2013;39(3) 3195-3204.

[121] Tomaszewski H, Boniecki M, Weglarz H. Toughness-curve behaviour of alumina-SiC and ZTA-SiC composites. Journal of the European Ceramic Society 2000;20(8) 1215-1224.

[122] Si T, Lou N, Zhang Q, You X. Thermal shock fatigue behavior of TiC/Al_2O_3 composite ceramics. Rare Metals 2008;27(3) 308-314.

[123] Laarz E, Carlsson M, Vivien B, Johnsson M, Nygren M, Bergström L. Colloidal processing of Al_2O_3-based composites reinforced with TiN and TiC particulates, whiskers and nanoparticles. Journal of the European Ceramic Society 2001;21(8) 1027-1035.

[124] Nassau K. Dr. A. V. L. Verneuil: The man and the method. Journal of Crystal Growth 1972;13–14 12-18.

[125] Ueltzen M. The Verneuil flame fusion process: substances. Journal of Crystal Growth 1993;132(1-2) 315-328.

[126] Arivuoli D. Principles of the Verneuil Growth Technique. Encyclopedia of Materials: Science and Technology (Second Edition); p.7854-7856; 2001. ISBN: 978-0-08-043152-9.

[127] R.R. Neurgaonkar. Single Crystal Processes. Encyclopedia of Materials: Science and Technology (Second Edition); p.8629-8635; 2001. ISBN: 978-0-08-043152-9.

[128] Tchamba AB, Melo UC, Lecomte-Nana GL, Kamseu E, Gault C, Yongue R, Njopwouo D. Use of bauxite from Cameroon for solid state sintering and characterization of calcium dialuminate (CaO $2Al_2O_3$) refractory cement. Ceramics International 2014;40(1) 1961-1970.

[129] Caballero A, Valle F., De Aza S, Castillo S. Constitution of calcined refractory-grade bauxites: An interpretation. Ceramics International 1985;11(2) 45-50.

[130] Caballero A, Requena J, De Aza S. Refractory bauxites. How processing can improve high temperature mechanical properties. Ceramics International 1986;12(3) 155-160.

[131] Abou-Sekkina MM, Abo-El-Enein SA, Khalil NM, Shalma OA. Phase composition of bauxite-based refractory castables. Ceramics International 2011;37(1) 411-418.

[132] Maciejewski M, Richarz W. Reduction of iron oxides in bauxite by hydrogen. Thermochimica Acta 1985;85 199-202.

[133] Milne DJ, Wibberley LJ. The removal of iron from bauxite using anhydrous hydrogen chloride. Light Metals 1977;2 125-145.

[134] Hussain SA, Jamal R. Evaluation of an HCl process for leaching of low-grade highly siliceous bauxite ore. Developments in Mineral Processing 2000;13 C6-8-C6-14.

[135] Zhao A.-ch, Liu Y, Zhang T.-a, Lü G.-z, Dou Z.-h. Thermodynamics study on leaching process of gibbsitic bauxite by hydrochloric acid. Transactions of Nonferrous Metals Society of China 2013;23(1) 266-270.

[136] Reddy BR, Mishra SK, Banerjee GN. Kinetics of leaching of a gibbsitic bauxite with hydrochloric acid. Hydrometallurgy 1999;51(1) 131-138.

[137] Zafar ZI. Determination of semi empirical kinetic model for dissolution of bauxite ore with sulfuric acid: Parametric cumulative effect on the Arrhenius parameters. Chemical Engineering Journal 2008;141(1–3) 233-241.

[138] Papassiopi N, Vaxevanidou K, Paspaliaris I. Effectiveness of iron reducing bacteria for the removal of iron from bauxite ores. Minerals Engineering 2010;23(1) 25-31.

[139] Vasan SS, Modak JM, Natarajan KA. Some recent advances in the bioprocessing of bauxite. International Journal of Mineral Processing 2001;62(1-4) 173-186.

[140] Yeh ChH, Zhang G. Stepwise carbothermal reduction of bauxite ores. International Journal of Mineral Processing 2013;124 1-7.

[141] Yu Ch, Yuan W, Deng Ch, Zhu H, Li J. Synthesis of hexagonal plate-like Al_4SiC_4 from calcined bauxite, silica and carbon black. Powder Technology 2013;247 76-80.

[142] Pontikes Y, Angelopoulos GN. Bauxite residue in cement and cementitious applications: Current status and a possible way forward. Resources, Conservation and Recycling 2013;73 53-63.

[143] Bayer KJ. Verfahren zur darstellung von thonerhydrat und alkalialuminat; 1892.

[144] Klauber C, Gräfe M, Power G. Bauxite residue issues: II. Options for residue utilization. Hydrometallurgy 2011;108(1-2) 11-32.

[145] Gräfe M., Power G, Klauber C. Bauxite residue issues: III. Alkalinity and associated chemistry. Hydrometallurgy 2011;108(1–2) 60-79.

[146] Piga L, Pochetti F, Stoppa L. Recovering metals from red mud generated during alumina production. JOM Journal of the Minerals, Metals and Materials Society 1993;45(11) 54-59.

[147] Valeton I. Bauxites. Amsterdam: Elsevier Publishing Company; p.226; 1972.

[148] Hao X, Leung K, Wang R, Sun W, Li Y. The geomicrobiology of bauxite deposits. Geoscience Frontiers 2010;1(1) 81-89.

[149] Smith P. The processing of high silica bauxites - Review of existing and potential processes. Hydrometallurgy 2009;98(1-2) 162-176.

[150] Bárdossy G. Classification of Bauxite Deposits; p.16-22. In: Developments in Economic Geology; Volume 14: Karst Bauxites Bauxite Deposits on Carbonate Rocks; 1882. ISBN: 978-0-444-99727-2.

[151] Gu J, Huang Z, Fan H, Jin Z, Yan Z, Zhang J. Mineralogy, geochemistry, and genesis of lateritic bauxite deposits in the Wuchuan–Zheng'an–Daozhen area, Northern Guizhou Province, China. Journal of Geochemical Exploration 2013;130 44-59.

[152] Bogatyrev BA, Zhukov VV, Tsekhovsky YG. Formation conditions and regularities of the distribution of large and superlarge bauxite deposits. Lithology and Mineral Resources 2009;44 135-151.

[153] Selim AQ, El-Midany AA, Abdel-Fattah AS, Ibrahim SS. Rationalization of the upgrading circuit of celestite for advanced applications. Powder Technology 2010;198(2) 233-239.

[154] El-Midany AA, Ibrahim SS. Does calcite content affect its separation from celestite by Falcon concentrator? Powder Technology 2011;213(1-3) 41-47.

[155] López-Valdivieso A, Robledo-Cabrera A, Uribe-Salas A. Flotation of celestite with the anionic collector sodium dodecyl sulfate. Effect of carbonate ions. International Journal of Mineral Processing 2000;60(2) 79-90.

[156] Hernáinz F, Calero M. The effect of the degree of grinding on the flotation of celestite ore. Advanced Powder Technology 2001;12(4) 481-491.

[157] Şener S, Bilgen S, Özbayoğlu G. Effect of heat treatment on grindabilities of celestite and gypsum and separation of heated mixture by differential grinding. Minerals Engineering 2004;17 473-475.

[158] Ozkan A, Ucbeyiay H, Aydogan S. Shear flocculation of celestite with anionic surfactants and effects of some inorganic dispersants. Colloids and Surfaces A: Physicochemical and Engineering Aspects 2006;281(1-3) 92-98.

[159] Ozkan A, Yekeler M. Shear flocculation of celestite with sodium oleate and tallow amine acetate: effects of cations. Journal of Colloid and Interface Science 2004;273(1) 170-174.

[160] Ozkan A, Yekeler M. Coagulation and flocculation characteristics of celestite with different inorganic salts and polymers. Chemical Engineering and Processing: Process Intensification 2004;43(7) 873-879.

[161] Laskowski JS. In: Colloid Chemistry in Mineral Processing; ed.: Laskowski JS, Ralston J., New York: Elsevier; 1992.

[162] Torres J, Mendez J, Sukiennik M. Transformation enthalpy of the alkali-earths sulfates ($SrSO_4$, $CaSO_4$, $MgSO_4$, $BaSO_4$). Thermochimica Acta 1999;334(1-2) 57-66.

[163] Erdemoğlu M. Carbothermic reduction of mechanically activated celestite. International Journal of Mineral Processing 2009;92(3-4) 144-152.

[164] Owusu G, Litz JE. Water leaching of SrS and precipitation of $SrCO_3$ using carbon dioxide as the precipitating agent. Hydrometallurgy 2000;57(1) 23-29.

[165] Erdemoğlu M, Canbazoğlu M. The leaching of SrS with water and the precipitation of $SrCO_3$ from leach solution by different carbonating agents. Hydrometallurgy 1998;49(1-2) 135-150.

[166] Bingöl D, Aydoğan S, Bozbaş SK. Production of $SrCO_3$ and $(NH_4)_2SO_4$ by the dry mechanochemical processing of celestite. Journal of Industrial and Engineering Chemistry 2012;18(2) 834-838.

[167] Zeppenfeld K. Crystallization kinetics of strontianite from $Sr(HCO_3)_2$ solutions. Chemie der Erde – Geochemistry 2006;66(4) 319-323.

[168] Setoudeh N, Welham NJ, Azami SM. Dry mechanochemical conversion of $SrSO_4$ to $SrCO_3$. Journal of Alloys and Compounds 2010;492(1-2) 389-391.

[169] Setoudeh N; Welham NJ. Ball milling induced reduction of $SrSO_4$ by Al. International Journal of Mineral Processing 2011;98(3-4) 214-218.

[170] Bingöl D, Aydogan S, Gultekin SS. Neural model for the leaching of celestite in sodium carbonate solution. Chemical Engineering Journal 2010;165(2) 617-624.

[171] Kocakusak S, Tolun R, Dogan H, Akcay K, Koroglu HJ, Yuzer H, Koral M, Isbilir F, Savascı OT, Ayok T. Patent No: TR2001/0326521.5; 2003.

[172] De Buda F. US Patent 4,666,688; 1987.

[173] Cheng Z, Jiang T. Production of strontium carbonate by ammonium bicarbonate method without removing barium. Huadong Huagong Xueyuan Xueba 1992;18 723-728.

[174] Di H, Wang Y, Zhang Y. Chine HC Patent 1.078.706; 1993.

[175] Liao F, Zhao L, Zhai Ch, Zhang Z, Ma X. Morphology and photoluminescence properties of $SrCO_3$ prepared by a simple solution method. Materials Letters 2014;122 331-333.

[176] Xue Y, Ren X, Zhai X, Yu M. Polyvinylpyrrolidone (PVP)-assisted solvothermal synthesis of flower-like $SrCO_3:Tb^{3+}$ phosphors. Materials Research Bulletin 2012;47(2) 393-399.

[177] Shi L, Du F. Solvothermal synthesis of $SrCO_3$ hexahedral ellipsoids. Materials Letters 2007;61(14-15) 3262-3264.

[178] Shi L, Du F. Solvothermal synthesis of fusiform hexagonal prism $SrCO_3$ microrods via ethylene glycol solution. Materials Research Bulletin 2007;42(8) 1550-1555.

[179] Yang J, Liu X, Li Ch, Quan Z, Kong D, Lin J. Hydrothermal synthesis of $SrCO_3:Eu^{3+}/Tb^{3+}$ microneedles and their luminescence properties. Journal of Crystal Growth 2007;303(2) 480-486.

[180] Li Sh, Zhang H, Xu J, Yang D. Hydrothermal synthesis of flower-like $SrCO_3$ nanostructures. Materials Letters 2005;59(4) 420-422.

[181] Zhu W, Liang Z, Liu X, Zhang H, Zheng Y, Piao X, Zhang Q. Soft-template self-assembly of hierarchical mesoporous $SrCO_3$ by low-temperature hydrothermal route and their application as adsorbents for methylene blue and heavy metal ions. Powder Technology 2012;226165-172.

[182] Ni Sh, Yang X, Li T. Hydrothermal synthesis and photoluminescence properties of $SrCO_3$. Materials Letters 2011;65(4) 766-768.

[183] Alavi MA, Morsali A. Syntheses and characterization of $Sr(OH)_2$ and $SrCO_3$ nanostructures by ultrasonic method. Ultrasonics Sonochemistry 2010;17(1) 132-138.

[184] Thongtem T, Tipcompor N, Phuruangrat A, Thongtem S. Characterization of $SrCO_3$ and $BaCO_3$ nanoparticles synthesized by sonochemical method. Materials Letters 2010;64(4) 510-512.

[185] Ma M.-G., Zhu Y.-J. Microwave synthesis of $SrCO_3$ one-dimensional nanostructures assembled from nanocrystals using ethylenediamine additive. Materials Letters 2008;62(16) 2512-2515.

[186] Tipcompor N, Thongtem T, Phuruangrat A, Thongtem S. Characterization of $SrCO_3$ and $BaCO_3$ nanoparticles synthesized by cyclic microwave radiation. Materials Letters 2012;87 153-156.

[187] Obut A, Baláž P, Girgin İ. Direct mechanochemical conversion of celestite to $SrCO_3$. Minerals Engineering 2006;19(11) 1185-1190.

[188] Li L, Lin R, Tong Z, Feng Q. Crystallization control of $SrCO_3$ nanostructure in imidazolium-based temperature ionic liquids. Materials Research Bulletin 2012;47(11) 3100-3106.

[189] Hind AR, Bhargava SK, Grocott SC. The surface chemistry of Bayer process solids: a review. Colloids and Surfaces A: Physicochemical and Engineering Aspects 1999;146(1-3) 359-374.

[190] Power G, Loh JSC, Vernon Ch. Organic compounds in the processing of lateritic bauxites to alumina Part 2: Effects of organics in the Bayer process. Hydrometallurgy 2012;127-128 125-149.

[191] Whittington BI, Fallows TM, Willing MJ. Tricalcium aluminate hexahydrate (TCA) filter aid in the Bayer industry: factors affecting TCA preparation and morphology. International Journal of Mineral Processing 1997;49(1-2) 1-29.

[192] Whittington BI, Cardile CM. The chemistry of tricalcium aluminate hexahydrate relating to the Bayer industry. International Journal of Mineral Processing 1996;48(1-2) 21-38.

[193] Zhang B, Li J, Chen Q, Chen G. Precipitation of $Al(OH)_3$ crystals from supersaturated sodium aluminate solution irradiated with ultrasonic sound. Minerals Engineering 2009;22(9-10) 853-858.

[194] Zhang Y, Zheng Sh, Du H, Xu H, Wang Sh, Zhang Y. Improved precipitation of gibbsite from sodium aluminate solution by adding methanol. Hydrometallurgy 2009;98(1-2) 38-44.

[195] Zeng J, Yin Z, Chen Q. Intensification of precipitation of gibbsite from seeded caustic sodium aluminate liquor by seed activation and addition of crown ether. Hydrometallurgy 2007;89(1-2) 107-116.

[196] Smith PG, Watling HR, Crew P. The effects of model organic compounds on gibbsite crystallization from alkaline aluminate solutions: polyols. Colloids and Surfaces A: Physicochemical and Engineering Aspects 1996;111(1-2) 119-130.

[197] Yin Z, Zeng J, Chen Q. Effect of oleic acid on gibbsite precipitation from seeded sodium aluminate liquors. International Journal of Mineral Processing 2009;92(3-4) 184-189.

[198] Watling H. Gibbsite crystallization inhibition: 2. Comparative effects of selected alditols and hydroxycarboxylic acids. Hydrometallurgy 2000;55(3) 289-309.

[199] Samal S, Ray AK, Bandopadhyay A. Proposal for resources, utilization and processes of red mud in India - A review. International Journal of Mineral Processing 2013;118 43-55.

[200] Liu W, Yang J, Xiao B. Application of Bayer red mud for iron recovery and building material production from alumosilicate residues. Journal of Hazardous Materials 2009;161(1) 474-478.

[201] Abhilash, Sinha Sh, Sinha MK, Pandey BD. Extraction of lanthanum and cerium from Indian red mud. International Journal of Mineral Processing 2014;12770-73.

[202] Couillard D. Use of red mud, a residue of alumina production by the Bayer process, in water treatment. Science of the Total Environment 1982;25(2) 181-191.

[203] He H, Yue Q, Qi Y, Gao B, Zhao Y, Yu H, Li J, Li Q, Wang Y. The effect of incorporation of red mud on the properties of clay ceramic bodies. Applied Clay Science 2012;70 67-73.

[204] Kumar A, Kumar S. Development of paving blocks from synergistic use of red mud and fly ash using geopolymerization. Construction and Building Materials 2013;38 865-871.

[205] Nishida T, Nakano H, Urabe K. Preparation and characterization of polycrystalline alumina with small grain size. Journal of the European Ceramic Society 1993;12(3) 197-203.

[206] Gu H, Tian P, Li Y, Hao X. Low-temperature electrolytic coloration and spectral property of ammonium alum crystals. Journal of Luminescence 2012;132(7) 1623-1626.

[207] Park HC, Park YJ, Stevens R. Synthesis of alumina from high purity alum derived from coal fly ash. Materials Science and Engineering: A 2004;367(1-2) 166-170.

[208] Lim AR, Moon H.-G, Chang J.-H. Nuclear magnetic resonance study of the phase transitions and local environments of α-alum $NH_4Al(SO_4)_2 \cdot 12H_2O$ single crystals. Chemical Physics 2010;371(1-3) 91-95.

[209] Petruševski VM, Ivanovski V, Šoptrajanov B, Zugik M. Vibrational spectra of hexaaqua complexes. X. Raman and IR studies of the sulfate group disorder in α-alums. Journal of Molecular Structure 2001;563-564 329-333.

[210] Wang LK, Yang JY. Total waste recycle system for water purification plant using alum as primary coagulant. Resource Recovery and Conservation 1975;1(1) 67-84.

[211] Gao S, Wang Ch, Pei Y. Comparison of different phosphate species adsorption by ferric and alum water treatment residuals. Journal of Environmental Sciences, Volume 25, Issue 5, 1 May 2013, Pages 986-992.

[212] Korchak YM, Kapustyanyk VB, Partyka MV, Rudyk VP. Temperature variation of the optical absorption edge for ammonium aluminum alum. Journal of Applied Spectroscopy 2007;74(2) 289-294.

[213] Lazarev VV, Blinov LM, Palto SP, Yudin SG. Electro-optical and ferroelectric switching of Langmuir–Blodgett films made of a chiral smectic-C*liquid crystalline compound. Thin Solid Films 2008;516(24) 8905-8908.

[214] Jotshi CK. Thermal storage in ammonium alum/ammonium nitrate eutectic for solars pace heating applications. Solar Energy Engineering 1998;120 20-24.

[215] Rathod MK, Banerjee J. Thermal stability of phase change materials used in latent heat energy storage systems: A review. Renewable and Sustainable Energy Reviews 2013;18 246-258.

[216] Aderemi BO, Hameed BH. Alum as a heterogeneous catalyst for the transesterification of palm oil. Applied Catalysis A: General 2009;370(1-2) 54-58.

[217] Sonar SS, Shelke KF, Kakade GK, Shingate BB, Shingare MS. Alum: An efficient catalyst for one-pot synthesis of α-aminophosphonates. Chinese Chemical Letters 2009;20(9) 1042-1046.

[218] Jandová J, Dvořák P, Formánek J, Vu HN. Recovery of rubidium and potassium alums from lithium-bearing minerals. Hydrometallurgy 2012;119-120 73-76.

[219] Wojciechowska R, Wojciechowski W, Kamiński J. Thermal decompositions of ammonium and potassium alums. Journal of thermal analysis 1988;33(2) 503-509.

[220] Toniolo JC, Lima MD, Takimi AS, Bergmann CP. Synthesis of alumina powders by the glycine–nitrate combustion proces. Materials Research Bulletin 2005;40(3) 561-571.

[221] Su X, Chen Sh, Zhou Z. Synthesis and characterization of monodisperse porous α-Al_2O_3 nanoparticles. Applied Surface Science 2012;258(15) 5712-5715.

[222] Kingsley JJ, Patil KC. A novel combustion process for the synthesis of fine particle α-alumina and related oxide materials. Materials Letters 1988;6(11-12) 427-432.

[223] Minnermann M, Grossmann HK, Pokhrel S, Thiel K, Hagelin-Weaver H, Bäumer M, Mädler L. Double flame spray pyrolysis as a novel technique to synthesize alumina-supported cobalt Fischer–Tropsch catalysts. Catalysis Today 2013;214 90-99.

[224] Kathirvel P, Chandrasekaran J, Manoharan D, Kumar S. Preparation and characterization of alpha alumina nanoparticles by in-flight oxidation of flame synthesis. Journal of Alloys and Compounds 2014;590 341-345.

[225] Aman Y, Rossignol C, Garnier V, Djurado E. Low temperature synthesis of ultrafine non vermicular α-alumina from aerosol decomposition of aluminum nitrates salts. Journal of the European Ceramic Society 2013;33(10) 1917-1928.

[226] Vallet-Regí M, Rodríguez-Lorenzo LM, Ragel CV, Salinas AJ, González-Calbet JM. Control of structural type and particle size in alumina synthesized by the spray pyrolysis method. Solid State Ionics 1997;101-103 197-203.

[227] Hu Y, Ding H, Li Ch. Preparation of hollow alumina nanospheres via surfactant-assisted flame spray pyrolysis. Particuology 2011;9(5) 528-532.

[228] Sharifi L, Beyhaghi M, Ebadzadeh T, Ghasemi E. Microwave-assisted sol–gel synthesis of alpha alumina nanopowder and study of the rheological behavior. Ceramics International 2013;39(2) 1227-1232.

[229] Cheng L.-T, Tsai M.-Y, Tseng WJ, Hsiang H.-I, Yen F.-S. Boehmite coating on θ-Al_2O_3 particles via a sol–gel route. Ceramics International 2008;34(2) 337-343.

[230] Liu H, Ning G, Gan Z, Lin Y. Emulsion-based synthesis of unaggregated, spherical alpha alumina. Materials Letters 2008;62(10-11) 1685-1688.

[231] Ponthieu E, Payen E, Grimblot J. Ultrafine alumina powders via a sol-emulsion-gel method. Journal of Non-Crystalline Solids 1992;147-148 598-605.

[232] Russell DS, Campbell JH, Bermaban SS. The spectrophotometric determination of strontium with murexide (ammonium purpurate). Analytica Chimica Acta 1961;25(1) 81-84.

[233] Fornaseri M, Grandi L. Flame photometric determination of strontium in silicates. Strontium content of the granite G-1 and the diabase W-1. Geochimica et Cosmochimica Acta 1960;19(3) 218-221.

[234] Andersen NR, Hume DN. Determination of barium and strontium in sea water. Analytica Chimica Acta 1968;40 207-220.

[235] Webb MSW, Wordingham ML. The direct flame photometric determination of Strontium: Calcium ratios in the ash of human bones and teeth. Analytica Chimica Acta 1963;28 450-456.

[236] Arslan Z, Tyson JF. Determination of calcium, magnesium and strontium in soils by flow injection flame atomic absorption spektrometry. Talanta 1999;50(5) 929-937.

[237] Höglund G. Determination of strontium in bone by X-ray fluorescence spectroscopy. Experimental Cell Research 1959;17(3) 565-569.

[238] Macka M, Paull B, Andersson P, Haddad PR. Determination of barium and strontium by capillary zone electrophoresis using an electrolyte containing sulfonazo III. Journal of Chromatography A 1997;767(1-2) 303-310.

[239] Gautier EA, Gettar RT, Servant RE. Simultaneous determination of lanthanum, strontium and copper in superconductor materials by ion chromatography. Analytica Chimica Acta 1993;283(1) 350-353.

[240] Loveridge BA, Webster RK, Morgan JW, Thomas AM, Smales AA. The determination of strontium in rocks and biological materials. Analytica Chimica Acta 1960;23 154-171.

[241] American Society for Testing and Materials (ASTM), Annual Book of ASTMStandards, vol. 04.01, C-114, Philadelphia, PA, USA, 2004.

[242] Vanhoyland G, Bourée F, Van Bael MK, Mullens J, Van Poucke LC. Structure Determination and Refinement of Acid Strontium Oxalate from X-Ray and Neutron Powder Diffraction. Journal of Solid State Chemistry 2001;157(2) 283-288.

[243] Brandenburg K. (Crystal Impact GbR.), &&DIAMOND: Visual Crystal Structure Information System," version 2.1c.

[244] Idriss KA, Sedaira H, Ahmed SS. Determination of strontium and simultaneous determination of strontium oxide, magnesium oxide and calcium oxide content of Portland cement by derivative ratio spectrophotometry. Talanta 2009;78(1) 81-87.

[245] Knaepen E, Mullens J, Yperman J, Van Poucke LC. Preparation and thermal decomposition of various forms of strontium oxalate. Thermochimica Acta 1996;284(1) 213-227.

[246] Yu J, Tang H, Cheng B. Morphological control of strontium oxalate particles by PSMA-mediated precipitation reaction. Materials Chemistry and Physics 2005;91(1) 134-139.

[247] Grahek Z, Macefat MR. Determination of radioactive strontium in seawater. Analytica Chimica Acta 2005;534 271-279.

[248] Jones P, Foulkes M, Paull B. Determination of barium and strontium in calcium-containing matrices using high-performance chelation ion chromatography.Journal of Chromatography A 1994;673(2) 173-179.

[249] Robinson P. Determination of calcium, magnesium, manganese, strontium, sodium and iron in the carbonate fraction of limestones and dolomites. Chemical Geology 1980;28 135-147.

[250] Patti F, Hernandez JA. Nouvelle methode de separation calcium-strontium appliquee a la preparation d'un sel de calcium de tres faible teneur en strontium. Analytica Chimica Acta 1971;55(2) 325-332.

[251] Sedaira H. Simultaneous determination of manganese and zinc in mixtures using first- and second-derivative spectrophotometry. Talanta 2000;51(1) 39-48.

[252] Agnihotri NK, Ratnani S, Singh VK, Singh HB. Simultaneous determination of gallium and indium with 2-(5-bromo-2-pyridylazo)-5-diethylaminophenol in cationic micellar medium using derivative spectrophotometry. Analytical Sciences 2003;9 1297-1301.

[253] Eskandari H, Saghseloo AG: Second and first-derivative spectrophotometry for efficient simultaneous and individual determination of palladium and cobalt using 1-(2-pyridylazo)-2-naphthol in sodium dodecylsulfate micellar media. Analytical Sciences 2003;19(11) 1513-1518.

[254] Bhalotra A, Puri BK. Simultaneous first derivative spectrophotometric determination of palladium and nickel using 2-(2-Thiazolylazo)-5-dimethylamino-benzoic acid as an analytical reagent. Microchimica Acta 2000;134(3-4) 139-143.

[255] Benamor M, Aguerssif N. Simultaneous determination of calcium and magnesium by derivative spectrophotometry in pharmaceutical products. Spectrochimica Acta Part A: Molecular and Biomolecular Spectroscopy 2008;69(2) 676-681.

[256] O'Haver TC, Fell AF, Smith G, Gans P, Sneddon J, Bezur L, Michel RG, Ottaway JM, Miller JN, Ahmad TA, Fell AF, Chadburn BP, Cottrell CT. Derivative spectroscopy and its applications in analysis. Analytical Proceedings 1982;19(1) 22-46.

[257] Ishii H, Satoh KZ. Determination of micro amounts of samarium and europium by analogue derivative spectrophotometryAnalytical Chemistry 1982;312(2) 114-120.

[258] Leviillain P, Fompeydie D. Spectrophotométrie dérivée: intérêt, limites et applications," Analysis, Analysis 1986;14 1-20.

[259] O'Haver TC. Derivative Spectroscopy: Theoretical Aspects.Analytical Proceedings 1982;19 22-28.

[260] Medinilla J, Ales F, Sanchez FG, Spectrophotometric and second-derivative spectrophotometric determination of mercury in organomercurials by means of benzyl 2-pyridyl ketone 2-quinolylhydrazone. Talanta 1986;33(4) 329-334.

[261] Sanchez FG, Lopez MH, Gomez and JCM. A graphical derivative approach to the photometric determination of lutetium and praseodymium in mixtures. Talanta 1987;34(7) 639-644.

[262] Salinas F, Nevado JJ, Espinosa-Mansilla A. A new spectrophotometric method for quantitative multicomponent analysis resolution of mixtures of salicylic and salicyluric acids. Talanta 1990;37(3) 347-351.

[263] Nevado JJB, Cabanillas CG, Salinas F. Spectrophotometric resolution of ternary mixtures of salicylaldehyde, 3-hydroxybenzaldehyde and 4-hydroxybenzaldehyde by the derivative ratio spectrum-zero crossing method. Talanta 1992;39(5) 547-533.

[264] Aggarwal J, Habicht-Mauche J, Juarez Ch. Application of heavy stable isotopes in forensic isotope geochemistry: A review. Applied Geochemistry 2008;23(9) 2658-2666.

[265] Goguel RL, St. John DA. Chemical identification of Portland cements in New Zealand concretes: I. Characteristics differences among New Zealand cements in minor and trace element chemistry. Cement and Concrete Research 1993;23(1) 59-68.

[266] I.J Graham, R.L Goguel, D.A St John. Use of strontium isotopes to determine the origin of cement in concretes: Case examples from New Zealand. Cement and Concrete Research 2000;30(7) 1105-1111.

[267] Redler L. Quantitative X- ray diffraction analysis of high alumina cements. Cement and Concrete Research 1991;21(5) 873-884.

[268] Midgley HG. Quantitative determination of phases in high alumina cement clinkers by X-ray diffraction. Cement and Concrete Research 1976;6(2) 217-223.

[269] Guirado F, Galí S, Chinchón S . Quantitative Rietveld analysis of aluminous cement clinker phases. Cement and Concrete Research 2000;30(7) 1023-1029.

[270] Wilson W, Krakowiak KJ, Ulm F.-J. Simultaneous assessment of phase chemistry, phase abundance and bulk chemistry with statistical electron probe micro-analyses: Application to cement clinkers. Cement and Concrete Research 2014;55 35-48.

[271] Haach VG, Vasconcelos G, Lourenço PB. Influence of aggregates grading and water/cement ratio in workability and hardened properties of mortars. nstruction and Building Materials 2011;25(6) 2980-2987.

[272] Neville AM. Properties of Concrete, Fourth Edition. Addison Wesley Longman Ltd.; 1995.

[273] Wong HS, Buenfeld NR. Determining the water–cement ratio, cement content, water content and degree of hydration of hardened cement paste: Method development and validation on paste samples. Cement and Concrete Research 2009;39(10) 957-965.

[274] Philippidis TP, Aggelis DG. An acousto-ultrasonic approach for the determination of water-to-cement ratio in concrete. Cement and Concrete Research 2003;33(4) 525-538.

[275] Sahu S, Badger S, Thaulow N, Lee RJ. Determination of water–cement ratio of hardened concrete by scanning electron microscopy. Cement and Concrete Composites 2004;26(8) 987-992.

[276] Beygi MHA, Kazemi MT, Nikbin IM, Amiri JV. The effect of water to cement ratio on fracture parameters and brittleness of self-compacting concrete. Materials & Design 2013;50 267-276.

[277] Bescher E, Sambol M, Rice EK, Mackenzie JD. Determination of water-to-cement ratio in freshly mixed rapid-setting calcium sulfoaluminate concrete using 2.45 GHz microwave radiation. Cement and Concrete Research 2004;34(5) 807-812.

[278] Yodsudjai W, Wang K. Chemical shrinkage behavior of pastes made with different types of cements. Construction and Building Materials 2013;40 854-862.

[279] Jankovic A, Valery W, Davis E. Cement grinding optimisation. Minerals Engineering 2004;17(11-12) 1075-1081.

[280] Austin LG, Bagga R, Celik M. Breakage properties of some materials in a laboratory ball mill. Powder Technology 1981;28 235-241.

[281] Bond FC. Crushing and Grinding Calculations Parts I and II. British Chemical Engineering 6 (6&8); 1961.

[282] Morrell S, Shi F, Tondo. Modelling and Scale-up of High Pressure Grinding Rolls. In: Proceedings of the XX International Mineral Processing Congress (IMPC). Aachen: Germany; September 1997.

[283] Magdalinovic N. Mathematical Model for Determination of an Optimal Crusher Product size. Aufbereitungs-Technik 1990;31(5) 277-279.

[284] Lynch AJ. Mineral Crushing and Grinding Circuits, Their Simulation, Optimization, Design and Control. Amsterdam: Elsevier Scientific Publishing Co.; 1977.

[285] Benzer H, Ergün L, Öner M, Lynch AJ. Simulation of open circuit clinker grinding. Minerals Engineering 2001;14(7) 701-710.

[286] Aldrich Ch. Consumption of steel grinding media in mills - A review. Minerals Engineering 2013;49 77-91.

[287] Katsioti M, Tsakiridis PE, Giannatos P, Tsibouki Z, Marinos J. Characterization of various cement grinding aids and their impact on grindability and cement performance. Construction and Building Materials 2009;23(5) 1954-1959.

[288] Reid KJ. A solution to the batch grinding equation. Chemical Engineering Science 1965;20 953-963.

[289] Deniz V. A study on the specific rate of breakage of cement materials in a laboratory ball mill. Cement and Concrete Research 2003;33 439-445.

[290] Austin LG, Luckie PT. Methods for determination of breakage distribution parameters. Powder Technology 1971;5 215-222.

[291] Kapur PC. Balling and Granulation. Advances in Chemical Engineering 1978;10 55-123.

[292] Adetayo AA, Litster JD, Cameron IT. Steady state modelling and simulation of a fertilizer granulation circuit. Computers & Chemical Engineering 1995;19 383-393.

[293] Cotabarren I, Schulz PG, Bucalá V, Piña J. Modelling of an industrial double-roll crusher of a urea granulation circuit. Powder Technology 2008;183(2) 224-230.

[294] Horio M. Binderless granulation - its potential, achievements and future issues. Powder Technology 2003;130(1-3) 1-7.

[295] Saidur R, Hossain MS, Islam MR, Fayaz H, Mohammed HA. A review on kiln system modeling. Renewable and Sustainable Energy Reviews 2011;15(5) 2487-2500.

[296] Mujumdar KS, Ranade VV. Simulation of Rotary Cement Kilns Using a One-Dimensional Model. Chemical Engineering Research and Design 2006;84(3) 165-177.

[297] Mujumdar KS, Ganesh KV, Kulkarni SB, Ranade VV. Rotary Cement Kiln Simulator (RoCKS): Integrated modeling of pre-heater, calciner, kiln and clinker cooler. Chemical Engineering Science 2007;62(9) 2590-2607.

[298] Rademaker O, Goessens LH, Voskamp JH, Debie ACP. Dynamics and control of a clinker cooler. Automatica 1970;6(2) 231-244.

[299] Ahamed JU, Madlool NA, Saidur R, Shahinuddin MI, Kamyar A, Masjuki HH. Assessment of energy and exergy efficiencies of a grate clinker cooling system through the optimization of its operational parameters. Energy 2012;46(1) 664-674.

[300] Stadler KS, Poland J, Gallestey E. Model predictive control of a rotary cement kiln. Control Engineering Practice 2011;19(1) 1-9.

[301] Yuko T, Ikabata T, Akiyama T, Yamamoto T, Kurumada N. New clinker formation process by the fluidized bed kiln system. Cement and Concrete Research 2000;301113-1120.

[302] Van Puyvelde DR. Modelling the hold up of lifters in rotary dryers. Chemical Engineering Research and Design 2009;87(2) 226-232.

[303] Barin I. Thermodynamic data of pure substances. Part I Ag – Kr; 1992. ISBN 3-527-28531-8.

[304] Barin I. Thermodynamic data of pure substances. Part II La – Zr; 1992. ISBN 3-527-28531-8.

[305] Rodríguez NH, Martínez-Ramírez S, Blanco-Varela MT, Donatello S, Guillem M, Puig J, Fos C, Larrotcha E, Flores J. The effect of using thermally dried sewage sludge as an alternative fuel on Portland cement clinker production. Journal of Cleaner Production 2013;52 94-102.

[306] Tsiliyannis ChA. Alternative fuels in cement manufacturing: Modeling for process optimization under direct and compound operation. Fuel 2012;99 20-39.

[307] Pipilikaki P, Katsioti M, Papageorgiou D, Fragoulis D, Chaniotakis E. Use of tire derived fuel in clinker burning. Cement and Concrete Composites 2005;27(7-8) 843-847.

[308] Valderrama C, Granados R, Cortina JL, Gasol CM, Guillem M, Josa A. Comparative LCA of sewage sludge valorisation as both fuel and raw material substitute in clinker production. Journal of Cleaner Production 2013;51 205-213.

[309] Nakano T, Ichitsubo K, Kurokawa D, Ichikawa M. The effect of cooling rate on the fluidity of mortar made from kiln clinker. Cement and Concrete Research 2008;38(5) 643-648.

[310] Ichikawa M, Ikeda S, Komukai Y. Effect of cooling rate and Na_2O content on the character of the interstitial materials in portland cement clinker. Cement and Concrete Research 1994;24(6) 1092-1096.

[311] Woodson RD. Concrete Materials Concrete Portable Handbook; 2012.

[312] Sverak TS, Baker CGJ, Kozdas O. Efficiency of grinding stabilizers in cement clinker processing. Minerals Engineering 2013;43-44 52-57.

[313] Touil D, Belaadi S, Frances C. The specific selection function effect on clinker grinding efficiency in a dry batch ball mill. International Journal of Mineral Processing 2008;87 141-145.

[314] Bensted J. Effects of the clinker - gypsum grinding temperature upon early hydration of Portland cement. Cement and Concrete Research 1982;12(3) 341-348.

[315] Marzouki A, Lecomte A, Beddey A, Diliberto C, Ouezdou MB. The effects of grinding on the properties of Portland-limestone cement. Construction and Building Materials 2013;481145-1155.

[316] Spitas V, Spitas C. Stochastic simulation of the power requirements of dry clinker pulverisation. International Journal of Mineral Processing 2012;106-10942-49.

[317] Shi F. Comparison of grinding media- cylpebs and balls, Minerals Engineering 2004;17 1259-1268.

[318] Ipek H. The effects of grinding media shape on breakage rate. Minerals Engineering 2006;19 91-93.

[319] Qian HY, Kong QG, Zhang BL. The effects of grinding media shapes on the grinding kinetics of cement clinker in ball mill. Powder Technology 2013;235422-425.

[320] Deniz V . The effect of mill speed on kinetic breakage parameters of clinker and limestone. Cement and Concrete Research 2004;34(8) 1365-1371.

[321] Wang J.-h, Chen Q.-r, Kuang Y.-l, Lynch AJ, Zhuo J.-w. Grinding process within vertical roller mills: experiment and simulation. Mining Science and Technology (China) 2009;19(1) 97-101.

[322] Sohoni S, Sridhar R, Mandal G. The effect of grinding aids on the fine grinding of limestone, quartz and Portland cement clinker. Powder Technology 1991;67(3) 277-286.

[323] Tsivilis S, Voglis N. Photou J. A study of the intergrinding of clinker and limestone. Minerals Engineering 1999;12(7) 837-840.

[324] Irassar EF, Violini D, Rahhal VF, Milanesi C, Trezza MA, Bonavetti VL. Influence of limestone content, gypsum content and fineness on early age properties of Portland limestone cement produced by inter-grinding. Cement and Concrete Composites 2011;33(2) 192-200.

[325] Karamalidis AK, Dzombak DA. Surface Complexation Modeling: Gibbsite. Wiley; 2011.

[326] Gallardo AR, Vegas A. Zeitschrift für Kristallographie. The cation array in aluminum oxides, hydroxides and oxihydroxides. 1996;211(5) 299-303.

[327] Megaw HD. Zeitschrift für Kristallographie, Mineralogie und Petrographie 1934;87 185-204.

[328] Kloprogge JT, Duong LV, Wood BJ, Frost RL. XPS study of the major minerals in bauxite: Gibbsite, bayerite and (pseudo-)boehmite.Journal of Colloid and Interface Science 2006;296(2) 572-576.

[329] Milligan WO, McAtee JL. Crystal Structure of γ-AlOOH and γ-ScOOH.Journal of Physical Chemisty 1956;60(3) 273-277.

[330] Christoph GG, Corbato CE, Hofmann A, Tettenhorst RT. The crystal structure of boehmite. Clays and Clay Minerals 1979;27(2) 81-86.

[331] Bosmans HJ. Unit cell and crystal structure of nordstrandite, $Al(OH)_3$. Acta Crystallographica, Section B 1970;26(5) 649-652.

[332] Hathaway JC, Schlanger SO. Nordsrandite ($Al_2O_3 \bullet 3H_2O$) from Guam. The American Mineralogis 1965;50 1029-1037.

[333] El-Katatny EA, Halawy SA, Mohamed MA, Zaki MI. A novel synthesis of high-area alumina via H_2O_2-precipitated boehmite from sodium aluminate solutions.Journal of Chemical Technology and Biotechnology1998;72(4) 320-328.

[334] Demichelis R, Civalleri B, Noel Y, Meyer A, Dovesi R. Structure and stability of aluminium trihydroxides bayerite and gibbsite: A quantum mechanical ab initio study with the crystal 06 code. Chemical Physics Letters 2008;465(4-6) 220-225.

[335] Sposito G. The Environmental Chemistry of Aluminum. Lewis Publishers is an imprint of CRC Press; 1996. ISBN: 1-56670-030-2.

[336] Dewey C. Geological section from Williamstown, Mass. to Troy, N. Y. on the Hudson. American Journal of Science 1820;2(2) 249.

[337] Freij SJ, Parkinson GM. Surface morphology and crystal growth mechanism of gibbsite in industrial Bayer liquors. Hydrometallurgy 2005;78(3-4) 246-255.

[338] Bhöm J. Über Aluminium- und Eisenoxide I. Zeitschrift für anorganische und allgemeine Chemie 1925;149 203-210.

[339] Cai W, Li H, Zhang G. An innovative approach for pseudoboehmite precipitation from seeded sodium aluminate solutions. Journal of Physics and Chemistry of Solids 2010;71(4) 515-518.

[340] Chao GY, Baker J. Sabina AP, Roberts AC. Doyleite, a new polymorph of Al(OH)r, and its relationship to bayerite, gibbsite, and nordstrandite. Canadian Mineralogist 1985;23 21-28.

[341] Van Nordstrad RA, Hettinger WP, Keith CD. A new alumina trihydrate. Nature 1956;177 713-714.

[342] Jodin M.-C, Gaboriaud F, Humbert B. Limitations of potenciometric studies to determine the surface charge of gibbsite γ-$Al(OH)_3$ particles. Journal of Collid and Interface Science 2005;287 581-591.

[343] Kosmulski M. pH-dependent surface charging and points of zero charge. IV. Update and new approach. Journal of Collid and interface Science 2009;337 439-448.

[344] Huang PM, Li Y, Sumner ME. Handbook of Soil Sciences: Properties and Processes, Second Edition. CRC Press; 2011. ISBN 978-1-4398-0305-9.

[345] Kosmulski M. Surface Charging and Points of Zero Charge, CRC, Taylor & Francis; Boca Raton; FL; London; 2009.

[346] Alwitt RS. The point of zero charge of pseudoboehmite. Journal of Colloids and Interface Science 1972;40 195-198.

[347] Liu Y, Naidu R, Ming H. Surface electrochemical properties of red mud (bauxite residue): Zeta potential and surface charge density. Journal of Colloid and Interface Science 2013;394 451-457.

[348] Kabengi NJ, Daroub SH, Rhue RD. Energetics of arsenate sorption on amorphous aluminum hydroxides studied using flow adsorption calorimetry.Journal of Colloid and Interface Science 2006;297(1) 86-94.

[349] Carre S, Gnep NS, Revel R, Magnoux P. Characterization of the acid-base properties of transition aluminas by model reaction. Applied Catalysis A: General 2008;348 71-78.

[350] Richardson JT. Principles of Catalyst Development; Plenum Press; London; 1989.

[351] Du X, Wang Y, Su X, Li J. Influences of pH value on the microstructure and phase transformation of aluminum hydroxide. Powder Technology 2009;192(1) 40-46.

[352] Sudworth JL, Tilley AR. The Sodium Sulphur Battery. London: Chapman & Hall; 1985.

[353] Mali A, Petric A. Synthesis of sodium β′′-alumina powder by sol-gel combustion. Journal of the European Ceramic Society 2012;32 1229-1234.

[354] Lu X, Xia G, Lemmon JP, Yang Z. Advanced materials for sodium-beta alumina batteries: Status, challenges and perspectives. Journal of Power Sources 2010;195 2431-2442.

[355] Bragg WL, Gottfried C, West J. The structure of β alumina. Zeitschrift für Kristallographie- Crystalline Materials 1931;77(1) 255-274.

[356] Beevers CA, Ross MAS. The crystal structure of "beta alumina" $Na_2O \cdot 11Al_2O_3$. Zeitschrift für Kristallographie 1937;97 59-66.

[357] Yamaguchi G, Suzuki K.Bulletin of the Chemical Society of Japan 1968;4193-99.

[358] Bettman M, Peters CR. The crystal structure of $Na_2O \cdot MgO \cdot 5Al_2O_3$ with reference to $Na_2O \cdot 5Al_2O_3$ and other isotypal compounds. Journal of Physical Chemistry 1969;73 1774-1780.

[359] Bourke MAM, Hooper A, Moseley PT, Taylor RG. Sodium-rich beta-alumina. Solid State Ionics 1980;1(5-6) 367-372.

[360] Koh J.-H, Weber N, Virkar AV. Synthesis of lithium-beta-alumina by various ion-exchange and conversion processes. Solid State Ionics 2012;220 32-38.

[361] Kalaignan GP, Seo DJ, Park SB. Characterization of Srβ-alumina prepared by sol–gel and spray pyrolysis methods. Materials Chemistry and Physics 2004;85(2-3)286-293.

[362] Yuji Masubuchi, Tomoyuki Hata, Teruki Motohashi, Shinichi Kikkawa. Crystal structure of Eu-doped magnetoplumbite-type lanthanum aluminum oxynitride with emission site splitting. Journal of Solid State Chemistry 2011;184(9) 2533-2537.

[363] Carrillo-Cabrera W, Thomas JO, Farrington GC. The structure of the lanthanide Gd^{3+}, Eu^{3+} and Nd^{3+} β″-aluminas. Solid State Ionics 1988;28–30 317-323.

[364] MacKenzie KJD, Temuujin J, Okada K. Thermal decomposition of mechanically activated gibbsite. Thermochimica Acta 1999;327(1-2) 103-108.

[365] Zhu B, Fang B, Li X. Dehydration reactions and kinetic parameters of gibbsite. Ceramics International 2010;36(8) 2493-2498.

[366] Li H, Shao T, Li D, Chen D. Nonisothermal reaction kinetics of diasporic bauxite. Thermochimica Acta 2005;427(1-2) 9-12.

[367] Whittington B, Ilievski D. Determination of the gibbsite dehydration reaction pathway at conditions relevant to Bayer refineries. Chemical Engineering Journal 2004;98(1-2) 89-97.

[368] Alphonse P, Courty M. Structure and thermal behavior of nanocrystalline boehmite. Thermochimica Acta 2005;425(1-2) 75-89.

[369] L'vov BV. Mechanism of thermal decomposition of alkaline-earth carbonates. Thermochimica Acta 1997;303 161-170.

[370] An L, Chan HM, Soni KK. Control of Calcium hexaluminate grain morphology in insitu toughened ceramic composites. Journal of Materials Science 1996;31(12) 3223–3229.

[371] Vishista K, Gnanam FD. Microstructural development of $SrAl_{12}O_{19}$ in alumina–strontia composites. Journal of the European Ceramic Society 2009;29(1) 77-83.

[372] Davar F, Salavati-Niasari M, Baskoutas S. Temperature controlled synthesis of SrCO3 nanorods via a facile solid-state decomposition rout starting from a novel inorganic precursor. Applied Surface Science 2011;257(9) 3872-3877.

[373] Miller F, Wilkins ChH. Infrared spectra and characteristic frequencies of inorganic ions. Their use in quantitative analysis. Analytical Chemistry 1952;24 1253-1274.

[374] Huang CK, Kerr PF. Infrared study of the carbonate minerals. The American Mineralogist 1960;45 311-324.

[375] Boumaza A, Favaro L, Lédion J, Sattonnay G, Brubach JB, Berthet P, Huntz AM, Roy P, Tétot R. Transition alumina phases induced by heat treatment of boehmite: An X-ray diffraction and infrared spectroscopy study. Journal of Solid State Chemistry 2009;182(5) 1171-1176.

[376] Boumaza A, Djelloul A, Guerrab F. Specific signatures of α-alumina powders prepared by calcination of boehmite or gibbsite. Powder Technology 2010;201(2) 177-180.

[377] Nag A, Kutty TRN. Role of B_2O_3 on the phase stability and long phosphorescence of $SrAl_2O_4$:Eu, Dy. Journal of Alloys and Compounds 2003;354(1-2) 221-231.

[378] Zhou T, Song Z, Bian L, Ren Q, Liu Q. Synthesis and luminescence properties of europium activated $Ca_3Al_2O_6$-$Sr_3Al_2O_6$ system. Journal of Rare Earths 2012;30(7) 632-636.

[379] Ptáček P, Šoukal F, Opravil T, Bartoníčková E, Zmrzlý M, Novotný R. Synthesis, hydration and thermal stability of hydrates in strontium-aluminate cement. Ceramics International 2014;40(7) 9971-9979.

[380] Hörkner W, Müller-Buschbaum HK. Zur kristallstruktur von $CaAl_2O_4$. Journal of Inorganic and Nuclear Chemistry 1976;38(5) 983-984.

[381] E. Ghiasvand, A.A. Ramezanianpour, A.M. Ramezanianpour. Effect of grinding method and particle size distribution on the properties of Portland-pozzolan cement. Construction and Building Materials 2014; 53 547-554.

[382] Bentz DP, Garboczi EJ, Haecker CJ, Jensen OM. Effects of cement particle size distribution on performance properties of Portland cement-based materials. Cement and Concrete Research 1999;29(10) 1663-1671.

[383] Zhang T, Yu Q, Wei J, Zhang P. Effects of size fraction on composition and fundamental properties of Portland cement. Construction and Building Materials 2011;25(7) 3038-3043.

[384] Frigione G, Marra S. Relationship between particle size distribution and compressive strength in portland cement. Cement and Concrete Research 1976;6(1) 113-127.

[385] Celik IB. The effects of particle size distribution and surface area upon cement strength development. Powder Technology 2009;188(3) 272-276.

[386] Vedalakshmi R, Raj AS, Srinivasan S, Babu KG. Quantification of hydrated cement products of blended cements in low and medium strength concrete using TG and DTA technique. Thermochimica Acta 2003;407(1-2) 49-60.

[387] Pane I, Hansen W. Investigation of blended cement hydration by isothermal calorimetry and thermal analysis. Cement and Concrete Research 2005;35(6) 1155-1164.

[388] Lilkov V, Petrov O, Tzvetanova Y, Savov P. Mössbauer, DTA and XRD study of Portland cement blended with fly ash and silica fume. Construction and Building Materials 2012;29 33-41.

[389] Hesse Ch, Goetz-Neunhoeffer F, Neubauer J. A new approach in quantitative in-situ XRD of cement pastes: Correlation of heat flow curves with early hydration reactions. Cement and Concrete Research 2011;41(1) 123-128.

[390] Pang X, Bentz DP, Meyer Ch, Funkhouser GP, Darbe R. A comparison study of Portland cement hydration kinetics as measured by chemical shrinkage and isothermal calorimetry. Cement and Concrete Composites 2013;3923-32.

[391] Xu Q, Hu J, Ruiz JM, Wang K, Ge Z. Isothermal calorimetry tests and modeling of cement hydration parameters. Thermochimica Acta 2010;499(1-2) 91-99.

[392] Gerstig M, Wadsö L. A method based on isothermal calorimetry to quantify the influence of moisture on the hydration rate of young cement pastes. Cement and Concrete Research 2010;40(6) 867-874.

[393] Mostafa NY, Brown PW. Heat of hydration of high reactive pozzolans in blended cements: Isothermal conduction calorimetry. Thermochimica Acta 2005;435(2) 162-167.

[394] Esteves LP. On the hydration of water-entrained cement–silica systems: Combined SEM, XRD and thermal analysis in cement pastes. Thermochimica Acta 2011;518(1-2) 27-35.

[395] Ylmén R, Jäglid U, Steenari B.-M, Panas I. Early hydration and setting of Portland cement monitored by IR, SEM and Vicat techniques. Cement and Concrete Research 2009;39(5) 433-439.

[396] Torréns-Martín D, Fernández-Carrasco L, Martínez-Ramírez S. Hydration of calcium aluminates and calcium sulfoaluminate studied by Raman spectroscopy. Cement and Concrete Research 2013;47 43-50.

[397] Corr DJ, Juenger MCG, Monteiro PJM, Bastacky J. Investigating entrained air voids and Portland cement hydration with low-temperature scanning electron microscopy. Cement and Concrete Composites 2004;26(8) 1007-1012.

[398] Gallucci E, Mathur P, Scrivener K. Microstructural development of early age hydration shells around cement grains. Cement and Concrete Research 2010;40(1) 4-13.

[399] Double DD. Some studies of the hydration of Portland cement using high voltage (1 MV) electron microscopy. Materials Science and Engineering 1973;12(1) 29-34.

[400] Goldschmidt A. About the hydration theory and the composition of the liquid phase of portland cement. Cement and Concrete Research 1982;12(6) 743-746.

[401] Michaux M, Fletcher P, Vidick B. Evolution at early hydration times of the chemical composition of liquid phase of oil-well cement pastes with and without additives. Part I. Additive free cement pastes. Cement and Concrete Research 1989;19(3) 443-456.

[402] Delmi MMY, A.-Mokhtar A, Amiri O. Modelling the coupled evolution of hydration and porosity of cement-based materials. Construction and Building Materials 2006;20(7) 504-514.

[403] Dongyu X, Shifeng H, Lei Q, Lingchao L, Xin C. Monitoring of cement hydration reaction process based on ultrasonic technique of piezoelectric composite transducer. Construction and Building Materials 2012;35 220-226.

[404] Smith A, Chotard T, Gimet-Breart N, Fargeot D. Correlation between hydration mechanism and ultrasonic measurements in an aluminous cement: effect of setting time and temperature on the early hydration. Journal of the European Ceramic Society 2002;22(12) 1947-1958.

[405] Cheng X, Qin L, Zhong QQ, Huang SF, Li ZJ. Temperature and boundary influence on cement hydration monitoring using embedded piezoelectric transducers. Ultrasonics 2013;53(2) 412-416.

[406] Chotard T, Gimet-Breart N, Smith A, Fargeot D, Bonnet JP, Gault C. Application of ultrasonic testing to describe the hydration of calcium aluminate cement at the early age. Cement and Concrete Research 2001;31(3) 405-412.

[407] McCarter WJ. A parametric study of the impedance characteristics of cement-aggregate systems during early hydration. Cement and Concrete Research 1994;24(6) 1097-1110.

[408] Dotelli G, Mari CM. The evolution of cement paste hydration process by impedance spectroscopy. Materials Science and Engineering: A 2001;303(1-2) 54-59.

[409] Zuo Y, Zi J, Wei X. Hydration of cement with retarder characterized via electrical resistivity measurements and computer simulation. Construction and Building Materials 2014;53411-418.

[410] Liao Y, Wei X, Li G. Early hydration of calcium sulfoaluminate cement through electrical resistivity measurement and microstructure investigations. Construction and Building Materials 2011;25(4) 1572-1579.

[411] Brooks JJ, Johari MAM, Mazloom M. Effect of admixtures on the setting times of high-strength concrete. Cement and Concrete Composites 2000;22(4) 293-301.

[412] Ramachandran VS, Malhotra VM. Superplasticizers: Properties and application in concrete; 1988. ISBN 0-660-17393-X.

[413] Janowska-Renkas E. The effect of superplasticizers' chemical structure on their efficiency in cement pastes. Construction and Building Materials 2013;38 1204-1210.

[414] Valcuende M, Marco E, Parra C, Serna P. Influence of limestone filler and viscosity-modifying admixture on the shrinkage of self-compacting concrete. Cement and Concrete Research 2012;42(4) 583-592.

[415] Lothenbach B, Le Saout G, Gallucci E, Scrivener K. Influence of limestone on the hydration of Portland cements. Cement and Concrete Research 2008;38(6) 848-860.

[416] Chakoumakos BC, Lager GA, Fernandez-Baca JA. Refinement of the structures of Sr3Al2O6 and the hydrogarnet $Sr_3Al_2(OD)_{12}$ by Rietveld analysis of neutron powder diffraction data. Acta Crystallographica Section C 1992;48 414–419.

[417] Mondal P, Jeffery JW. The crystal structure of tricalcium aluminate, $Ca_3Al_2O_6$. Acta Crystallographica Section B 31 (1975) 689–697.

[418] Handy NC, Carter S, Yamaguchi Y, Li S, Turney JM, Schaefer HF. Rovibrational energy levels for the electronic ground state of AlOH. Chemical Physics Letters 2006;427(1-3) 14-17.

[419] Hirota F, Tanimoto M, Tokiwa H. Ab initio study on the AlOH/HAlO system. Chemical Physics Letters 1993;208(1-2) 115-119.

[420] Tarte P. Infra-red spectra of inorganic aluminates and characteristic vibrational frequencies of AlO_4 tetrahedra and AlO_6 octahedra. Spectrochimica Acta Part A: Molecular Spectroscopy 1967;23(7) 2127-2143.

[421] Preudhomme J, Tarte P. Infrared studied of spinel-III: the normal II–III spinel. Spectrochimica Acta Part A 1971;27(9) 1817–1835.

[422] Li X, Wang D, Zhou Q, Liu G, Peng Z. Concentration variation of aluminate ions during the seeded precipitation process of gibbsite from sodium aluminate solution. Hydrometallurgy 2011;106(1-2) 93-98.

[423] Andrews A, Nsiah-Baafi E, Gawu SKY, Olubambi PA. Synthesis of high alumina refractories from lithomargic clay. Ceramics International 2014;40(4) 6071-6075.

[424] Galusek D, Ghillányová K. Ceramic oxides. In: Ceramics Science and Technology, Volume 2, Materials and Properties. Ed.: Riedel R and Chen I-Wei; Wiley-VCH; 2010. ISBN: 978-3-527-31156-9.

[425] Amrane B, Ouedraogo E, Mamen B, Djaknoun S, Mesrati N. Experimental study of the thermo-mechanical behaviour of alumina-silicate refractory materials based on a mixture of Algerian kaolinitic clays. Ceramics International 2011;37(8) 3217-3227.

[426] Lavat AE, Grasselli MC, Lovecchio EG. Effect of α and γ polymorphs of alumina on the preparation of $MgAl_2O_4$-spinel-containing refractory cements. Ceramics International 2010;36(1) 15-21.

[427] Reddy NK. Reaction-bonded silicon carbide refractories. Materials Chemistry and Physics 2002;76(1) 78-81.

[428] Pierson HO. Applications of Refractory Carbides and Nitrides. Handbook of Refractory Carbides and Nitrides; 1996.

[429] Schmalzried C, Schwetz K. Silicon Carbide- and Boron Carbide-Based Hard materials. In: Ceramics Science and Technology, Volume 2, Materials and Properties. Ed.: Riedel R and Chen I.-W; Wiley-VCH; 2010. ISBN: 978-3-527-31156-9.

[430] Rao MPLN, Gupta GS, Manjunath ., Kumar S, Suri AK, Krishnamurthy N, Subramanian C. Temperature measurements in the boron carbide manufacturing process – A hot model study. International Journal of Refractory Metals and Hard Materials 2009;27(3) 621-628.

[431] Liu G, Li J, Chen K. Combustion synthesis of refractory and hard materials: A review. International Journal of Refractory Metals and Hard Materials 2013;39 90-102.

[432] Lihrmann J.-M. Thermodynamics of the Al_2O_3–Al_4C_3 system: I. Thermochemical functions of Al oxide, carbide and oxycarbides between 298 and 2100 K. Journal of the European Ceramic Society 2008;28(3) 633-642.

[433] Chen K, Huang Z, Liu Y.-g, Fang M, Huang J, Xu Y. Synthesis of β-Si_3N_4 powder from quartz via carbothermal reduction nitridation. Powder Technology 2013;235 728-734.

[434] Jones MI, Valecillos M.-C, Hirao K, Yamauchi Y. Grain growth in microwave sintered Si_3N_4 ceramics sintered from different starting powders. Journal of the European Ceramic Society 2002;22(16) 2981-2988.

[435] Yu J, Yang J, Li H, Xi X, Huang Y. Study on particle-stabilized Si_3N_4 ceramic foams. Materials Letters 2011;65(12) 1801-1804.

[436] Eichler J, Lesniak Ch. Boron nitride (BN) and BN composites for high-temperature applications. Journal of the European Ceramic Society 2008;28(5) 1105-1109.

[437] Wang Q, He D, Peng F, Xiong , Wang J, Wang P, Xu Ch, Liu J. Compression behavior of nanocrystalline TiN. Solid State Communications 2014;182 26-29.

[438] Kartavykh AV, Tcherdyntsev VV, Zollinger J. TiAl–Nb melt interaction with AlN refractory crucibles. Materials Chemistry and Physics 2009;116(1) 300-304.

[439] Šajgalík P, Lenčéš Z, Hnatko M. Nitrides. In: Ceramics Science and Technology, Volume 2, Materials and Properties. Ed.: Riedel R and Chen I.-W; Wiley-VCH, 2010. ISBN: 978-3-527-31156-9.

[440] Taleghani PR, Bakhshi SR, Erfanmanesh M, Borhani GH, Vafaei R. Improvement of $MoSi_2$ oxidation resistance via boron addition: Fabrication of $MoB/MoSi_2$ composite by mechanical alloying and subsequent reactive sintering. Powder Technology 2014;254241-247.

[441] Raju GB, Basu B, Suri AK. Thermal and electrical properties of TiB_2–$MoSi_2$. International Journal of Refractory Metals and Hard Materials 2010;28(2) 174-179.

[442] Luo M, Li Y, Jin S, Sang S, Zhao L, Wang Q, Li Y. Microstructure and mechanical properties of multi-walled carbon nanotubes containing Al_2O_3–C refractories with addition of polycarbosilane. Ceramics International 2013;39(5) 4831-4838.

[443] Fan H, Li Y, Sang S. Microstructures and mechanical properties of Al_2O_3–C refractories with silicon additive using different carbon sources. Materials Science and Engineering: A 2011;528(7-8) 3177-3185.

[444] Musante L, Martorello LF, Galliano PG, Cavalieri AL, Martinez AGT. Mechanical behaviour of MgO–C refractory bricks evaluated by stress–strain curves. Ceramics International 2012;38(5) 4035-4047.

[445] Musante L, Muñoz V, Labadie MH, Martinez AGT. High temperature mechanical behavior of Al_2O_3–MgO–C refractories for steelmaking use. Ceramics International 2011;37(5) 1473-1483.

[446] Muñoz V, Martinez AGT. Thermal Evolution of Al_2O_3-MgO-C Refractories. Procedia Materials Science 2012;1 410-417.

[447] Sadeghbeigi R. Refractory Lining Systems. Fluid Catalytic Cracking Handbook (Third Edition); 2012.

[448] Stamenković I, Simičić V, Sigulinski F, Martinović P, Stefanović R. Properties of Al_2O_3-SiO_2 heat insulating refractories. Ceramurgia International 1977;3(4) 168.

[449] Schmitt N, Hernandez J.-F, Lamour V, Berthaud Y, Meunier P, Poirier J. Coupling between kinetics of dehydration, physical and mechanical behaviour for high alumina castable. Cement and Concrete Research 2000;30(10) 1597-1607.

[450] Nouri-Khezrabad M, Braulio MAL, Pandolfelli VC, Golestani-Fard F, Rezaie HR. Nano-bonded refractory castables. Ceramics International 2013;39(4) 3479-3497.

[451] Zhou N, Hu S, Zhang S. Advances in modern refractory castables, CN-Refractories 2004;13(2) 3-12.

[452] Zhou N. New castables and their role in advancements in monolithic refractories. Interceram 2006;55(1) 24-26.

[453] Deville HS.-C. On the compositions of different alloys. Annales des Chimie et des Physique 1856;46(3) 418-420.

[454] Soc J, de Lafarge PA. French patent applications 320290; 39454; 1908.

[455] Parr Ch, Bier TA, Bunt NE, Spreafico E. Calcium aluminate cement (CAC) based castables for demanding applications. Technical Paper (Ref.TP-GB-RE-LAF-007) presented at the 1st Monolithics Conference; Teheran; Iran; 1997.

[456] Guirado F, Galí S, Chinchón JS. Thermal Decomposition of Hydrated Alumina Cement (CAH_{10}). Cement and Concrete Research 1998;28(3) 381-390.°

[457] Mostafa NY, Zaki ZI, Elkader OHA. Chemical activation of calcium aluminate cement composites cured at elevated temperature. Cement and Concrete Composites 2012;34(10) 1187-1193.

[458] Luz AP, Pandolfelli VC. $CaCO_3$ addition effect on the hydration and mechanical strength evolution of calcium aluminate cement for endodontic applications. Ceramics International 2012;38(2) 1417-1425.

[459] Antonovič V, Kerienė J, Boris R, Aleknevičius M. The Effect of Temperature on the Formation of the Hydrated Calcium Aluminate Cement Structure. Procedia Engineering 2013;57 99-106.

[460] Kopanda JE, Maczura G. Production Processes, Properties, and Applications for Calcium Aluminate Cements. Aluminum Company of America; Pittsburg 1987;15212 171-183.

[461] Cardoso FA, Innocentini MDM, Akiyoshi MM, Pandolfelli VC. Effect of curing time on the properties of CAC bonded refractory castables. Journal of the European Ceramic Society 2004;24(7) 2073-2078.

[462] Gungor A, Celikcioglu O, Sahin S. The physical and mechanical properties of alumina-based ultralow cement castable refractories. Ceramics International 2012;38(5) 4189-4194.

[463] Karadeniz E, Gurcan C, Ozgen S, Aydin S. Properties of alumina based low-cement self flowing castable refractories. Journal of the European Ceramic Society 2007;27(2-3) 1849-1853.

[464] Karamian E, Monshi A. Influence of additives on nano-SiC whisker formation in alumina silicate–SiC–C monolithic refractories. Ceramics International 2010;36(2) 811-816.

[465] Auvray JM, Gault C, Huger M. Microstructural changes and evolutions of elastic properties versus temperature of alumina and alumina–magnesia refractory castables. Journal of the European Ceramic Society 2008;28(10) 1953-1960.

[466] Braulio MAL, Morbioli GG, Milanez DH, Pandolfelli VC. Calcium aluminate cement source evaluation for Al_2O_3–MgO refractory castables. Ceramics International 2011;37(1) 215-221.

[467] Braulio MAL,Tontrup C,Medeiros J,Pandolfelli VC. Colloidal alumina as anovel castable bonding system, Refractories World-forum 2011;3(3)136-141.

[468] Hongo Y.ρ-alumina bonded castable refractories. Taikabutsu Overseas 1988;40(4)226–229.

[469] Ma W, Brown PW. Mechanisms of reaction of hydratable aluminas. Journal of the American Ceramic Society 1999;82(2) 453-456.

[470] Cardoso FA, Innocentini MDM, Miranda MFS, Valenzuela FAO, Pandolfelli VC. Drying behavior of hydratable alumina-bonded refractory castables. Journal of the European Ceramic Society 2004;24(5) 797-802.

[471] Miśta W, Wrzyszcz J. Rehydration of transition aluminas obtained by flash calcination of gibbsite. Thermochimica Acta 1999;331(1) 67-72.

[472] Violante A, Huang PM. Formation Mechanism of Aluminum Hydroxide Polymorphs. Clays and Clay Minerals 1993;41 590-597.

[473] S. Mukhopadhyay, S. Ghosh, M.K. Mahapatra, R. Mazumder, P. Barick, S. Gupta, S. Chakraborty. Easy-to-use mullite and spinel sols as bonding agents in a high-alumina based ultra low cement castable. Ceramics International 2002;28(7) 719-729.

[474] Mukhopadhyay S, Sen S, Maiti T, Mukherjee M, Nandy RN, Sinhamahapatra BK. In situ spinel bonded refractory castable in relation to co-precipitation and sol-gel derived spinel forming agents. Ceramics International 2003;29(8) 857-868.

[475] Risbud SH, Pask JA. SiO_2-Al_2O_3 metastable phase equilibrium diagram without mullite. Journal of Materials Science 1978;132449–2454.

[476] Abo-El-Enein SA, Abou-Sekkina MM, Khalil Nagy M, Osama A. Shalma. Microstructure and refractory properties of spinel containing castables. Ceramics International 2010;36(5) 1711-1717.

[477] Ahari KG, Sharp JH, Lee WE. Hydration of refractory oxides in castable bond systems-II: alumina–silica and magnesia–silica mixtures. Journal of the European Ceramic Society 2003;23(16) 3071-3077.

[478] Braulio MAL, Bittencourt LRM, Pandolfelli VC. Selection of binders for in situ spinel refractory castables. Journal of the European Ceramic Society 2009;29(13) 2727-2735.

[479] Salomão R, Pandolfelli VC. The role of hydraulic binders on magnesia containing refractory castables: Calcium aluminate cement and hydratable alumina. Ceramics International 2009;35(8) 3117-3124.

[480] Otroj S, Bahrevar MA, Mostarzadeh F, Nilforoshan MR. The effect of deflocculants on the self-flow characteristics of ultra low-cement castables in Al_2O_3–SiC–C systém. Ceramics International 2005;31(5) 647-653.

[481] Evangelista PC, Parr C. Control of formulation and optimization of self-flow castables based on pure calcium aluminates. Refractories Applications and News 2002;7 14–18.

[482] Pinto DG, Silva AP, Segadães AM, Devezas TC. Thermomechanical evaluation of self-flowing refractory castables with and without the addition of aluminate cement. Ceramics International 2012;38(4) 3483-3488.

[483] Ukrainczyk N, Matusinovic T, Kurajica S, Zimmermann B, Sipusic J. Dehydration of a layered double hydroxide-C_2AH_8. Thermochim Acta 2007;464(1-2) 7-15.

[484] Luz AP, Neto ABS, Santos Jr. T, Medeiros J, Pandolfelli VC. Mullite-based refractory castable engineering for the petrochemical industry. Ceramics International 2013;39(8) 9063-9070.

[485] Dinger DR, Funk JE. Particle packing II: review of packing of polydisperse particle systems. Interceramic 1992;41 95-97.

[486] Silva AP, Segadães AM, Pinto DG, Oliveira LA, Devezas TC. Effect of particle size distribution and calcium aluminate cement on the rheological behaviour of all-alumina refractory castables. Powder Technology 2012;226 107-113.

[487] Lange F, Mörtel H, Rudert V. Dense packing of cement pastes and resulting consequences on mortar properties. Cement and Concrete Research 1997;27(10) 1481-1488.

[488] Oliveira IR, Ortega FS, Pandolfelli VC. Hydration of CAC cement in a castable refractory matrix containing processing additives. Ceramics International 2009;35(4) 1545-1552.

[489] Zhou X, Sankaranarayanane K, Rigaud M. Design of bauxite-based low-cement pumpable castables: a rheological approach. Ceramics International 2004;30(1) 47-55.

[490] Andrews A, Adam J, Gawu SKY. Development of fireclay aluminosilicate refractory from lithomargic clay deposits. Ceramics International 2013;39(1) 779-783.

[491] Tseng T.-T, Wu H.-M, Chen Ch.-N, Cheng Ch.-Ch, Uan J.-Y, Wu W, Tseng W.J. Refractory filler sands with core–shell composite structure for the taphole nozzle in slide-gate system of steel ladles. Ceramics International 2012;38(2) 967-971.

[492] Merzouki T, Blond E, Schmitt N, Bouchetou M-L, Cutard T, Gasser A. Modelling of the swelling induced by oxidation in SiC-based refractory castables. Mechanics of Materials 2014;68253-266.

[493] Luz AP, Miglioli MM, Souza TM, Hashimoto S, Zhang S, Pandolfelli VC. Effect of Al_4SiC_4 on the Al_2O_3 single bond SiC single bond SiO_2 single bond C refractory castables performance. Ceramics International 2012;38(5) 3791-3800.

[494] Aksel C. The influence of zircon on the mechanical properties and thermal shock behaviour of slip-cast alumina–mullite refractories. Materials Letters 2002;57(4) 992-997.

[495] Aksel C, Dexet M, Logen N, Porte F, Riley FL, Konieczny F. The influence of zircon in a model aluminosilicate glass tank forehearth refraktory. Journal of the European Ceramic Society 2003;23(12) 2083-2088.

[496] Zargar HR, Oprea C, Oprea G, Troczynski T. The effect of nano-Cr_2O_3 on solid-solution assisted sintering of MgO refractories. Ceramics International 2012;38(8) 6235-6241.

[497] Kim T, Kim D, Kang S. Effect of additives on the sintering of $MgAl_2O_4$. Journal of Alloys and Compounds 2014;587 594-599.

[498] Sutcu M, Akkurt S, Bayram A, Uluca U. Production of anorthite refractory insulating firebrick from mixtures of clay and recycled paper waste with sawdust addition. Ceramics International 2012;38(2) 1033-1041.

[499] Demir I, Baspınar MS, Orhan M. Utilization of kraft pulp production residues in clay brick production. Building and Environment 2005;40(11) 1533–1537.

[500] Sutcu M, Akkurt S. The use of recycled paper processing residues in making porous brick with reduced thermal conductivity. Ceramics International 2009;35(7) 2625–2631.

[501] Sutcu M, Akkurt S. Utilization of recycled paper processing residues and clay of different sources for the production of porous anorthite ceramics. Journal of the European Ceramic Society 2010;30(8) 1785-1793.

[502] ASTM C155-97; Standard Classification of Insulating Firebrick. ASTM International, West Conshohocken, PA; 2002.

[503] ASTM Standard C134-95; Standard Test Methods for Size, Dimensional Measurements, and Bulk Density of Refractory Brick and Insulating Firebrick. ASTM International, West Conshohocken, PA; 2005.

[504] ASTM Standard C20; Standard Test Methods for Apparent Porosity, Water Absorption, Apparent Specific Gravity, and Bulk Density of Burned Refractory Brick and Shapes by Boiling Water. ASTM International, West Conshohocken, PA; 2005.

[505] ASTM C133-97; Standard Test Methods for Cold Crushing Strength and Modulus of Rupture of Refractories. ASTM International, West Conshohocken, PA; 2003.

[506] ASTM C1171-96; Standard Test Method for Quantitatively Measuring the Effect of Thermal Shock and Thermal Cycling on Refractories. ASTM International, West Conshohocken, PA; 2003.

[507] Chudzik S. Measurement of thermal parameters of a heat insulating material using infrared thermography. Infrared Physics & Technology 2012;55(1) 73-83.

[508] Gong L, Wang Y, Cheng X, Zhang R, Zhang H. A novel effective medium theory for modelling the thermal conductivity of porous materials. International Journal of Heat and Mass Transfer 2014;68 295-298.

[509] Hirao K, Yhou Y. Thermal conductivity. In: Ceramics Science and Technology, Volume 2, Materials and Properties. Ed.: Riedel R, Chen I.-W; Wiley-VCH; 2010. ISBN: 978-3-527-31156-9.

[510] Bart GCJ, Hoogendoorn CJ, Schaareman PBJ. Stationary and transient heat conduction in a non homogeneous material. Wärme - und Stoffübertragung 1986;20(4) 269-272.

[511] Tritt TM. Thermal Conductivity, Theory, Properties, and Applications. Springer; 2004. ISBN 978-0-306-48327-1.

[512] Carson JK, Lovatt SJ, Tanner DJ, Cleland AC. Thermal conductivity bounds for isotropic, porous materials. International Journal of Heat and Mass Transfer 2005;48(11) 2150-2158.

[513] Carson JK. Review of effective thermal conductivity models for foods.International Journal of Refrigeration 2006;296)958-967.

[514] Wang J, Carson JK, North MF, Cleland DJ. A new structural model of effective thermal conductivity for heterogeneous materials with co-continuous phases. International Journal of Heat and Mass Transfer 2008;51(9-10) 2389-2397.

[515] Francl J, Kingery WD. Thermal conductivity: IX, Experimental investigation of effect of porosity on thermal conductivity. Journal of the American Ceramic Society 1954;37 99-107.

[516] Han Y, Li C, Bian Ch, Li S, Wang Ch.-A. Porous anorthite ceramics with ultra-low thermal conductivity. Journal of the European Ceramic Society 2013;33(13-14) 2573-2578.

[517] Barea R, Osendi MI, Ferreira JMF, Miranzo P. Thermal conductivity of highly porous mullite material.Acta Materialia 2005;53(11) 3313-3318.

[518] Progelhof RC, Throne JL, Ruetsch RR. Methods for predicting the thermal conductivity of composite systems: a review. Polymer Engineering & Science 1976;16(9) 615-625.

[519] Cheng P, Hsu CT. The effective stagnant thermal conductivity of porous media with periodic structures. Journal of Porous Media 1999;2(1) 19-38.

[520] Brailsford AD, Major KG. The thermal conductivity of aggregates of several phases including porous materials.British Journal of Applied Physics 1964;15 313-319.

[521] Dondero M, Cisilino AP, Carella JM, Pablo TJ. Effective thermal conductivity of functionally graded random micro-heterogeneous materials using representative volume

element and BEM. International Journal of Heat and Mass Transfer 2011;54(17-18) 3874-3881.

[522] Wang J, Carson JK, North MF, Cleland DJ. A new approach to modelling the effective thermal conductivity of heterogeneous materials. International Journal of Heat and Mass Transfer 2006;49(17-18) 3075-3083.

[523] Kirkpatrick S. Percolation and conduction. Reviews of Modern Physics 1973;45 574-588.

[524] Fricke H. A mathematical treatment of the electric conductivity and capacity of disperse systems I. The electric conductivity of a suspension of homogeneous spheroids. Physical Review 1924;24(5) 575-587.

[525] Hamilton RL, Crosser OK. Thermal conductivity of heterogeneous two-component systems. Industrial & Engineering Chemistry Fundamentals 1962;1 187-191.

[526] Levy FL. A modified Maxwell-Eucken equation for calculating the thermal conductivity of two-component solutions or mixtures. International Journal of Refrigeration 1981;4(4) 223-225.

[527] Gong L, Wang Y, Cheng X, Zhang R, Zhang H. Thermal conductivity of highly porous mullite materials. International Journal of Heat and Mass Transfer 2013;67253-259.

[528] Smith BH, Szyniszewski S, Hajjar JF, Schafer BW, Arwade SR. Steel foam for structures: A review of applications, manufacturing and material properties. Journal of Constructional Steel Research 2012;71 1-10.

[529] Friberg SE. Foams from non-aqueous systems. Current Opinion in Colloid & Interface Science 2010;15 359-364.

[530] Jang W.-Y, Kraynik AM, Kyriakides S. On the microstructure of open-cell foams and its effect on elastic properties. International Journal of Solids and Structures 2008;45 1845-1875.

[531] Luyten J, Mullens S, Cooymans J, De Wilde AM, Thijs I, Kemps R. Different methods to synthesize ceramic foams. Journal of the European Ceramic Society 2009;29 829-832.

[532] Carey E, Stubenrauch C. Free drainage of aqueous foams stabilized by mixtures of a non-ionic (C12DMPO) and an ionic (C12TAB) surfactant. Colloids and Surfaces A: Physicochemical and Engineering Aspects 2013;419 7-14.

[533] Malysa K, Lunkenheimer K. Foams under dynamic conditions. Current Opinion in Colloid & Interface Science 2008;13 150-162.

[534] Grassia P, Neethling SJ, Cervantes C, Lee HT. The growth, drainage and bursting of foams. Colloids and Surfaces A: Physicochemical and Engineering Aspects 2006;274110-124.

[535] Kim H, Lee S, Han Y, Park JK. Control of pore size in ceramic foams: Influence of surfactant concentration. Materials Chemistry and Physics 2009;113 441-444.

[536] Wang M, Du H, Guo A, Hao R, Hou Z. Microstructure control in ceramic foams via mixed cationic/anionic surfactant. Materials Letters 2012;88(1) 97-100.

[537] Twigg MV, Richardson JT. Theory and Applications of Ceramic Foam Catalysts. Chemical Engineering Research and Design 2002;80 183-189.

[538] Juettner T, Moertel H, Svinka V, Svinka R. Structure of kaoline–alumina based foam ceramics for high temperature applications. Journal of the European Ceramic Society 2007;27(2-3) 1435-1441.

[539] Lee HT, Neethling SJ, Cilliers JJ. Particle and liquid dispersion in foams. Colloids and Surfaces A: Physicochemical and Engineering Aspects 2005;263 320-329.

[540] Drenckhan W, Langevin D. Monodisperse foams in one to three dimensions. Current Opinion in Colloid & Interface Science 2010;15 341-358.

[541] Kruglyakov PM, Karakashev SI, Nguyen AV, Vilkova NG. Foam drainage. Current Opinion in Colloid & Interface Science 2008;13 163-170.

[542] Britan A, Liverts M, Ben-Dor G, Koehler SA, Bennani N. The effect of fine particles on the drainage and coarsening of foam. Colloids and Surfaces A: Physicochemical and Engineering Aspects 2009;344 15-23.

[543] Kaptay G. Interfacial criteria for stabilization of liquid foams by solid particles. Colloids and Surfaces A: Physicochemical and Engineering Aspects 2003;230 67-80.

[544] Hunter TN, Pugh RJ, Franks GV, Jameson GJ. The role of particles in stabilising foams and emulsions. Advances in Colloid and Interface Science 2008;137 57-81.

[545] Zhang S, Sun D, Dong X, Li C, Xu J. Aqueous foams stabilized with particles and nonionic surfactants. Colloids and Surfaces A: Physicochemical and Engineering Aspects 2008;324 1-8.

[546] Horozov TS. Foams and foam films stabilised by solid particles. Current Opinion in Colloid & Interface Science 2008;13 134-140.

[547] Peng HX, Fan Z, Evans JRG. Factors affecting the microstructure of a fine ceramic foam. Ceramics International 2000;26 887-895.

[548] Yu J, Sun X, Li Q, Li X. Preparation of Al_2O_3 and Al_2O_3–ZrO_2 ceramic foams with adjustable cell structure by centrifugal slip casting. Materials Science and Engineering: A 2008;476 274-280.

[549] Montanaro L, Jorand Y, Fantozzi G, Negro A. Ceramic foams by powder processing. Journal of the European Ceramic Society 1998;18 1339-1350.

[550] Nor MAAM, Hong LCh, Ahmad ZA, Akil HM. Preparation and characterization of ceramic foam produced via polymeric foam replication method. Journal of Materials Processing Technology 2008;207 235-239.

[551] Dressler M, Reinsch S, Schadrack R, Benemann S. Burnout behavior of ceramic coated open cell polyurethane (PU) sponges. Journal of the European Ceramic Society 2009;29 3333-3339.

[552] Akpinar S, Kusoglu IM, Ertugrul O, Onel K. Silicon carbide particle reinforced mullite composite foams. Ceramics International 2012;38 6163-6169.

[553] de Sousa E, Rambo CR, Hotza D, de Oliveira APN, Fey T, Greil P. Microstructure and properties of LZSA glass-ceramic foams. Materials Science and Engineering: A 2008;476 89-97.

[554] Guo X, Zhou Z, Ma G, Wang S, Zhao S, Zhang Q. Effect of forming process on the integrity of pore-gradient Al_2O_3 ceramic foams by gelcasting. Ceramics International 2012;38 713-719.

[555] Bartuli C, Bemporad E, Tulliani JM, Tirillò J, Pulci G, Sebastiani M. Mechanical properties of cellular ceramics obtained by gel casting: Characterization and modeling. Journal of the European Ceramic Society 2009;29 2979-2989.

[556] Potoczek M, Zima A, Paszkiewicz Z, Ślósarczyk A. Manufacturing of highly porous calcium phosphate bioceramics via gel-casting using agarose. Ceramics International 2009;35 2249-2254.

[557] Prabhakaran K, Ananthakumar S, Pavithran C. Gel Casting of Alumina using Boehmite as a Binder. Journal of the European Ceramic Society 1999;19 2875-2881.

[558] Sundermann E, Viedt J. Method of manufacturing ceramic foam bodies. US Pat no. 3 745 201; 1973.

[559] Xu Ch, Wang S, Flodström K, Mao X, Guo J. Cellular silica-based ceramics prepared by direct foaming at high temperature. Ceramics International 2010;6(3) 923-927.

[560] Nambiar EKK, Ramamurthy K. Air void characterisation of foam concrete. Cement and Concrete Research2007;37(2) 221-230.

[561] Ranjani GIS, Ramamurthy K. Behaviour of foam concrete under sulphate environments. Cement and Concrete Composites 2012;34(7) 825-834.

[562] Huang J.-S, Cheng Ch.-K. Fracture toughness variability of foamed alumina cements. Cement and Concrete Research 2004;34(5) 883-888.

[563] Huang J.-S, Lin J.-Y, Jang M-J. Stress relaxation of foamed high-alumina cement paste. Cement and Concrete Research 2005;35(8) 1503-1509.

[564] Jitchaiyaphum K, Sinsiri T, Chindaprasirt P. Cellular Lightweight Concrete Containing Pozzolan Materials. Procedia Engineering 2011;14 1157-1164.

[565] Chen X, Yan Y, Liu Y, Hu Z. Utilization of circulating fluidized bed fly ash for the preparation of foam concret. Construction and Building Materials 2014;54137-146.

[566] Li G, Muthyala VD. Acement based syntactic foam. Materials Science and Engineering: A 2008;478(1-2) 77-86.

[567] Akthar FK, Evans JRG. High porosity (> 90%) cementitious foams. Cement and Concrete Research 2010;40(2) 352-358.

[568] Just A, Middendorf B. Microstructure of high-strength foam concrete. Materials Characterization 2009;60(7) 741-748.

[569] Panesar DK. Cellular concrete properties and the effect of synthetic and protein foaming agents. Construction and Building Materials 2013;44 575-584.

[570] Valore RC. Cellular concrete part 1, Composition and methods of production. published in American Concrete Institute Jornal 1954;50 773-796.

[571] Ramamurthy K, Nambiar EKK, Ranjani GIS. A classification of studies on properties of foam concrete. Cement and Concrete Composites 2009;31(6) 388-396.

[572] Nora KH, Natascha S, Regine VK. Effects of oppositely charged surfactants on the stability of foam films. Colloids and Surfaces A: Physicochemical and Engineering Aspects 2011;382 165-173.

[573] Simjoo M, Rezaei T, Andrianov A, Zitha PLJ. Foam stability in the presence of oil: Effect of surfactant concentration and oil type. Colloids and Surfaces A: Physicochemical and Engineering Aspects 2013;438 148-158.

[574] Stevenson P. Inter-bubble gas diffusion in liquid foam. Current Opinion in Colloid & Interface Science 2010;15(5) 374-381.

[575] Vivaldini DO, Salvini VR, Luz AP, Pandolfelli VC. Road maps for processing foams containing particles. Ceramics International 2013;39(6) 6153-6163.

[576] Haw M. Colloidal suspensions, Brownian motion, molecular reality: a short history, Journal of Physics: Condensed Matter 2002;14 7769-7779.

[577] Ptáček P, Lang K, Šoukal F, Opravil T, Bartoníčková E, Tvrdík L. Preparation and properties of enstatite ceramic foam from talc. Journal of the European Ceramic Society 2014;34(2) 515-522.

[578] Ptáček P, Lang K, Šoukal F, Opravil T, Tvrdík L, Novotný R. Preparation and properties of nanostructured ceramic foam from kaolinite. Powder Technology 2014;253 29-34.

[579] Vaou V, Panias D. Thermal insulating foamy geopolymers from perlite. Minerals Engineering 2010;23(14) 1146-1151.

[580] Manu KM, Joseph T, Sebastian MT. Temperature compensated $Sr_2Al_2SiO_7$ ceramic for microwave applications. Materials Chemistry and Physics 2012;133(1) 21-23.

[581] Dear PS. Isomorphism of åkermanite and strontio-gehlenite. Lithos 1970;3(1) 13-16.

[582] De Aza AH, Turrillas X, Rodriguez MA, Duran T, Pena P. Time-resolved powder neutron diffraction study of the phase transformation sequence of kaolinite to mullite. Journal of the European Ceramic Society 2014;34(5) 1409-1421.

[583] Zhou HM, Qiao XC, Yu JG. Influences of quartz and muscovite on the formation of mullite from kaolinite. Applied Clay Science 2013;80-81 176-181.

[584] Kobayashi Y, Inagaki M. Preparation of reactive Sr-celsian powders by solid-state reaction and their sintering. Journal of the European Ceramic Society 2004;24(2) 399-404.

[585] Ptáček P, Šoukal F, Opravil T, Havlica J, Brandštetr J. Crystallization of spinel phase from metakaoline: The nonisothermal thermodilatometric CRH study. Powder Technology 243;2013 40-45.

[586] Ptáček P, Křečková M, Šoukal F, Opravil T, Havlica J, Brandštetr J. The kinetics and mechanism of kaolin powder sintering I. The dilatometric CRH study of sinter-crystallization of mullite and cristobalite. Powder Technology 2012;232 24-30.

[587] Ptáček P, Šoukal F, Opravil T, Nosková M, Havlica J, Brandštetr J. Mid-infrared spectroscopic study of crystallization of cubic spinel phase from metakaolin. Journal of Solid State Chemistry 2011;184(10) 2661-2667.

[588] Jin X.-H, Gao L, Guo J.-K. The structural change of diphasic mullite gel studied by XRD and IR spectrum analysis. Journal of the European Ceramic Society 2002;22(8) 1307-1311.

[589] Voll D, Angerer P, Beran A, Schneider H. A new assignment of IR vibrational modes in mullite. Vibrational Spectroscopy 2002;30(2) 237-243.

[590] Birchall JD, Howard AJ, Kendall K. Flexural strength and porosity of cements. Nature 1981;289 388-389.

[591] Drabik M, Mojumdar SC, Slade RCT. Prospects of novel macro-defect-free cements for the new millenium. Ceramics – Silikaty 2001;46(2) 68-73.

[592] Donatello S, Tyrer M, Cheeseman CR. Recent developments in macro-defect-free (MDF) cements. Construction and Building Materials 2009;23(5) 1761-1767.

[593] Šoukal F, Másilko J, Havlica J, Ptáček P, Opravil T. Temperature and moisture effect of macrodefect-free composite structure and properties. Chemické Listy 2008;102 s265–s1309.

[594] Alfani R, Colombet P, D´amore A, Rizzo N, Nicolais L. Effect of temperature on thermo-mechanical properties of macro-defect-free cement–polymer composite. Journal of Materials Science1999;34 5683-5687.

[595] Pushpalal GKD, Kobayashi T, Hasegawa M. High alumina cement-phenol resin composite: water resistivity and effect of post hydration of unreacted cement on durability. Cement and Concrete Research 1997;27(9) 1393-1405.

[596] Pushpalal GKD, Kobayashi T, Kawano T, Maeda N. The processing, properties, and applications of calcium aluminate–phenol resin composite. Cement and Concrete Research 1999;29(1) 121-132.

[597] McHugh AJ, Walberer JA. Rheology and structuring in organo-ceramic composites. Composites Part A: Applied Science and Manufacturing 2001;32(8) 1085-1093.

[598] Pushpalal GKD, Kawano T, Kobayashi T, Hasegawa M. Chemical Characterization of Calcium Aluminate-Phenol Resin Composite. Advanced Cement Based Materials 1997;6(2) 45-52.

[599] Hewlett PC, Lea´s Chemistry of Cement and Concrete. Fourth edition; Elsevier Ltd.; 1988. ISBN: 978-0-7506-6256-7.

[600] Wu Z, Naik TR. Properties of concrete produced from multicomponent blended cements. Cement and Concrete Research 2002;32(12) 1937-1942.

[601] Antiohos SK, Papadakis VG, Chaniotakis E, Tsimas S. Improving the performance of ternary blended cements by mixing different types of fly ashes. Cement and Concrete Research 2007;37(6) 877-885.

[602] Pal SC, Mukherjee A, Pathak SR. Investigation of hydraulic activity of ground granulated blast furnace slag in concrete. Cement and Concrete Research 2003;33(9) 1481-1486.

[603] Mostafa NY, El-Hemaly SAS, Al-Wakeel EI, El-Korashy SA, Brown PW. Characterization and evaluation of the hydraulic activity of water-cooled slag and air-cooled slag. Cement and Concrete Research 2001;31(6) 899-904.

[604] Mostafa NY, El-Hemaly SAS, Al-Wakeel EI, El-Korashy SA, Brown PW. Hydraulic activity of water-cooled slag and air-cooled slag at different temperatures. Cement and Concrete Research 2001;31(3) 475-484.

[605] Singh NB, Bhattacharjee KN, Shukla AK. Effect of alkali bypass dust on the hydration of granulated blast furnace slag blended cement. Cement and Concrete Research 1995;25(4) 883-892.

[606] Bijen J. Benefits of slag and fly ash. Construction and Building Materials 1996;10(5) 309-314.

[607] Erdem TK, Meral Ç, Tokyay M, Erdoğan TY. Use of perlite as a pozzolanic addition in producing blended cements. Cement and Concrete Composites 2007;29(1) 13-21.

[608] Uzal B, Turanli L. Studies on blended cements containing a high volume of natural pozzolans. Cement and Concrete Research 2003;33(11) 1777-1781.

[609] Hossain KMA. Blended cement using volcanic ash and pumice. Cement and Concrete Research 2003;33(10) 1601-1605.

[610] Rasheeduzzafar, Hussain ES, Al-Gahtani AS. Pore solution composition and reinforcement corrosion characteristics of microsilica blended cement concrete. Cement and Concrete Research 1991;21(6) 1035-1048.

[611] Malhotra VM, Hammings RT. Blended cements in North America - A review. Cement and Concrete Composites 1995;17(1) 23-35.

[612] Ramachandran VS, Paroli RM, Beaudoin JJ, Delgado AH. Handbook of Thermal Analysis of Construction Materials. William Andrew; 2002. ISBN: 0-8155-1487-5

[613] Rivas-Mercury JM, Pena P, de Aza AH, Turrillas X. Dehydration of $Ca_3Al_2(SiO_4)_y(OH)_{4(3-y)}$ ($0 < y < 0.176$) studied by neutron thermodiffractometry. Journal of the European Ceramic Society 2008;28(9) 1737-1748.

[614] Maitra S, Bose S, Bandyopadhyay N, Roychoudhury A. Dehydration kinetics of calcium aluminate cement hydrate under non-isothermal conditions. Ceramics International 2005;31(3) 371-374.

[615] Carlson ET. A study of some strontium aluminate and calcium - strontium aluminate solid solutions, Journal of research of the National Bureau of Standards 1955;54 2334-2595.

[616] Desmoulins H, Malo S, Boudin S, Caignaert V, Hervieu M. Polymorphism of the iron doped strontium aluminate $SrAl_{1.5}Fe_{0.5}O_4$. Journal of Solid State Chemistry 2009;182(7) 1806-1820.

[617] Macphee DE, Lachowski EE. Cement Components and Their Phase Relations. In Lea ′s Chemistry of Cement and Concrete; ed. Hewltett PC; 1998. ISBN: 978-0-7506-6256-7.

[618] Stöber S, Redhammer G, Schorr S, Prokhnenko O, Pöllmann H. Structure refinements of members in the brownmillerite solid solution series $Ca_2Al_x(Fe0.5Mn_{0.5})_{2-}xO_{5+\delta}$ with $1/2 \leq x \leq 4/3$. Journal of Solid State Chemistry 2013;197420-428.

[619] Sullivan E, Greaves C. Fluorine insertion reactions of the brownmillerite materials $Sr_2Fe_2O_5$, Sr_2CoFeO_5, and $Sr_2Co_2O_5$. Materials Research Bulletin 2012;47(9) 2541-2546.

[620] Yang Y, Cao Z, Jiang Y, Liu L, Sun Y. Photoinduced structural transformation of $SrFeO_3$ and $Ca_2Fe_2O_5$ during photodegradation of methyl orange. Materials Science and Engineering: B 2006;132(3) 311-314.

[621] Mizusaki J, Okayasu M, Yamauchi S, Fueki K. Nonstoichiometry and phase relationship of the $SrFeO_{2.5}SrFeO_3$ system at high temperature. Journal of Solid State Chemistry 1992;99(1) 166-172.

[622] Auckett JE, Studer AJ, Sharma N, Ling ChD. Floating-zone growth of brownmillerite $Sr_2Fe_2O_5$ and the observation of a chain-ordered superstructure by single-crystal neutron diffraction. Solid State Ionics 2012;225(4) 432-436.

[623] Kendall KR, Navas C, Thomas JK, zurLoye HC. Structural and Magnetic Studies of $Bi_2Fe_{4-x}Al_xO_9$. Solid State Ionics 1995;82 215-223.

[624] Casey PS, Barker D, Hayward MA. Charge and structural ordering in the brownmillerite phases: $La_{1-x}Sr_xMnO_{2.5}$ ($0.2<x<0.4$). Journal of Solid State Chemistry 2006;179(5) 1375-1382.

[625] Greaves C, Jacobson AJ, Tofield BC, Fender BEF. A powder neutron diffraction investigation of the nuclear and magnetic structure of $Sr_2Fe_2O_5$. Acta Crystallographica B 1975;31 641-646.

[626] Schmidt M, Campbell SJ. Crystal and Magnetic Structures of $Sr_2Fe_2O_5$ at Elevated Temperature. Journal of Solid State Chemistry 156 (2001) 292-304.

[627] D'Hondt H, Abakumov AM, Hadermann J, Kalyuzhnaya AS, Rozova MG, Antipov EV, Van Tendeloo G. Tetrahedral Chain Order in the $Sr_2Fe_2O_5$ Brownmillerite. Chemistry of Materials 2008;20 7188-7194.

[628] Mahboub MS, Zeroual S, Boudjada A. Synthesis of homogeneous $Ca_{0.5}Sr_{0.5}FeO_{2.5+\delta}$ compound using a mirror furnace method. Materials Research Bulletin 2012;47(2) 370-374.

[629] Prado F, Grunbaum N, Caneiro A, Manthiram A. Effect of La^{3+} doping on the perovskite-to-brownmillerite transformation in $Sr_{1-x}La_xCo_{0.8}Fe_{0.2}O_{3-\delta}$ ($0\leq x\leq 0.4$). Solid State Ionics 167(2004) 147-154.

[630] Andreas Heyden, Shuguo Ma, Fanglin Chen. Synthesis and characterization of Mo-doped SrFeO3−δ as cathode materials for solid oxide fuel cells. Journal of Power Sources 2012;202 63-69.

[631] Roth G, Adelmann P, Knitter R, Massing S, Wolf Th. The crystal structure of RESrGaCuO$_5$ (RE: La, Pr, Nd), a gallate-cuprate with strong structural similarities to superconducting cuprates. Journal of Solid State Chemistry 1992;99 376-387.

[632] Hadermann J, van Tendeloo G, Abakumov AM, Pavlyuk BPh, Rozova M, Antipov EV. Structural transformation in fluorinated $LaACuGaO_5$ (A = Ca, Sr) brownmilleritesInternational Journal of Inorganic Materials2000;2493-502.

[633] Luo K, Hayward MA. The synthesis and characterisation of $LaCa_2Fe_2GaO_8$. Journal of Solid State Chemistry 2013;198203-209.

[634] Wright AJ, Palmer HM, Anderson PA, Greaves C. Synthesis and structure of Sr_2MnGaO_5, a new layered manganese oxide. Journal Of Materials Chemistry 2001;11(5) 1324-1326.

[635] Abakumov AM, Rozova MG, Ph Pavlyuk B, Lobanov MV, Antipov EV, Lebedev OI, van Tendeloo G, Ignatchik OL, Ovtchenkov EA, Koksharov YA, Vasil'ev AN. Syn-

thesis, Crystal Structure, and Magnetic Properties of a Novel Layered Manganese Oxide $Sr_2MnGaO_{5+\delta}$. Journal of Solid State Chemistry 2001;160 353-361.

[636] Abakumov AM, Alekseeva AM, Rozova MG, Antipov EV, Lebedev OI, Van Tendeloo G. Ordering of tetrahedral chains in the Sr_2MnGaO_5 brownmillerite. Journal of Solid State Chemistry 2003;174(2) 319-328.

[637] Antipov EV, Lebedev OI, Van Tendeloo G. Synthesis and structure of $Sr_2MnGaO_{5+\delta}$ brownmillerites with variable oxygen content. Solid State Sciences 2003;5(6) 871-882.

[638] D'Hondt H, Hadermann J, Abakumov AM, Kalyuzhnaya AS, Rozova MG, Tsirlin AA, Nath R, Tan H, Verbeeck J, Antipov EV, Tendeloo GV. Synthesis, crystal structure and magnetic properties of the $Sr_2Al_{0.78}Mn_{1.22}O_{5.2}$ anion-deficient layered perovskite. Journal of Solid State Chemistry 2009;182(2) 356-363.

[639] Zhao YM, Zhou PF. Metal–insulator transition in helical $SrFeO_{3-\delta}$ antiferromagnet. Journal of Magnetism and Magnetic Materials 2004;281(2-3)214-220.

[640] Augustin CO, Berchmans LJ, Selvan RK. Structural, electrical and electrochemical properties of co-precipitated $SrFeO_{3-\delta}$. Materials Letters 2004;58(7-8) 1260-1266.

[641] Leonidov IA, Patrakeev MV, Bahteeva JA, Mitberg EB, Kozhevnikov VL, Colomban P, Poeppelmeier KR. High-temperature phase equilibria in the oxide systems $SrFe_{1-x}Ga_xO_{2.5}$-$SrFe_{1-x}Ga_xO_3$ (x = 0, 0.1, 0.2). Journal of Solid State Chemistry 2006;179(4) 1093-1099.

[642] Shaula AL, Pivak YV, Waerenborgh JC, Gaczyñski P, Yaremchenko AA, Kharton VV.Ionic conductivity of brownmillerite-type calcium ferrite under oxidizing conditions. Solid State Ionics 2006;177(33-34) 2923-2930.

[643] Zhang GB, Smyth DM. Protonic conduction in $Ba_2In_2O_5$. Solid State Ionics 1995;82(3-4)153-160.

[644] Schober T, Friedrich J, Krug F. Phase transition in the oxygen and proton conductor $Ba_2In_2O_5$ in humid atmospheres below 300°C.Solid State Ionics 1997;999-13.

[645] Schober T, Friedrich J. The oxygen and proton conductor $Ba_2In_2O_5$: Thermogravimetry of proton uptake. Solid State Ionics 1998;113-115 369-375.

[646] Fischer W, Reckand G, Schober T. Structural transformation of the oxygen and proton conductor $Ba_2In_2O_5$ in humid air: an in-situ X-ray powder diffraction study. Solid State Ionics 1999;116(3-4) 211-215.

[647] Omata T, Kita M, Goto Y, Okura T, Otsuka-Yao-Matsuo S. Formation and Thermal Stability of Hydrate-Like Compounds of$Ba_2(In_{1-x}M^{III}{}_x)_2O_5$ $n$$H_2O$ (M^{III} = Ga, Sc, Lu, and Y).Journal of The Electrochemical Society 2005;152(6) A1068-1072.

[648] Omata T, Fuke T, Otsuka-Yao-Matsuo S. Hydration behavior of $Ba_2Sc_2O_5$ with an oxygen-deficient perovskite structure. Solid State Ionics 2006;177(26-32) 2447-2451.

[649] Muñoz R, Masó N, Julián B, Márquez F, Beltrán H, Escribano P, Cordoncillo E. Environmental study of Cr_2O_3-Al_2O_3 green ceramic pigment synthesis. Journal of the European Ceramic Society 2004;24(7) 2087-2094.

[650] Shibata K, Yoshinaka M, Hirota K, Yamaguchi O. Fabrication and mechanical properties of Cr_2O_3 solid solution ceramics in the system Cr_2O_3-Al_2O_3. Materials Research Bulletin 1997;32(5) 627-632.

[651] Cho S.-A, Arenas FJ, Ochoa J. Densification and hardness of Al_2O_3-Cr_2O_3 system with and without Ti addition. Ceramics International 1990;16(5) 301-309.

[652] Riu D.-H, Kong Y.-M, Kim H.-E. Effect of Cr_2O_3 addition on microstructural evolution and mechanical properties of Al_2O_3. Journal of the European Ceramic Society 2000;20(10) 1475-1481.

[653] Hirata T, Akiyama K, Yamamoto H. Sintering behavior of Cr_2O_3-Al_2O_3 ceramics. Journal of the European Ceramic Society 2000;20(2) 195-199.

[654] Kitaoka Y, Nakamura K, Akiyama T, Ito T. Structural stability and electronic properties in Al_2O_3-Cr_2O_3 mixed crystal. Journal of Crystal Growth 2013;362(1) 42-44.

[655] Mitra NK, Maitra S, Gnanabharathi D, Parya TK, Dey R. Effect of Cr_2O_3 on the sintering of aluminosilicate precursor leading to mullite formation. Ceramics International 2001;27(3) 277-282.

[656] Sarkar R, Das SK, Banerjee G. Effect of addition of Cr_2O_3 on the properties of reaction sintered MgO-Al_2O_3 spinels. Journal of the European Ceramic Society 2002;22(8) 1243-1250.

[657] Pakhomov NA, Kashkin VN, Nemykina EI. Molchanov VV, Nadtochiy VI, Noskov AS. Dehydrogenation of C3–C4 paraffins on Cr_2O_3/Al_2O_3 catalysts in fluidized and fixed bed reactors. Chemical Engineering Journal 2009;154(1-3) 185-188.

[658] Gascón J, Téllez C, Herguido J, Menéndez M. Propane dehydrogenation over a Cr_2O_3/Al_2O_3 catalyst: transient kinetic modeling of propene and coke formation. Applied Catalysis A: General 2003;248(1-2) 105-116.

[659] Shee D, Sayari A. Light alkane dehydrogenation over mesoporous Cr_2O_3/Al_2O_3 catalysts. Applied Catalysis A: General 2010;389(1-2) 155-164.

[660] Tang Y, Elzinga EJ, Lee YJ, Reeder RJ. Coprecipitation of chromate with calcite: Batch experiments and X-ray absorption spectroscopy. Geochimica et Cosmochimica Acta 2007;71(6) 1480-1493.

[661] Areán CO, Carbonell CM, Platero EE, Macía EM, Zecchina A, Geobaldo F. Thermolysis of mixed alums: a route to mesoporous chromia-alumina. Materials Chemistry and Physics 1993;34(3-4) 214-218.

[662] Peters D, Hummel FA. Phase studies in the systems CaO Al_2O_3 $CaCrO_4$ and SrO Al_2O_3 $SrCrO_4$. Cement and Concrete Research 1979;9(2) 259-268.

[663] Chang Y.-L, Hsiang H.-I, Liang M.-T. Characterizations of Eu, Dy co-doped $SrAl_2O_4$ phosphors prepared by the solid-state reaction with B_2O_3 addition. Journal of Alloys and Compounds 2008;461(1-2) 598-603.

[664] Neumair SC, Kaindl R, Huppertz H. The new high-pressure borate $Co_7B_{24}O_{42}(OH)_2 \cdot 2 H_2O$ - Formation of edge-sharing BO_4 tetrahedra in a hydrated borate. Journal of Solid State Chemistry 2012;185 1-9.

[665] Goh K.-H, Lim T.-T, Dong Z. Application of layered double hydroxides for removal of oxyanions: A review. Water Research 2008;42(6-7) 1343-1368.

[666] Liang X, Zang Y, Xu Y, Tan X, Hou W, Wang L, Sun Y. Sorption of metal cations on layered double hydroxides. Colloids and Surfaces A: Physicochemical and Engineering Aspects 2013;433 122-131.

[667] Yang K, Yan L.-g, Yang Y.-m, Yu Sh.-j, Shan R.-r, Yu H.-q, Zhu B.-c, Du B. Adsorptive removal of phosphate by Mg–Al and Zn–Al layered double hydroxides: Kinetics, isotherms and mechanisms. Separation and Purification Technology 2014;12436-42.

[668] Heredia AC, Oliva MI, Agú U, Zandalazini CI, Marchetti SG, Herrero ER, Crivello ME. Synthesis, characterization and magnetic behavior of Mg–Fe–Al mixed oxides based on layered double hydroxide. Journal of Magnetism and Magnetic Materials 2013;342 38-46.

[669] Cui G, Evans DG, Li D. Synthesis and UV absorption properties of 5, 5′-thiodisalicylic acid intercalated Zn-Al layered double hydroxides. Polymer Degradation and Stability 2010;95(10) 2082-2087.

[670] Pérez MR, Barriga C, Fernández JM, Rives V, Ulibarri MA. Synthesis of Cd/(Al+Fe) layered double hydroxides and characterization of the calcination products. Journal of Solid State Chemistry 2007;180(12) 3434-3442.

[671] Mishra G, Dash B, Pandey S, Mohanty PP. Antibacterial actions of silver nanoparticles incorporated Zn-Al layered double hydroxide and its spinel. Journal of Environmental Chemical Engineering 2013;1(4) 1124-1130.

[672] Zhang X, Li Sh. Mechanochemical approach for synthesis of layered double hydroxides. Applied Surface Science 2013;274 158-163.

[673] Wittmann FH. Creep and shrinkage mechanisms. Part II in creep and shrinkage. In: Concrete structures; Ed. Bazant ZP, Wittmann FH. John Wiley & Sons; 1982.

[674] Zhang T, Gao P, Luo R, Guo Y, Wei J, Yu Q. Measurement of chemical shrinkage of cement paste: Comparison study of ASTM C 1608 and an improved method. Construction and Building Materials 2013;48 662-669.

[675] Bouasker M, Mounanga P, Turcry P, Loukili A, Khelidj A. Chemical shrinkage of cement pastes and mortars at very early age: Effect of limestone filler and granular inclusions. Cement and Concrete Composites 2008;30(1) 13-22.

[676] Barcelo L, Moranville M, Clavaud B. Autogenous shrinkage of concrete: a balance between autogenous swelling and self-desiccation. Cement and Concrete Research 2005;35(1) 177-183.

[677] Tazawa E.-i, Miyazawa Sh. Influence of cement and admixture on autogenous shrinkage of cement paste. Cement and Concrete Research 1995;25(2) 281-287.

[678] ASTM C 1608. Standard test method for chemical shrinkage of hydraulic cement paste. American society for testing and materials. West Conshohocken (PA): ASTM International; 2007.

[679] Li Y, Bao J, Guo Y. The relationship between autogenous shrinkage and pore structure of cement paste with mineral admixtures. Construction and Building Materials 2010;24(10) 1855-1860.

[680] Geiker M, Knudsen T. Chemical shrinkage of portland cement pastes. Cement and Concrete Research 1982;12(5) 603-610.

[681] Tongaroonsri S, Tangtermsirikul S. Effect of mineral admixtures and curing periods on shrinkage and cracking age under restrained condition. Construction and Building Materials 2009;23(2) 1050-1056.

[682] Bissonnette B, Pierre P, Pigeon M. Influence of key parameters on drying shrinkage of cementitious materials. Cement and Concrete Research 1999;29(10) 1655–1662.

[683] Mo L, Deng M, Tang M. Effects of calcination condition on expansion property of MgO-type expansive agent used in cement-based materials. Cement and Concrete Research 2010;40(3) 437-446.

[684] Mo L, Deng M, Wang A. Effects of MgO-based expansive additive on compensating the shrinkage of cement paste under non-wet curing conditions. Cement and Concrete Composites 2012;34(3) 377-383.

[685] Zheng L, Xuehua C, Mingshu T. Hydration and setting time of MgO-type expansive cement Original. Cement and Concrete Research 1992;22(1) 1-5.

[686] Zheng L, Xuehua C, Mingshu T. MgO-type delayed expansive cement. Cement and Concrete Research 1991;21(6) 1049-1057.

[687] Gao P, Lu X, Geng F, Li X, Hou J, Lin H, Shi N. Production of MgO-type expansive agent in dam concrete by use of industrial by-products. Building and Environment 2008;43(4) 453-457.

[688] Appah D, Reichetseder P. Practical improvements in CaO-swelling cements. Journal of Petroleum Science and Engineering 2002;36(1-2) 61-70.

[689] Opravil T, Ptáček P, Šoukal F, Havlica J, Brandštetr J. The synthesis and characterization of an expansive admixture for M-type cements I. The influence of free CaO to the formation of ettringite. Journal of Thermal Analysis and Calorimetry 2013;111(1) 517-526.

[690] Mo L, Deng M, Tang M, Al-Tabbaa A. MgO expansive cement and concrete in China: Past, present and future. Cement and Concrete Research 2014;57 1-12.

[691] Pavoine A, Brunetaud X, Divet L. The impact of cement parameters on Delayed Ettringite Formation. Cement and Concrete Composites 2012;34(4) 521-528.

[692] ACI Committee 223: Expansive cement concretes - Present state of knowledge. ACI Journal (1970) 583-610.

[693] Cohen MD. Theories of expansion in sulfoaluminate-type expansive cements: schools of thought. Cements and Concrete Research 1983;13 809-818.

[694] Kasselouri V, Tsakiridis P, Malami Ch, Georgali B, Alexandridou C. A study on the hydration products of a non-expansive sulfoaluminate cement. Cement and Concrete Research 1995;25 1726-1736.

[695] Lu Y, Su M, Wang Y. Microstructural study of the interfacial zone between expansive sulphoaluminate cement pastes and limestone aggregates. Cement and Concrete Research 1996;26805-812.

[696] Havlica J, Sahu S. Formation and expansion of ettringite crystals. Cement and Concrete Research 1992;22(4) 671-677.

[697] Nagataki S, Gomi H. Expansive admixtures (mainly ettringite). Cement and Concrete Composites 1998;20(2-3) 163-170.

[698] Brown PW, LaCroix P. The kinetics of ettringite formation. Cement and Concrete Research 1989;19 879-884.

[699] Evju C, Hansen S. The kinetics of ettringite formation and dilatation in a blended cement with β-hemihydrate and anhydrite as calcium sulfate. Cement and Concrete Research 2005;35 2310-2321.

[700] Palou M, Majling J, Dovál◎ M, Kozanková J, Mojmudar SCh. Formation and stability of crystallohydrates in the non-equilibrium system during hydration of SAB cements. Ceramics – Silikáty 2005;49 230-236.

[701] Gabrišová A, Havlica J, Sahu S. Stability of calcium sulphoaluminate hydrates in water solutions with various pH values Cement and Concrete Research 1991;21(1) 1023-1027.

[702] Palou MT, Majling J. Effect of sulphate, calcium and aluminium ions upon the hydration of sulphoaluminate belite cement. Journal of Thermal Analysis 1996;46 549-556.

[703] Palou MT, Majling J. Hydration in the system C_4AS_3-CSH_2-CH-H. Journal of Thermal Analysis 1996;46557-563.

[704] Gaines RV, Skinner HCW, Foord EF, Mason B, Rosenzweig A. Dana's New Mineralogy; 8th ed.; John Wiley; New York; 1998.

[705] Zhou Q, Lachowski EE, Glasser FP. Metaettringite, a decomposition product of ettringite. Cement and Concrete Research 2004;34(4) 703-710.

[706] Gougar MLD, Scheetz BE, Roy DM. Ettringite and C-S-H Portland cement phases for waste ion immobilization: A review. Waste Management 1996;16(4) 295-303.

[707] Frost RL, López A, Xi Y, Scholz R, da Costa GM, Lima RMF, Granja A. The spectroscopic characterization of the sulphate mineral ettringite from Kuruman manganese deposits, South Africa. Vibrational Spectroscopy 2013;68 266-271.

[708] Möschner G, Lothenbach B, Winnefeld F, Ulrich A, Figi R, Kretzschmar R. Solid solution between Al-ettringite and Fe-ettringite ($Ca_6[Al_{1-x}Fe_{x(}OH)_6]_2(SO_4)_3 \cdot 26H_2O$). Cement and Concrete Research 2009;39(6) 482-489.

[709] Clark SM, Colas B, Kunz M, Speziale S. Monteiro PJM. Effect of pressure on the crystal structure of ettringite. Cement and Concrete Research 2008;38(1) 19-26.

[710] Kumarathasan P, McCarthy GJ, Hassett DJ, Pflughoeft-Hasset DF. Oxyanion substituted ettringites: synthesis and characterization; and their potential role in immobilization of As, B, Cr, Se and V. Materials Research Society Symposium Proceedings 1990;178 83-104.

[711] Poellmann H, St. Auer H-J, Wenda KR. Solid solution of ettringites: Part II: Incorporation of $B(OH)_4-$ and CrO_4^{2-} in $3CaO \cdot Al_2O_3 \cdot 3CaSO_4 \cdot 32H_2O$. Cement and Concrete Research 1993;23(2) 422-430.

[712] Baur I, Johnson CA. The solubility of selenate-AFt ($3CaO \cdot Al_2O_3 \cdot 3CaSeO_4 \cdot 37.5H_2O$) and selenate-AFm ($3CaO \cdot Al_2O_3 \cdot CaSeO_4 \cdot xH_2O$). Cement and Concrete Research 2003;33(11) 1741-1748.

[713] Kadiyski M, Armbruster T, Günther D, Reusser E, Peretti A. Johachidolite, $CaAl[B_3O_7]$, a mineralogical and structural peculiarity. European Journal of Mineralogy 2008, 20, 965-973.

[714] Champenois J.-B, Mesbah A, Coumes CCD, Renaudin G, Leroux F, Mercier C, Revel B, Damidot D. Crystal structures of Boro-AFm and sBoro-AFt phases. Cement and Concrete Research 2012;42(10) 1362-1370.

[715] Perkins RB, Palmer CD. Solubility of $Ca_6[Al(OH)_6]_2(CrO_4)_3 \cdot 26H_2O$, the chromate analog of ettringite at 5–75 °C.Applied Geochemistry 2000;15 1203-1218.

[716] Sharp JH, Milestone NB, Hill J, Miller EW. Cementitious systems for encapsulation of intermediate level waste; in: The 9th International Conference on Radioactive Waste Management and Environmental Remediation; Oxford; UK; 2003.

[717] Hassett DJ, McCarthy GJ, Kumarathasan P, Pflughoeft-Hassett D. Synthesis and characterization of selenate and sulfate-selenate ettringite structure phases. Materials Research Bulletin 1990;25(11) 1347-1354.

[718] OECD, Nuclear Energy Agency. Chemical Thermodynamics Chemical Thermodynamics of Solid Solutions of Interest in Radioactive Waste Management. Chemical Thermodynamics. OECD Publishing; 2007. ISBN: 9789264033191.

[719] Barnett SJ, Adam CD, Jackson ARW. An XRPD profile fitting investigation of the solid solution between ettringite, $Ca_6Al_2(SO_4)_3(OH)_{12}{\cdot}26H_2O$, and carbonate ettringite, $Ca_6Al_2(CO_3)_3(OH)_{12}{\cdot}26H_2O$. Cement and Concrete Research 2001;31(1) 13-17.

[720] Hampson CJ, Bailey JE. On the structure of some precipitated calcium alumino-sulfate hydrates Journal of Materials Science 1982;17 3341-3346.

[721] Damidot D, Glasser FP. Thermodynamic investigation of the CaO-Al_2O_3-$CaSO_4$-H_2O system at 25°C and the influence of Na_2O. Cement and Concrete Research 1993;23(1) 221-238.

[722] Damidot D, Glasser FP. Thermodynamic investigation of the CaO-Al_2O_3-$CaSO_4H_2O$ system at 50°C and 85°C. Cement and Concrete Research 1992;22(6) 1179-1191.

[723] Damidot D, Glasser FP. Thermodynamic investigation of the CaO-Al_2O_3-$CaSO_4$-K_2O-H_2O system at 25°C. Cement and Concrete Research 1993;23(5) 1195-1204.

[724] Pajares I, Martínez-Ramírez S, Blanco-Varela MT. Evolution of ettringite in presence of carbonate, and silicate ions. Cement and Concrete Composites 2003;25(8) 861-865.

[725] Lachowski EE, Barnett SJ, Macphee DE. Transmission electron optical study of ettringite and thaumasite. Cement and Concrete Composites 2003;25(8) 819-822.

[726] Ramezanianpour AM, Hooton RD. Thaumasite sulfate attack in Portland and Portland-limestone cement mortars exposed to sulfate solution. Construction and Building Materials 2013;40 162-173.

[727] Skaropoulou A, Kakali G, Tsivilis S. Thaumasite form of sulfate attack in limestone cement concrete: The effect of cement composition, sand type and exposure temperature. Construction and Building Materials 2012;36 527-533.

[728] Schmidt T, Lothenbach B, Romer M, Scrivener K, Rentsch D, Figi R. A thermodynamic and experimental study of the conditions of thaumasite formation. Cement and Concrete Research 2008;38(3) 337-349.

[729] Thomas MDA, Rogers CA, Bleszynski RF. Occurrences of thaumasite in laboratory and field concrete. Cement and Concrete Composites 2003;25(8) 1045-1050.

[730] Brown PW. Thaumasite formation and other forms of sulfate attack. Cement and Concrete Composites 2002;24(3-4) 301-303.

[731] Carlson ET, Berman HA. Some observation on the calcium aluminate carbonate hydrates. Journal of Research, National Bureau of Standards 1960;64A(4) 333-341.

[732] Chrysochoou M, Dermatas D. Evaluation of ettringite and hydrocalumite formation for heavy metal immobilization: Literature review and experimental study. Journal of Hazardous Materials 2006;136(1) 20-33.

[733] Hall C, Barnes P, Billimore AD, Jupe AC, Turrillas X. Thermal decomposition of ettringite $Ca_6[Al(OH)_6]_2(SO_4)_3 \cdot 26H_2O$. Journal of the Chemical Society, Faraday Transactions 1996;92 2125-2129.

[734] Vladu CM, Hall Ch, Maitland GC. Flow properties of freshly prepared ettringite suspensions in water at 25 °C. Journal of Colloid and Interface Science 2006;294(2) 466-472.

[735] Álvarez-Ayuso E, Nugteren HW. Synthesis of ettringite: a way to deal with the acid wastewaters of aluminium anodising industry. Water Research 2005;39(1) 65-72.

[736] Perkins RB, Palmer CD. Solubility of chromate hydrocalumite ($3CaO \cdot Al_2O_3 \cdot CaCrO_4 \cdot nH_2O$) 5-75°C. Cement and Concrete Research 2001;31(7) 983-992.

[737] Bothe JV Jr, Brown PW. Phase formation in the system CaO-Al_2O_3-B_2O_3-H_2O at 23±1°C. Journal of Hazardous Materials 1998;63(2-3) 199-210.

[738] Taylor HFW, Famy C, Scrivener KL. Delayed ettringite formation. Cement and Concrete Research 2001;31(5) 683-693.

[739] Collepardi M. A state-of-the-art review on delayed ettringite attack on concrete. Cement and Concrete Composites 2003;25(4-5) 401-407.

[740] Diamond S. Delayed ettringite formation - Processes and problems. Cement and Concrete Composites 1996;18(3) 205-215.

[741] Odler I. Ettringite nomenclature. Cement and Concrete Research 1997;27(3) 473-474.

[742] Ouhadi VR, Yong RN. Ettringite formation and behaviour in clayey soils. Applied Clay Science 2008;42(1-2) 258-265.

[743] Aldaood A, Bouasker M, Al-Mukhtar M. Geotechnical properties of lime-treated gypseous soils. Applied Clay Science 2014;88-8939-48.

[744] Vlachou P.-V, Piau J.-M. Physicochemical study of the hydration process of an oil well cement slurry before setting. Cement and Concrete Research 1999;29(1) 27-36.

[745] García-Maté M, De la Torre AG, León-Reina L, Aranda MAG, Santacruz I. Hydration studies of calcium sulfoaluminate cements blended with fly ash. Cement and Concrete Research 2013;54 12-20.

[746] Narayanan PS, Lakshmanan BR. Infrared and raman spectra fo witherite ($BaCO_3$) and strontianite ($SrCO_3$). Journal of the Indian Institute of Science 1957;40(1) 1-12.

[747] Pasierb P, Komornicki S, Rokita M, Rekas M. Structural properties of Li_2CO_3-$BaCO_3$ system derived from IR and Raman spectroscopy. Journal of Molecular Structure 2001;596 151-156.

[748] Turianicová E, Obut A, Zorkovská A, Baláž P, Matik M, Briančin J. The effects of LiOH and NaOH on the carbonation of $SrSO_4$ by dry high-energy milling. Minerals Engineering 2013;49 98-102.

[749] Tang T.-P, Lee Ch.-M, Yen F.-Ch. The photoluminescence of $SrAl_2O_4$:Sm phosphors. Ceramics International 2006;32(6) 665-671.

[750] Derjaguin B, Landau L. Theory of the stability of strongly charged lyophobic sols and of the adhesion of strongly charged particles in solutions of electrolytes. Acta Physico Chemica URSS 1941;14 633-662.

[751] Pechini MP. US Patent No. 3330697; 1967.

[752] Zaki T, Kabel KI, Hassan H. Preparation of high pure α-Al_2O_3 nanoparticles at low temperatures using Pechini method. Ceramics International 2012;38(3) 2021-2026.

[753] Escribano P, Marchal M, Sanjuán ML, Alonso-Gutiérrez P, Julián B, Cordoncillo E. Low-temperature synthesis of $SrAl_2O_4$ by a modified sol-gel route: XRD and Raman characterization. Journal of Solid State Chemistry 2005;178(6) 1978-1987.

[754] Xu Y, Peng W, Wang Sh, Xiang X, Lu P. Synthesis of $SrAl_2O_4$ and $SrAl_{12}O_{19}$ via ethylenediaminetetraacetic acid precursor. Materials Chemistry and Physics 2006;98(1) 51-54.

[755] Peng T, Huajun L, Yang H, Yan Ch. Synthesis of $SrAl_2O_4$:Eu, Dy phosphor nanometer powders by sol-gel processes and its optical properties. Materials Chemistry and Physics 2004;85 68-72.

[756] Xiang Ying Chen, Zhao Li, Shi Ping Bao, Ping Ting Ji. Porous MAl_2O_4:Eu^{2+} (Eu^{3+}), Dy^{3+} (M=Sr, Ca, Ba) phosphors prepared by Pechini-type sol-gel method: The effects of solvents. Optical Materials 34 (2011) 48-55.

[757] Qiu X, Xu Y, Qiao X. Synthesis of $SrAl_2O_4$ from a mixed-metal citrate precursors. Materials Letters 2007;61 2731-2734.

[758] Misevicius M, Scit O, Grigoraviciute-Puroniene I, Degutis G, Bogdanoviciene I, Karciva A. Sol-gel synthesis and investigation of un-doped and Ce-doped strontium aluminates. Ceramics International 2012;385915-5924.

[759] Lu Y, Li Y, Xiong Y, Wang D, Yin Q. $SrAl_2O_4$: Eu^{2+}, Dy^{3+} phosphors derived from a new sol-gel route. Microelectronics Journal 2004;35 379-382.

[760] Chen I.-Ch, Chen K.-K, Chen H.-S, Du J.-k, Lin T.-J, Lin S.-Sh, Chen T.-M, Shieh T.-Y. Investigation into thermoluminescence and afterglow characterization of strontium aluminates with boron-modification and reductions via sol-gel route. Journal of Rare Earths 2012;30 972-978.

[761] Tang Z, Zhang F, Zhang Z, Huang Ch, Lin Y. Luminescent properties of $SrAl_2O_4$: Eu, Dy material prepared by the gel method. Journal of the European Ceramics Society 2000;20 2129-2132.

[762] Luo X, Cao W, Xiao Z. Investigation on the distribution of rare earth ions in strontium aluminate phosphors. Journal of Alloys and Compounds 2006;416(1-2) 250-255.

[763] Yang P, Lü MK, Song ChF, Xu D, Yuan DR, Cao DX, Gu F. Effect of Cu^{2+} ions on the photoluminescence characteristics of Ce^{4+}-doped $SrAl_2O_4$. Optical Materials 2002;20(2) 135-140.

[764] Kovalevsky AV, Kharton VV, Snijkers FMM, Cooymans JFC, Luyten JJ, Marques FMB. Oxygen transport and stability of asymmetric $SrFe(Al)O_{3-\delta}$-$SrAl_2O_4$ composite membranes. Journal of Membrane Science 2007;301(1-2) 238-244.

[765] Cheng Y, Zhao Y, Zhang Y, Cao X. Preparation of $SrAl_2O_4$: Eu^{2+}, Dy^{3+} fibers by electrospinning combined with sol-gel process. Journal of Colloid and Interface Science 2010;344 321-326.

[766] Chang Ch, Yuan Z, Mao D. Eu^{2+} activated long persistent strontium aluminate nano scaled phosphor prepared by precipitation method. Journal of Alloys and Compounds 2006;415(1-2) 220-224.

[767] Azorin J. Preparation methods of thermoluminescent materials for dosimetric applications: An overview. Applied Radiation and Isotopes 2014;83(Part C) 187-191.

[768] Xue Z, Deng S, Liu Y, Lei B, Xiao Y, Zheng M. Synthesis and luminescence properties of $SrAl_2O_4$:Eu^{2+},Dy^{3}+ hollow microspheres via a solvothermal co-precipitation method. Journal of Rare Earths 2013;31(3) 241-246.

[769] Xiao Q, Xiao L, Liu Y, Chen X, Li Y. Synthesis and luminescence properties of needle-like $SrAl_2O_4$:Eu, Dy phosphor via a hydrothermal co-precipitation method. Journal of Physics and Chemistry of Solids 2010;71(7) 1026-1030.

[770] Yu X, Zhou Ch, He X, Peng Z, Yang S.-P. The influence of some processing conditions on luminescence of $SrAl_2O_4$:Eu^{2+} nanoparticles produced by combustion method. Materials Letters 2004;581087-1091.

[771] Peng T, Yang H, Pu X, Hu B, Jiang Z, Yan Ch. Combustion synthesis and photoluminescence of $SrAl_2O_4$: Eu, Dy phosphor nanoparticles. Materials Letters 2004;58 359-356.

[772] Shafia E, Bodaghi M, Esposite S, Aghaei A. A critical role of pH in the combustion synthesis of nano-sized $SrAl_2O_4$: Eu^{2+}, Dy^{3+} phosphor. Ceramics International 2014;40(3) 4697-4706.

[773] Patil KC, Aruna ST, Mimani T. Combustion synthesis: an update. Current Opinion in Solid State and Materials Science 2002;6(6) 507-512.

[774] Li X, Qu Y, Xie X, Wang Z, Li R. Preparation of $SrAl_2O_4$: Eu_{2+}, Dy_{3+} nanometer phosphors by detonation method. Materials Letters 2006;60(29-30) 3673-3677.

[775] Li X, Qu Y, Sun G, Jiang D, Ouyang X. Study on the lattice distortion of the As-prepared nanosized TiO_2 particles via detonation method. Journal of Physics and Chemistry of Solids 2007;68(12) 2405-2410.

[776] Aruna ST, Mukasyan AS. Combustion synthesis of nanomaterials. Current Opinion in Solid State and Materials Science 2008;12 44-50.

[777] Xu R, Su Q. Chapter 2 - High-temperature Synthesis. Modern Inorganic Synthetic Chemistry; 2011.

[778] Yeh CL, Chuang HC, Liu EW, Chang YC. Effects of dilution and preheating on SHS of vanadium nitride. Ceramics International 2005;31(1) 95-104.

[779] Merzhanov AG. History and recent developments in SHS. Ceramics International 1995;21(5) 371-379.

[780] Esharghawi A, Penot C, Nardou F. Elaboration of porous mullite-based materials via SHS reaction. Ceramics International 2010;36(1) 231-239.

[781] Russias J, Cardinal S, Fontaine J, Fantozzi G, Esnouf C, Bienvenu K. Bulk titanium nitride material obtained from SHS starting powder: Densification, mechanical characterization and tribological approach. International Journal of Refractory Metals and Hard Materials 2005;23(4-6) 344-349.

[782] Qian Q, Wang J, Gu Y, Li J, Zhao G, Zhang L, Pan X. Convenient synthesis of Fe-filled boron nitride nanotubes by SHS method. Materials Letters 2011;65(5) 866-868.

[783] Cano IG, Rodríguez MA. Synthesis of β-Silicon nitride by SHS: fibber growth. Scripta Materialia 2004;50(3) 383-386.

[784] Bermudo J, Osendi MI. Study of AlN and Si_3N_4 powders synthesized by SHS reactions. Ceramics International 1999;25(7) 607-612.

[785] Pradeilles N, Record MC, Granier D, Marin-Ayral RM. Synthesis of β-SiAlON: A combined method using sol–gel and SHS processes. Ceramics International 2008;34(5) 1189-1194.

[786] Lotfi B, Shipway PH, McCartney DG, Edris H. Abrasive wear behaviour of Ni(Cr)–TiB_2 coatings deposited by HVOF spraying of SHS-derived cermet powders. Wear 2003;254(3-4) 340-349.

[787] Contreras L, Turrillas X, Mas-Guindal MJ, Vaughan GBM, Kvick Å, Rodríguez MA. Synchrotron diffraction studies of TiC/FeTi cermets obtained by SHS. Journal of Solid State Chemistry 2005;178(5) 1595-1600.

[788] Mas-Guindal MJ, Benko E, Rodríguez MA. Nanostructured metastable cermets of Ti–Al_2O_3 through activated SHS reaction. Journal of Alloys and Compounds 2008;454(1-2) 352-358.

[789] Córdoba JM, Alcalá MD, Avilés MA, Sayagués MJ, Gotor FJ. New production of $TiC_xN1{-}x$-based cermets by one step mechanically induced self-sustaining reaction:

Powder synthesis and pressureless sintering. Journal of the European Ceramic Society 2008;28(10) 2085-2098.

[790] Sathaporn T, Niyomwas S. Synthesis and Characterization of MAl_2O_4 (M = Ba, Ca, Sr) Phosphor by Self-propagating Hingh Temperature Synthesis. Energy Procedia 2011;9 410-417.

[791] Zhang Q, Saito F. A review on mechanochemical syntheses of functional materials. Advanced Powder Technology 2012;23(5) 523-531.

[792] Chen G, Niu D, Liu X. Preparation of $SrAl^2O^4$ from an oxide mixture via a high-energy ball milling. Journal of Alloys and Compounds 2005;399(1-2) 280-283.

[793] Garcés RS, Torres JT, Valdés AF. Synthesis of $SrAl_2O_4$ and $Sr_3Al_2O_6$ at high temperature, starting from mechanically activated $SrCO_3$ and Al_2O_3 in blends of 3:1 molar ratio. Ceramics International 2012;38(2) 889-894.

[794] Zollfrank C, Gruber S, Batentschuk M, Osvet A, Goetz-Neunhoeffer F, Dittrich S, Grabow J, Kurland H.-D, Müller FA. Synthesis of Eu-doped $SrAl_2O_4$ nanophosphors by CO_2 laser vaporization. Acta Materialia 2013;61(19) 7133-7141.

[795] Aroz R, Lennikov V, Cases R, Sanjuán ML, de la Fuente GF, Muñoz E. Laser synthesis and luminescence properties of $SrAl_2O_4$:Eu^{2+}, Dy^{3+} phosphors. Journal of the European Ceramic Society 2012;32(16) 4363-4369.

[796] Katsumata T, Nabae T, Sasajima K, Matsuzawa T. Growth and characteristics of long persistent $SrAl_2O_4$- and $CaAl_2O_4$-based phosphor crystals by a floating zone technique. Journal of Crystal Growth 1998;183(3) 361-365.

[797] Jia W, Yuan H, Holmstrom S, Liu H, Yen WM. Photo-stimulated luminescence in $SrAl_2O_4$: Eu^{2+},Dy^{3+} single crystal fibers. Journal of Luminescence 1999;83-84 465-469.

[798] Yen WM. Synthesis, Characterization and Applications of Shaped Single Crystals. Physics of the Solid State 1999;41(5) 693-696.

[799] Nsimama PD, Ntwaeaborwa OM, Swart HC. The effect of different gas atmospheres on luminescent properties of pulsed lase ablated $SrAl_2O_4$: Eu^{2+}, Dy^{3+} thin films. Journal of Luminescence 2011;131 119-125.

[800] Swart HC, Ntwaeaborva OM, Nsimama PD, Terblans JJ. Surface characterization and luminescent properties of $SrAl_2O_4$: Eu^{2+}, Dy^{3+} nano thin films. Physica B 2012;4071660-1663.

[801] Nsimama PD, Ntwaeaborwa OM, Coetsee E, Svart HC. The influence of the number of pulses on the morphological and photoluminescence properties of $SrAl_2O_4$: Eu^{2+}, Dy^{3+} thin films prepared by pulsed laser deposition. Physica B 2009;404 4489-4492.

[802] Nsimama PD, Ntwaeaborwa OM, Coetsee E, Svart HC. The effect of substrate temperature on the structure, morphology and photoluminescence properties of pulsed laser deposited $SrAl_2O_4$: Eu^{2+}, Dy^{3+} thin film. Physica B 2009;404 4436-4439.

[803] Sato K, Komuro S, Morikawa T, Aizawa H, Katsumata T, Harako S, Zhao X. Long afterglow characteristics of thin film phosphor fabricate by laser ablation. Journal of Crystal Growth 2005;275 e1137-e1141.

[804] Lee JS, Kim YJ. The deposition and the photoluminescence of $SrAl_2O_4$: Eu^{2+} thin films. Thin Solid Films 2010;518 e149-e151.

[805] Marla D, Bhandarkar UV, Joshi SS. Modeling nanosecond pulsed laser ablation: A focus on temperature dependence of material properties. Manufacturing Letters 2014;2(2) 13-16.

[806] Suematsu H, Sengiku M, Kato K, Mitome M, Kimuto K, Matsui Y, Jiang W, Yatsui K. Photoluminescence properties of crystallized strontium aluminate thin films prepared by ion-beam evaporation. Thin solid films 2002;407(1-2) 136-138.

[807] Ji Z, Tian S, Chen W, Kong Z, Wu J. Enhanced long lasting persistent luminescent $SrAl_2O_4$: Eu, Dy ceramics prepared by electron beam bombardment. Radiation Measurements 2013;59 210-213.

[808] Chang Ch, Mao D. Long lasting phosphorescence of $Sr_4Al_{14}O_{25}$: Eu^{2+},Dy^{3+} thin films by magnetron sputtering. Thin Solid Films 2004;46048-52.

[809] Kato K, Tsutai I, Kamimura T, Kaneko F, Shinbo K, Ohta M, Kawakami T. Thermoluminescence properties of $SrAl_2O_4$: Eu sputtered films with long phosphorescence. Journal of Luminescence 1999;82(3) 213-220.

[810] Haranath D, Sharma P, Chander H, Ali A, Bhalla N, Halder SK. Role of boric acid in synthesis and tailoring the properties of calcium aluminate phosphor. Materials Chemistry and Physics 2007;101(1) 163-169.

[811] Mothudi BM, Ntwaeaborwa OM, Botha JR, Swart HC. Photoluminescence and phosphorescence properties of MAl_2O_4:Eu^{2+}, Dy^{3+} (M=Ca, Ba, Sr) phosphors prepared at an initiating combustion temperature of 500°C. Physica B: Condensed Matter 2009;404(22) 4440-4444.

[812] Singh V, Natarajan V, Zhu J. J. Luminescence and EPR investigations of Mn activated calcium aluminate prepared via combustion method. Optical Materials 2007;30(3) 468-472.

[813] Zhao Ch, Chen D. Synthesis of $CaAl_2O_4$: Eu,Nd long persistent phosphor by combustion processes and its optical properties. Materials Letters 2007;61(17) 3673-3675.

[814] Sharma SK, Pitale SS, Malik MM, Qureshi MS, Dubey RN. Spectral and kinetic characterization of orange-red emitting $Sr_3Al_2O_6$:Eu^{3+}/Sm^{3+} phosphor. Journal of Alloys and Compounds 2009;482(1-2) 468-475.

[815] Kim SJ, Won HI, Hayk N, Won ChW, Jeon DY, Kirakosyan AG. Preparation and characterization of $Sr_4Al_2O_7$:Eu^{3+}, Eu^{2+} phosphors. Materials Science and Engineering: B 2011;176(18) 1521-1525.

[816] Yerpude AN, Dhoble SJ. Luminescence in trivalent rare earth activated $Sr_4Al_2O_7$ phosphor. Optik - International Journal for Light and Electron Optics 2013;124(18) 3567-3570.

[817] Huang P, Zhang Q, Cui C.-e, Li J. Influence of excitation wavelengths on luminescent properties of $Sr_3Al_2O_6:Eu^{2+}$, Dy^{3+} phosphors prepared by sol-gel-combustion processing. Optical Materials 2011;33(8) 1252-1257.

[818] Chen XY, Bao SP, Wu YCh. Controlled synthesis and luminescent properties of Eu^{2+} (Eu^{3+}), Dy^{3+}-doped $Sr_3Al_2O_6$ phosphors by hydrothermal treatment and postannealing approach. Journal of Solid State Chemistry 2010;183(9) 2004-2011.

[819] Zhang P, Xu M.-x, Zheng Z.-t, Sun B, Zhang Y.-h. Microwave synthesis and characterization of new red long afterglow phosphor Sr_3Al_2O6:Eu. Transactions of Nonferrous Metals Society of China 2006;16(s1) s423-s425.

[820] Li G, Lai Y, Cui T, Yu H, Liu D, Gan S. Luminescence properties and charge compensation of $Sr_3Al_2O_6$ doped with Ce^{3+} and alkali metal ions. Materials Chemistry and Physics 2010;124(2-3) 1094-1099.

[821] Feng W.-L. Preparation and luminescent properties of green $SrAl_2O_4$: Eu^{2+} and blue $SrAl_2O_4$: Eu^{2+}, Gd^{3+} phosphors. Materials Letters 2013;110 91-93.

[822] Wu ZC, Shi JX, Wang J, Wu H, SuQ, Gong ML. Synthesis and luminescent properties of $SrAl_2O_4:Eu^{2+}$ green-emitting phosphor for white LEDs. Materials Letters 2006;60(29-30) 3499-3501.

[823] Ayvacıklı M, Ege A, Can N. Radioluminescence of $SrAl_2O_4:Ln^{3+}$ (Ln = Eu, Sm, Dy) phosphor ceramic. Optical Materials 2011;34(1) 138-142.

[824] Shafia E, Bodaghi M, Tahriri M. The influence of some processing conditions on host crystal structure and phosphorescence properties of $SrAl_2O_4:Eu^{2+}$, Dy^{3+} nanoparticle pigments synthesized by combustion technique. Current Applied Physics 2010;10(2) 596-600.

[825] Sharma SK, Pitale SS, Malik MM, Rao TKG, Chawla S, Qureshi MS, Dubey RN. Spectral and defect analysis of Cu-doped combustion synthesized new $SrAl_4O_7$ phosphor. Journal of Luminescence 2010;130(2) 240-248.

[826] Singh V, Rao TKG, Zhu J.-J. Preparation, luminescence and defect studies of Eu2+-activated strontium hexa-aluminate phosphor prepared via combustion method. Journal of Solid State Chemistry 2006;179(8) 2589-2594.

[827] Blasse G. Energy transfer between inequivalent Eu^{2+} ions. Journal of Solid State Chemistry 1986;62(2) 207-211.

[828] Lu Z, Weng L, Song S, Zhang P, Luo X. Hydrothermal synthesis, morphology and photoluminescence of hexagonal $SrSiO_3:Eu^{2+}$ micro-octahedrons and prism-like hollow microstructures. Materials Chemistry and Physics 2012;132(2-3) 800-807.

[829] Cui Z, Ye R, Deng D, Hua Y, Zhao S, Jia G, Li Ch, Xu S. Eu^{2+}/Sm^{3+} ions co-doped white light luminescence $SrSiO_3$ glass-ceramics phosphor for White LED. Journal of Alloys and Compounds 2011;509(8) 3553-3558.

[830] Kuang J, Liu Y, Zhang J. White-light-emitting long-lasting phosphorescence in Dy^{3+}-doped $SrSiO_3$. Journal of Solid State Chemistry 2006;179(1) 266-269.

[831] Liu H, Wang Y, Yang J, Li L, Su W, Guan Z, Yu B. The structure and luminescence characteristics of $SrSiO_3:Eu^{3+}:Bi^{3+}$ synthesized at a high pressure and high temperature. Journal of Alloys and Compounds 1993;191(1) 1-4.

[832] Tshabalala MA, Dejene FB, Pitale SS, Swart HC, Ntwaeaborwa OM. Generation of white-light from Dy^{3+} doped Sr_2SiO_4 phosphor. Physica B: Condensed Matter 2014;439 126-129.

[833] Saradhi MP, Lakshminarasimhan N, Boudin S, Gupta KVK, Varadaraju UV, Raveau B. Enhanced luminescence of $Sr_2SiO_4:Dy^{3+}$ by sensitization (Ce^{3+}/Eu^{2+}) and fabrication of white light-emitting-diodes. Materials Letters 2014;117 302-304.

[834] Gupta SK, Kumar M, Natarajan V, Godbole SV. Optical properties of sol-gel derived $Sr_2SiO_4:Dy^{3+}$ - Photo and thermally stimulated luminescence. Optical Materials 2013;35(12) 2320-2328.

[835] Yang R.-Y, Chen H.-Y, Chang S.-J, Yang Y.-K. Effect of Eu^{3+} concentration on microstructure and photoluminescence of $Sr^2SiO^4:Eu^{3+}$ phosphors prepared by microwave assisted sintering. Journal of Luminescence 2012;132(3) 780-783.

[836] Zhang L, Han P, Wang K, Lu Z, Wang L, Zhu Y, Zhang Q. Enhanced luminescence of $Sr_2SiO_4:Dy^{3+}$ by sensitization (Ce^{3+}/Bi^{3+}) and its composition-induced phase transition. Journal of Alloys and Compounds 2012;541 54-59.

[837] Qiao Y, Zhang X, Ye X, Chen Y, Guo H. Photoluminescent properties of $Sr_2SiO_4:Eu^{3+}$ and $Sr_2SiO_4:Eu^{2+}$ phosphors prepared by solid-state reaction method. Journal of Rare Earths 2009;27(2) 323-326.

[838] Dutczak D, Milbrat A, Katelnikovas A, Meijerink A, Ronda C, Jüstel T. Yellow persistent luminescence of $Sr_2SiO_4:Eu^{2+},Dy^{3+}$. Journal of Luminescence 2012;132(9) 2398-2403.

[839] Qiua J, Kawasaki M, Tanaka K, Shimizugawa Y, Hirao K . Phenomenon and mechanism of long-lasting phosphorescence in Eu2+-doped aluminosilicate glasses. Journal of Physics and Chemistry of Solids 1998;59(9) 1521-1525.

[840] Guan Y, Wei Z, Huang Y, Maalej R, Seo HJ. 1.55 μm emission and upconversion luminescence of Er^{3+}-doped strontium borate glasses. Ceramics International 2013;39(6) 7023-7027.

[841] Rajesh D, Ratnakaram YC, Seshadri M, Balakrishna A, Krishna TS. Structural and luminescence properties of Dy^{3+} ion in strontium lithium bismuth borate glasses. Journal of Luminescence 2012;132(3) 841-849.

[842] Li Y, Niu P, Tang Ch, Hu L. Blue-excited luminescence of Eu-doped strontium boroaluminate glasses. Journal of Luminescence 2008;128(2) 273-276.

[843] Rajesh D, Balakrishna A, Ratnakaram YC. Luminescence, structural and dielectric properties of Sm^{3+} impurities in strontium lithium bismuth borate glasses. Optical Materials 2012;35(2) 108-116.

[844] Jiang Ch, Huang Y, Park S, Jang K, Seo HJ. Luminescence and spectral hole burning of Sm^{2+} doped in Li_2O-SrO-B_2O_3 glass-ceramics. Spectrochimica Acta Part A: Molecular and Biomolecular Spectroscopy 2009;72(2) 412-416.

[845] Aleksandrovsky AS, Krylov AS, Malakhovskii AV, Potseluyko AM, Zaitsev AI, Zamkov AV. Europium doped strontium borate glasses and their optical properties. Journal of Physics and Chemistry of Solids 2005;66(1) 75-79.

[846] Wu B, Qiu J, Wu E, Zeng H. Broadband near-infrared luminescence from transparent glass–ceramics containing Ni^{2+}-doped $SrTiO_3$ nanocrystals. Optical Materials 2013;35(5) 983-987.

[847] Martín LL, Martín IR, Haro-González P. Transfer and backtransfer processes in Yb^{3+}-Er^{3+} codoped Strontium Barium Niobate glass-ceramics. Journal of Luminescence 2011;131(12) 2446-2450.

[848] Jarý V, Boháček P, Mihóková E, Havlák L, Trunda B, Nikl M. Photoluminescence properties of non-stoichiometric strontium zirconate powder phosphor. Optical Materials 2013;35(5) 1019-1022.

[849] Wang Zh, Zhang J, Zheng G, Peng X, Dai H. Violet-blue afterglow luminescence properties of non-doped $SrZrO_3$ material. Journal of Luminescence 2013;144 30-33.

[850] Zhang A, Lü M, Zhou G, Zhou Y, Qiu Z, Ma Q. Synthesis, characterization and luminescence of Eu^{3+}-doped $SrZrO_3$ nanocrystals. Journal of Alloys and Compounds 2009;468(1-2) L17-L20.

[851] Rétot H, Bessière A, Kahn-Harari A, Viana B. Synthesis and optical characterization of $SrHfO_3$:Ce and $SrZrO_3$:Ce nanoparticles. Optical Materials 2008;30(7) 1109-1114.

[852] Nikl M, Bohacek P, Trunda B, Jary V, Fabeni P, Studnicka V, Kucerkova R, Beitlerova A. $SrHfO_3$-based phosphors and scintillators. Optical Materials 2011;34(2) 433-438.

[853] Yamamoto H, Mikami M, Shimomura Y, Oguri Y. Host-to-activator energy transfer in a new blue-emitting phosphor $SrHfO_3$: Tm^{3+}. Journal of Luminescence 2000;87-89 1079-1082.

[854] Höllriegl V, München HZ. Strontium in the Environment and Possible Human Health Effects. Reference Module in Earth Systems and Environmental Sciences, from Encyclopedia of Environmental Health; 2011.

[855] Wohl GR, Chettle DR, Pejović-Milić A, Druchok Ch, Webber CE, Adachi JD, Beattie KA. Accumulation of bone strontium measured by in vivo XRF in rats supplemented with strontium citrate and strontium ranelate. Bone 2013;52(1) 63-69.

[856] Harrigan TP, Kareh JA, O'Connor DO, Burke DW, Harris WH. A finite element study of the initiation of failure of fixation in cemented femoral total hip components. Journal of Orthopaedic Research 1992;10 134-144.

[857] Wang JS, Franzen H, Toksvig-Larsen S, Lidgren L. Does vacuum mixing of bone cement affect heat generation? Analysis of four cement brands. Journal of Applied Biomaterials 1995;6 105-108.

[858] Sudarsanan K, Young RA. Structure of strontium hydroxide phosphate, $Sr_5(PO_4)_3OH$. Acta Crystallographica Section B 1972;B28(12) 3668-3670.

[859] Cox SC, Jamshidi P, Grover LM, Mallick KK. Preparation and characterisation of nanophase Sr, Mg, and Zn substituted hydroxyapatite by aqueous precipitation. Materials Science and Engineering: C 2014;35106-114.

[860] Elliott JC. Structure and Chemistry of the Apatites and Other Calcium Orthophosphates; Amsterdam: Elsevier; 1994.

[861] Bigi A, Boanini E, Capuccini Ch, Gazzano M. Strontium-substituted hydroxyapatite nanocrystals. Inorganica Chimica Acta 2007;360(3) 1009-1016.

[862] Meunier PJ, Roux C, Seeman E, Ortolani S, Badurski J, Spector TD, Cannata J, Balogh A, Lemmel EM, Pors-Nielsen S, Rizzoli R, Genant HK, Reqinster JY. The effects of strontium ranelate on the risk of vertebral fracture in women with postmenopausal osteoporosis. The New England Journal of Medicine 2004;350(5) 459-468.

[863] Zhaoyang Li, Ning Yuan, Raymond Wing Moon Lam, Zhenduo Cui, Xianjin Yang, Lu WW. Preclinical evaluation of strontium-containing bioactive bone cement. Materials Science and Engineering: C 2013;33(8) 5100-5104.

[864] Li YW, Leong JCY, Lu WW, Luk KDK, Cheung KMC, Chiu KY, Chow SP.A novel injectable bioactive bone cement for spinal surgery: a developmental and preclinical study. Journal of Biomedical Materials Research 2000;52(1) 164-170.

[865] Lu WW, Cheung KM, Li YW, Luk KDK, Holmes AD, Zhu QA, Leong JC. Bioactive bone cement as a principal fixture for spinal burst fracture: an in vitro biomechanical and morphologic study. Spine 2001;26 2684-2690.

[866] Ni GX, Chiu KY, Lu WW, Wang Y, Zhang YG, Hao LB, Li ZY, Lam WM, Lu SB, Luk KDK. Strontium-containing hydroxyapatite bioactive bone cement in revision hip arthroplasty. Biomaterials 2006;27(24) 4348-4355.

[867] Ni GX, Lu WW, Xu B, Chiu KY, Yang C, Li ZY, Lam WM, Luk KDK. Interfacial behaviour of strontium-containing hydroxyapatite cement with cancellous and cortical bone. Biomaterials 2006;27(29) 5127-5133.

[868] Aina V, Bergandi L, Lusvardi G, Malavasi G, Imrie FE, Gibson IR, Cerrato G, Ghigo D. Sr-containing hydroxyapatite: morphologies of HA crystals and bioactivity on osteoblast cells. Materials Science and Engineering: C 2013;33(3) 1132-1142.

[869] Zhu K, Yanagisawa K, Shimanouchi R, Onda A, Kajiyoshi K, Qiu J. Synthesis and crystallographic study of Pb–Sr hydroxyapatite solid solutions by high temperature mixing method under hydrothermal conditions. Materials Research Bulletin 2009;44(6) 1392-1396.

[870] Tan S.-H, Chen X.-G, Ye Y, Sun J, Dai L.-Q, Ding Q. Hydrothermal removal of Sr^{2+} in aqueous solution via formation of Sr-substituted hydroxyapatite. Journal of Hazardous Materials 2010;179(1-3) 559-563.

[871] Ning Z, Chang Z, Li W, Sun Ch, Zhang J, Liu Y. Solvothermal synthesis and optical performance of one-dimensional Strontium Hydroxyapatite nanorod. Chinese Journal of Chemical Engineering 2012;20(1) 89-94.

[872] Grover LM, Gbureck U, Wright AJ, Barralet JE. Cement formulations in the calcium phosphate H_2O-H_3PO_4-$H_4P_2O_7$ system. Journal of the American Ceramic Society 2005;88 3096-3103.

[873] Hofmann MP, Mohammed AR, Perrie Y, Gbureck U, Barralet JE. High-strength resorbable brushite bone cement with controlled drug-releasing capabilities. Acta Biomaterialia 2009;5 43-49.

[874] Alkhraisat MH, Rueda C, Cabrejos-Azama J, Lucas-Aparicio J, Mariño FT, García-Denche JT, Jerez LB, Gbureck U, Cabarcos EL. Loading and release of doxycycline hyclate from strontium-substituted calcium phosphate cement. Acta Biomaterialia 2010;6(4) 1522-1528.

[875] Ohura K, Bohner M, Hardouin P, Lemaitre J, Pasquier G, Flautre B. Resorption of, and bone formation from, new beta-tricalcium phosphate-monocalcium phosphate cements: an in vivo study. Journal of Biomedical Materials Research 1996;30(2) 193-200.

[876] Frost RL, Palmer SJ. Thermal stability of the 'cave' mineral brushite $CaHPO_4 \cdot 2H_2O$-Mechanism of formation and decomposition. Thermochimica Acta 2011;521(1-2) 14-17.

[877] Francis MD, Webb NC. Hydroxyapatite formation for a hydrated calcium monohydrogen phospahte precursor.Calcified Tissue International 1971;6 335-342.

[878] Abbona F, Christensson F, Angela MF, Madsen HEL. Crystal habit and growth conditions of brushite, $CaHPO_4 \cdot 2H_2O$. Journal of Crystal Growth 1993;131(3-4) 331-346.

[879] Vereeche G, Lemaitre J. Calculation of the solubility diagrams in the system $Ca(OH)_2$-H_3PO_4-KOH-HNO_3-CO_2-H_2O. Journal of Crystal Growth 1990;104 820-832.

[880] Grover LM, Gbureck U, Young AM, Wright AJ, Barralet JE. Temperature dependent setting kinetics and mechanical properties of β-TCP-pyrophosphoric acid bone cement. Journal of Materials Chemistry 2005;15(46) 4955-4962.

[881] Alkhraisat MH, Mariño FT, Rodríguez CR, Jerez LB, Cabarcos EL. Combined effect of strontium and pyrophosphate on the properties of brushite cements. Acta Biomaterialia 2008;4(3) 664-670.

[882] Kondo N, Ogose A, Tokunaga K, Umezu H, Arai K, Kudo N, Hoshino M, Inoue H, Irie H, Kuroda K, Mera H. Osteoinduction with highly purified beta-tricalcium phosphate in dogdorsal muscles and the proliferation of osteoclasts before heterotopic bone formation. Biomaterials 2006;27(25) 4419-4427.

[883] Hench LL. Bioceramics: from concept to clinic. Journal of the American Ceramic Society 1991;74(7) 1487-1510.

[884] Yatongchai Ch, Wren AW, Curran DJ, Hornez J.-Ch, Towler MR. Comparison of the Weibull characteristics of hydroxyapatite and strontium doped hydroxyapatite. Journal of the Mechanical Behavior of Biomedical Materials 2013;21 95-108.

[885] Tamimi F, Le Nihouannen D, Eimar H, Sheikh Z, Komarova S, Barralet J. The effect of autoclaving on the physical and biological properties of dicalcium phosphate dihydrate bioceramics: Brushite vs. Monetite. Acta Biomaterialia 2012;8(8) 3161-3169.

[886] Njema H, Debbichi M, Boughzala K, Said M, Bouzouita K. Structural, electronic and thermodynamic properties of britholites $Ca_{10-x}La_x(PO_4)_{6-x}(SiO_4)_xF_2$ ($0 \le x \le 6$): Experiment and theory. Materials Research Bulletin 2014;51 210-216.

[887] Njema H, Boughzala K, Boughzala H, Bouzouita K. Structural analysis by Rietveld refinement of calcium and lanthanum phosphosilicate apatites. Journal of Rare Earths 2013;31(9) 897-904.

[888] Wang X, Gan J, Huang Y, Seo HJ. The doping concentration dependent tunable yellow luminescence of $Sr_5(PO_4)_2(SiO_4)$:Eu^{2+}. Ceramics International 2012;38(1) 701-706.

[889] Ouenzerfi RE, Cohen-Adad M.-T, Goutaudier Ch, Panczer G. Uranium-doped britholites $Ca_xLa_y(SiO_4)_{6-u}(PO_4)_uO_t$:U synthesis, characterization and preliminary study of uranium diffusion. Solid State Ionics 2005;176(1-2) 225-231.

[890] Gmati N, Boughzala K, Abdellaoui M, Bouzouita K. Mechanochemical synthesis of strontium britholites: Reaction mechanism. Comptes Rendus Chimie 2011;14(10) 896-903.

[891] Gmati N, Boughzala K, Chaabène A, Fattah N, Bouzouita K. Préparation par mécanosynthèse d'apatites strontiques dopées au lanthane et au césium. Comptes Rendus Chimie 2013;16(8) 712-720.

[892] Krzmanc MM, Valant M, Suvorov D. The synthesis and microwave dielectric properties of $Sr_xBa_{1-x}Al_2Si_2O_8$ and $Ca_yBa_{1-y}Al_2Si_2O_8$ ceramics. Journal of the European Ceramic Society 2007;27(2-3) 1181-1185.

[893] Ferone C, Liguori B, Marocco A, Anaclerio S, Pansini M, Colella C. Monoclinic (Ba, Sr)-celsian by thermal treatment of (Ba, Sr)-exchanged zeolite A. Microporous and Mesoporous Materials 2010;134(1-3) 65-71.

[894] Sung YM, Kim S. Sintering and crystallization of off-stoichiometric $SrO.Al_2O_3.2SiO_2$ glasses. Journal of Materials Science 2000;35 4293-4299.

[895] Fu Y.-P, Chang Ch.-Ch, Lin Ch.-H, Chin T-S. Solid-state synthesis of ceramics in the $BaO-SrO-Al_2O_3-SiO_2$ system. Ceramics International 2004;30(1) 41-45.

[896] Guillem MC, Guillem C. Kinetics and mechanism of formation of celsian from barium cabonate and kaolin. Transactions and Journal of the British Ceramic Society 1984;83 150-154.

[897] Bansal NP, Setlock JA. Fabrication of fiber-reinforced celsian matrix composites. Composites Part A: Applied Science and Manufacturing 2001;32(8) 1021-1029.

[898] BP. Mechanical properties of Hi-Nicalon fiber-reinforced celsian composites after high-temperature exposures in air. Journal of the European Ceramic Society 2009;29(3) 525-535.

[899] Limeng L, Feng Y, Haijiao Z, Jie Y, Zhiguo Z. Celsian formation in Si_3N_4-$Ba_{0.75}Sr_{0.25}Si_2Al_2O_8$ composites. Scripta Materialia 2009;60(6) 463-466.

[900] Strnad Z. Glass-Ceramic Materials, Glass Science and Technology; Volume 8; Amsterdam: Elsevier; 1986.

[901] Barbeeri L, Corradi AB, Leonelli C, Manfredini T, Romagnoli M, Siligardi C. The microstructure and mechanical properties of sintered celsian and strontium-celsian glass-ceramics. Materials Research Bulletin 1995;30(1) 27-41.

[902] Beall GH. Refractory glass-ceramics based on alkaline earth aluminosilicates. Journal of the European Ceramic Society 2009;29(7) 1211-1219.

[903] Salman SM, Salama S.N, Abo-Mosallam HA. The role of strontium and potassium on crystallization and bioactivity of $Na_2O-CaO-P_2O_5-SiO_2$. Ceramics International 2012;38(1) 55-63.

[904] Fredholm YC, Karpukhina N, Law RV, Hill RG. Strontium containing bioactive glasses: Glass structure and physical properties. Journal of Non-Crystalline Solids 2010;356(44-49) 2546-2551.

[905] O'Donnell MD, Hill RG. Influence of strontium and the importance of glass chemistry and structure when designing bioactive glasses for bone regeneration. Acta Biomaterialia 2010;6(7) 2382-2385.

[906] Xiao Li, Xiao Q, Liu Y, Ai P, Li Y, Wang H. A transparent surface-crystallized Eu^{2+}, Dy^{3+} co-doped strontium aluminate long-lasting phosphorescent glass-ceramic. Journal of Alloys and Compounds 2010;495(1) 72-75.

[907] Qiua J, Shimizugawa Y, Kojima K, Tanaka K, Hirao K. Relaxation of ultraviolet-radiation-induced structure and long-lasting phosphorescence in Eu^{2+}-doped strontium aluminosilicate glasses. Journal of Materials Research 2001;16(1) 88-92.

[908] Lan Z, Chengyu L, Qiang S. Long lasting phosphorescence in Eu^{2+} and Ce^{3+} co-coped strontium borate glasses. Journal of Rare Earths 2006;24(1) 196-198.

[909] Huang A, Chen Y, Liu Q, Wang N, Jiang J, Caro J. Synthesis of highly hydrophobic and permselective metal-organic framework $Zn(BDC)(TED)_{0.5}$ membranes for H_2/CO_2 separation. Journal of Membrane Science 2014;454 126-132.

[910] Chen X, Tong Y, Han M.-M, Cao K.-L, Feng Y.-L. Two luminescent metal - organic frameworks constructed by unsymmetric tricarboxylate. Inorganic Chemistry Communications 2014;40 62-65.

[911] Lo S.-H, Liu H.-K, Zhan J.-X, Lin W.-Ch, Kao Ch.-Ch, Lin Ch.-H, Zima V. Assembly of a water-insoluble strontium metal–organic framework with luminescent properties. Inorganic Chemistry Communications 2011;14(10) 1602-1605.

[912] Shen H, Song Y, Gu H, Wang P, Xi Y. A high-permittivity $SrTiO_3$-based grain boundary barrier layer capacitor material single-fired under low temperature. Materials Letters 2002;56(5) 802-805.

[913] Liu CY, Tseng TY. Correlation between deep depletion and current-voltage saturation of $SrTiO_3$ gate dielectric capacitor. Ceramics International 2004;30(7) 1101-1106.

[914] Hu Y, Tan OK, Pan JS, Huang H, Cao W. The effects of annealing temperature on the sensing properties of low temperature nano-sized $SrTiO_3$ oxygen gas sensor. Sensors and Actuators B: Chemical 2005;108(1-2) 244-249.

[915] Hu Y, Tan OK, Cao W, Zhu W. A low temperature nano-structured $SrTiO_3$ thick film oxygen gas sensor. Ceramics International 2004;30(7) 1819-1822.

[916] Schultz AM, Zhu Y, Bojarski SA, Rohrer GS, Salvador PA. Eutaxial growth of hematite Fe_2O_3 films on perovskite $SrTiO_3$ polycrystalline substrates. Thin Solid Films 2013;548 220-224.

[917] Nishio K, Ohnishi T, Akatsuka K, Takada K. Crystal orientation of epitaxial $LiCoO_2$ films grown on $SrTiO_3$ substrates. Journal of Power Sources 2014;247 687-691.

[918] Chen L, Zhang S, Wang L, Xue D, Yin S. Preparation and photocatalytic properties of strontium titanate powders via sol–gel proces. Journal of Crystal Growth 2009;311(3) 746-748.

[919] Huang X, Zhao H, Qiu W, Wu W, Li X. Performances of planar solid oxide fuel cells with doped strontium titanate as anode materials. Energy Conversion and Management 2007;48(5) 1678-1682.

[920] Park B.-K, Lee J.-W, Lee S.-B, Lim T.-H, Park S.-J, Song R.-H, Im WB, Shin D.-R. La-doped $SrTiO_3$ interconnect materials for anode-supported flat-tubular solid oxide fuel cells. International Journal of Hydrogen Energy 2012;37(5) 4319-4327.

[921] McCarthy GJ, White WB, Roy R. Phase Equilibria in the 1375°C Isotherm of the System Sr-Ti-O. Journal of the American Ceramic Society 1969;52(9) 463-467.

[922] Jacob KT, Rajitha G. Thermodynamic properties of strontium titanates: Sr_2TiO_4, $Sr_3Ti_2O_7$, $Sr_4Ti_3O_{10}$, and $SrTiO_3$. The Journal of Chemical Thermodynamics 2011;43(1) 51-57.

[923] Selmi F, Guerin F, Yu XD, Varadan VK, Varadan VV, Komarneni S. Microwave calcination and sintering of barium strontium titanate. Materials Letters 1992;12(6) 424-428.

[924] Mohammadi MR, Fray DJ. Sol-gel derived nanocrystalline and mesoporous barium strontium titanate prepared at room temperature. Particuology 2011;9(3) 235-242.

[925] Lahiry S, Mansingh A. Dielectric properties of sol-gel derived barium strontium titanate thin films. Thin Solid Films 2008;516(8) 1656-1662.

[926] Makarova M, Dejneka A, Franc J, Drahokoupil J, Jastrabik L, Trepakov V. Soft chemistry preparation methods and properties of strontium titanate nanoparticles. Optical Materials 2010;32(8) 803-806.

[927] Fujishiro F, Arakawa T, Hashimoto T. Substitution site and photoluminescence spectra of Eu^{3+}-substituted $SrTiO_3$ prepared by Pechini method. Materials Letters 2011;65(12) 1819-1821.

[928] Choi K.-M, Kil H.-S, Lee Y.-S, Lim D.-Y, Cho S.-B, Lee BW. Preparation and luminescence properties of $SrTiO^3$:Pr^{3+},Al^{3+} phosphor from the glycolate method. Journal of Luminescence 2011;131(5) 894-899.

[929] Lu X, Pine TS, Mumm DR, Brouwer J. Modified Pechini synthesis and characterization of Y-doped strontium titanate perovskite. Solid State Ionics 2007;178(19-20) 1195-1199.

[930] Burnat D, Heel A, Holzer L, Kata D, Lis J, Graule T. Synthesis and performance of A-site deficient lanthanum-doped strontium titanate by nanoparticle based spray pyrolysis. Journal of Power Sources 2012;201 26-36.

[931] Monteiro JF, Ferreira AAL, Antunes I, Fagg DP, Frade JR. Thermodynamic restrictions on mechanosynthesis of strontium titanate. Journal of Solid State Chemistry 2012;185 143-149.

[932] Sekhar KC, Hong KP, Key SH, Han ChS, Kim JCh, Kim DS, Park JCh, Cho YS. Enhanced dielectric and tunable characteristics of K-doped $Ba_{0.5}Sr_{0.5}TiO_3$ thin films prepared by pulsed laser deposition. Current Applied Physics 2012;12(3) 654-658.

[933] Ioachim A, Toacsan MI, Banciu MG, Nedelcu L, Dutu A, Antohe S, Berbecaru C, Georgescu L, Stoica G, Alexandru HV. Transitions of barium strontium titanate ferroelectric ceramics for different strontium content. Thin Solid Films 2007;515(16) 6289-6293.

[934] Hu T, Jantunen H, Uusimäki A, Leppävuori S. $Ba_{0.7}Sr_{0.3}TiO_3$ powders with B_2O_3 additive prepared by the sol–gel method for use as microwave material. Materials Science in Semiconductor Processing 2002;5(2-3) 215-221.

[935] Jain M, Majumder SB, Guo R, Bhalla AS, Katiyar RS. Synthesis and characterization of lead strontium titanate thin films by sol–gel technique. Materials Letters 2002;56(5) 692-697.

[936] Balasubramaniam KR, Cao Y, Patel N, Havelia S, Cox PJ, Devlin EC, Yu EP, Close BJ, Woodward PM, Salvador PA. Phase and structural characterization of $Sr_2Nb_2O_7$ and $SrNbO_3$ thin films grown via pulsed laser ablation in O_2 or N_2 atmospheres. Journal of Solid State Chemistry 2008;181(4) 705-714.

[937] Liu D, Yao X, Smyth DM. Structure intergrowth in ceramic $SrNbO_{3.5-x}$. Materials Research Bulletin 1992;27(3) 387-392.

[938] Lichtenberg F, Herrnberger A, Wiedenmann K, Mannhart J. Synthesis of perovskite-related layered $A_nB_nO3_{n+2}$ = ABOX type niobates and titanates and study of their structural, electric and magnetic properties. Progress in Solid State Chemistry 2001;29(1-2) 1-70.

[939] Dahl PI, Haugsrud R, Lein HL, Grande T, Norby T, Einarsrud M.-A. Synthesis, densification and electrical properties of strontium cerate ceramics. Journal of the European Ceramic Society 2007;27(16) 4461-4471.

[940] Aksenova TI, Khromushin IV, Zhotabaev ZR, Bukenov KD, Berdauletov AK, Medvedeva ZV. Thermodesorption study of barium and strontium cerates. Solid State Ionics 2003;162-16331-36.

[941] Liu S, Tan X, Li K, Hughes R. Synthesis of strontium cerates-based perovskite ceramics via water-soluble complex precursor routes. Ceramics International 2002;28(3) 327-335.

[942] Wei X, Lin YS. Protonic and electronic conductivities of terbium doped strontium cerates. Solid State Ionics 2008;178(35-36) 1804-1810.

[943] Matsuka M, Sakai T, Matsumoto H, Braddock RD, Agranovski IE, Ishihara T. Effects of hydrogen on phase stability of ytterbium doped strontium cerates. Materials Letters 2010;64(7) 833-835.

[944] Zhu Z.-L, Gu J.-H, Jia Y, Hu X. A comparative study of electronic structure and magnetic properties of $SrCrO_3$ and $SrMoO_3$. Physica B: Condensed Matter 2012;407(12) 1990-1994.

[945] Chamberland BL. Preparation and properties of $SrCrO_3$. Solid State Communications 1967;5(8) 663-666.

[946] Agarwal R, Singh Z, Venugopal V. Calorimetric investigations of $SrMoO_3$ and $BaMoO_3$ compounds. Journal of Alloys and Compounds 1999;282(1-2) 231-235.

[947] Macquart RB, Kennedy BJ, Avdeev M. Neutron diffraction study of phase transitions in perovskite-type strontium molybdate $SrMoO_3$. Journal of Solid State Chemistry 2010;183(1) 250-255.

[948] Yoo Y.-Z, Chmaissem O, Song J.-H. RHEED study on continuously repeated step flow and layer-by-layer growth modes in $SrRuO_3$/$SrMnO_3$ superlattice. Current Applied Physics 2014;14(3) 378-382.

[949] Doroftei C, Popa PD, Rezlescu E, Rezlescu N. Nanocrystalline $SrMnO_3$ powder as catalyst for hydrocarbon combustion. Journal of Alloys and Compounds 2014;584 195-198.

[950] Yaremchenko AA, Kharton VV, Valente AA, Shaula AL, Marques FMB, Rocha J. Mixed conductivity and electrocatalytic performance of $SrFeO_{3-\delta}$–$SrAl_2O_4$ composite membranes. Solid State Ionics 2006;177(26-32) 2285-2289.

[951] Hancock CA, Slade RCT, Varcoe JR, Slater PR. Synthesis, structure and conductivity of sulfate and phosphate doped $SrCoO_3$. Journal of Solid State Chemistry 2011;184(11) 2972-2977.

[952] Pyatnitsky YI, Ilchenko NI, Raevskaya LN, Dolgikh LY, Pavlenko NV. Methane coupling over $SrCoO_3$-based perovskites in the absence of gas-phase oxygen. Studies in Surface Science and Catalysis 2000;130 707-712.

[953] Rafferty A, Prescott T, Brabazon D. Sintering behaviour of cobalt ferrite ceramic. Ceramics International 2008;34 15-21.

[954] Onreabroy W, Papato K, Rujijanagul G, Pengpat K, Tunkasiri T. Study of strontium ferrites substituted by lanthanum on the structural and magnetic properties. Ceramics International 2012;38(Supplement 1) S415-S419.

[955] Stanica N, Cimpoesu F, Dobrescu G, Munteanu G, Suh S.-H. Monte Carlo simulation of magnetic ordering in the $Gd_3Fe_5O_{12}$ Ising ferrite with garnet structure. Journal of Magnetism and Magnetic Materials 2008;320(17) 2149-2154.

[956] Liu X, Zhong W, Yang S, Yu Z, Gu B, Du Y. Influences of La^{3+} substitution on the structure and magnetic properties of M-type strontium ferrites. Journal of Magnetism and Magnetic Materials 2002;238(2-3) 207-214.

[957] Ketov SV, Yagodkin YD, Menushenkov VP. Structure and magnetic properties of strontium ferrite anisotropic powder with nanocrystalline structure. Journal of Alloys and Compounds 2011;509(3) 1065-1068.

[958] Liu X, Hernández-Gómez P, Huang K, Zhou S, Wang Y, Cai X, Sun H, Ma B. Research on La^{3+}-Co^{2+}-substituted strontium ferrite magnets for high intrinsic coercive force. Journal of Magnetism and Magnetic Materials 2006;305(2) 524-528.

[959] Veverka P, Knížek K, Pollert E, Boháček J, Vasseur S, Duguet E, Portier J. Strontium ferrite nanoparticles synthesized in presence of polyvinylalcohol: Phase composition, microstructural and magnetic properties. Journal of Magnetism and Magnetic Materials 2007;309(1) 106-112.

[960] Ghasemi A, Morisako A, Liu X. Magnetic properties of hexagonal strontium ferrite thick film synthesized by sol-gel processing using SrM nanoparticles.Journal of Magnetism and Magnetic Materials 2008;320(18) 2300-2304.

[961] Teh GB, Wong YCh, Tilley RD. Effect of annealing temperature on the structural, photoluminescence and magnetic properties of sol–gel derived Magnetoplumbite-type (M-type) hexagonal strontium ferrite. Journal of Magnetism and Magnetic Materials 2011;323(17) 2318-2322.

[962] Lu HF, Hong RY, Li HZ. Influence of surfactants on co-precipitation synthesis of strontium ferrite. Journal of Alloys and Compounds 2011;509(41) 10127-10131.

[963] Anis-ur-Rehman M, Asghar G. Variation in structural and dielectric properties of co-precipitated nanoparticles strontium ferrites due to value of pH. Journal of Alloys and Compounds 2011;509(2) 435-439.

[964] Shirtcliffe NJ, Thompson S, O'Keefe ES, Appleton S, Perry CC. Highly aluminium doped barium and strontium ferrite nanoparticles prepared by citrate auto-combustion synthesis. Materials Research Bulletin 2007;42(2) 281-287.

[965] Burtfoot JC. Ferroelectrics: An Introduction to the Physical Principles. London: Van Nostrand-Reinbold; 1967.

[966] Molla J, Gonzalez M, Villa R, Ibara A. Effect of humidity on microwave dielectric losses of porous alumina. Journal of Applied Physics 1999;85 1727-1730.

[967] Bednorz JG, Müller KA. Possible high T_c superconductivity in the Ba-La-Cu-O system Zeitschrift für Physik B 1986;64 189-193.

[968] Kamimura H, Ushio H, Matsuno S. Theory of lanthanum copper oxide (LSCO) superconductors. Physica C: Superconductivity 2007;460-462 991-992.

[969] Anis-ur-Rehman M, Mubeen M. Synthesis and enhancement of current density in cerium doped Bi(Pb)Sr(Ba)-2223 high Tc superconductor. Synthetic Metals 2012;162(19-20) 1769-1774.

[970] Li S, Hu QY, Liu HK, Dou SX, Gao W. The grain alignment of Bi2223, Bi2212 and Bi2223 + Bi2212 phases in mechanical deformation and annealing processes. Physica C: Superconductivity 1997;279(3-4) 265-276.

[971] On DT, Sato O, Fujishima A, Hashimoto K. Change of the critical temperature of high-Tc single (2223) phase Bi–Pb–Sr–Ca–Cu–O superconductors by intercalation process Journal of Physics and Chemistry of Solids 1999;60(7) 883-890.

[972] Anis-ur-Rehman M, Maqsood A. Study of the thermal behaviour of Bi(Pb)Sr(Ba)-2223 high-Tc granular superconductors. Physica C: Superconductivity 2005;418(3-4) 121-130.

[973] Tampieri A, Calestani G, Celotti G, Masini R, Lesca S. Multi-step process to prepare bulk BSCCO (2223) superconductor with improved transport properties. Physica C: Superconductivity 1998;306(1-2) 21-33.

[974] Kandyel E, Wu X.-J, Adachi S, Tajima S. New Tl–Sr–Ca–Cu–O superconductor with 2223-type structure stabilized by mercury under high-pressure. Physica C: Superconductivity 1999;328(1-2) 44-52.

[975] Sundaresan A, Asada H, Crisan A, Nie JC, Kito H, Iyo A, Tanaka Y, Kusunoki M, Ohshima S. Preparation of Tl-2212 and -1223 superconductor thin films and their microwave properties. Physica C: Superconductivity 2003;388-389 473-474.

[976] Liu RS, Hu SH, Jefferson DA, Edwards PP.Superconductivity at 124 K in $(Tl_{0.5}Pb_{0.5})Sr_2Ca_2Cu_3O_9$. Physica C 1992;198 318-322.

[977] Marcos MD, Attfield JP. Crystal structure of $Tl_{0.5}Pb_{0.5}Sr_2Ca_2Cu_3O_9$ at 300 K and around Tc (118 K). Physica C: Superconductivity 1996;270(3-4) 267-273.

[978] Hayri EA, Greenblatt M. Superconductivity in $Tl_2Ba_{2-x}S_{rx}CaCu_2O_8$ solid solutions: T_c≈44 K for a composition with x=2.0. Physica C 1988;156(5) 775-780.

[979] Singh B, Gupta S, Sharma N, Goyal SC. Higher order elastic constants of $La_{1.85}Sr_{0.15}CuO_4$ high temperature superconductor. Physica C: Superconductivity 2005;419(1-2) 1-6.

[980] Akhter S, Paul DP. Estimation of nucleation thermodynamical parameters of $La_2Sr_2CuO_4$ (LSCO) crystallizing from high temperature solution. Materials Chemistry and Physics 2004;88(1) 41-45.

[981] Jayachandran KP, Menon CS. Mode Grüneisen parameters and the low temperature thermal expansion of high-Tc superconductor $La_{1.8}Sr_{0.2}CuO_4$. Physica C: Superconductivity 2002;383(1-2) 159-168.

[982] Hardy V, Martin C, Damay F, André G. Magnetic couplings in the quasi-2D triangular Heisenberg antiferromagnets α-ACr_2O_4 (A=Ca, Sr, Ba). Journal of Magnetism and Magnetic Materials 2013;330 111-118.

[983] Zhao L, Wang K.-J, Wen M.-H, Wu M.-K. Floating-zone growth and characterization of triangular lattice antiferromagnetic α-$SrCr_2O_4$ crystals. Journal of Crystal Growth 2014;392 81-86.

[984] Müller-Buschbaum H. The crystal chemistry of AM_2O_4 oxometallates. Alloys and Compounds 2003;349(1-2) 49-104.

[985] Moissan H. Nouvelles expériences sur la reproduction du diamant. Comptes rendus de l'Académie des sciences 1894;118 320-326.

[986] Moissan H.Bulletin de la Societe Chimique de France 1894;11 1002.

[987] Moissan H.Annales de Chimie et de Physique 1896;9 247.

[988] Ruschewitz U. Binary and ternary carbides of alkali and alkaline-earth metals. Coordination Chemistry Reviews 2003;244(1-2) 115-136.

[989] Vohn V, Knapp M, Ruschewitz U. Synthesis and Crystal Structure of SrC_2. Journal of Solid State Chemistry 2000;151(1) 111-116.

[990] Xiao B, Feng J, Chen JC, Yu L. Crystal structures and electronic properties of MC_2 (M = Mg, Ca, Sr, Ba) by comparative studies based on ab-initio calculations. Chemical Physics Letters 2007;448(1-3) 35-40.

[991] Inumaru K, Kuroda Y, Sakamoto K, Murashima M, Yamanaka S. Synthesis and high metallic conductivity of layer-structured Sr_2N thin film deposited onto MgO(001) substrate. Journal of Alloys and Compounds 2004;372(1-2) L1-L3.

[992] Reckeweg O, DiSalvo FJ. Alkaline earth metal nitride compounds with the composition M_2NX (M=Ca, Sr, Ba; X=□, H, Cl or Br). Solid State Sciences 2002;4(5) 575-584.

[993] Bowman A, Smith RI, Gregory DH. Synthesis and structure of the ternary and quaternary strontium nitride halides, $Sr_2N(X, X')$ (X, X'=Cl, Br, I). Journal of Solid State Chemistry 2006;179(1) 130-139.

[994] Alahmed ZA, Reshak AH. DFT calculation of the electronic structure and optical properties of two strontium germanium nitrides: α-Sr_2GeN_2 and β-Sr_2GeN_2. Journal of Alloys and Compounds 2013;559 181-187.

[995] Bowman A, Gregory DH. Synthesis and characterisation of the ternary nitride, Sr_2TaN_3. Journal of Alloys and Compounds 2003;348(1-2) 80-87.

[996] Balducci G, Brutti S, Ciccioli A, Gigli G, Trionfetti G, Palenzona A, Pani M. Vapor pressures and thermodynamic properties of strontium silicides. Intermetallics 2006;14(5) 578-583.

[997] Shein IR, Ivanovskii AL. Electronic band structure of pseudo-binary AlB_2-like hexagonal silicides $SrNi_xSi_{2-x}$ as novel low-TC superconductors. Physica B: Condensed Matter 2012;407(23) 4592-4594.

[998] Garay-Tapia AM, Romero AH, Trapaga G, Arróyave R. First-principles investigation of the Al–Si–Sr ternary system: Ground state determination and mechanical properties. Intermetallics 2012;21(1) 31-44.

[999] Kauzlarich SM, Condron CL, Wassei JK, Ikeda T, Snyder GJ. Structure and high-temperature thermoelectric properties of $SrAl_2Si_2$. Journal of Solid State Chemistry 2009;182(2) 240-245. D.Moser,D.

[1000] Lee MH, Sankey OF, Björling T, Noréus D,Parker SF, Häussermann U. Vibrational Properties of Polyanionic Hydrides $SrAl_2H_2$ and SrAlSiH: New Insights into Al-H Bonding Interactions Inorganic Chemistry200;74669876991.

[1001] Kranak VF, Evans MJ, Daemen LL, Proffen T, Lee MH, Sankey OF, Häussermann U. Structural and dynamic properties of the polyanionic hydrides SrAlGeH and BaAlGeH. Solid State Sciences 2009;11(11) 1847-1853.

[1002] Evans NDM. Binding mechanisms of radionuclides to cement. Cement and Concrete Research 2008;38(4) 543-553.

[1003] Omotoso OE, Ivey G, Mikula R. Containment mechanism of trivalent chromium in tricalcium silicate. Journal of Hazardous Materials 1998;60 1-28.

[1004] Atkins M, Glasser FP. Application of Portland cement-based materials to radioactive waste immobilization. Waste Management 1992;12 105-131.

[1005] Rahman ROA, El Abidin DHAZ, Abou-Shady H. Assessment of strontium immobilization in cement–bentonite matrices. Chemical Engineering Journal 2013;228 772-780.

[1006] El-Kamash AM, El-Naggar MR, El-Dessouky MI. Immobilization of cesium and strontium radionuclides in zeolite-cement blends. Journal of Hazardous Materials 2006;136(2) 310-316.

[1007] Bensted J, Varma SP. Ettringite and its derivatives. Cement Technology 1971;2 73-76.

[1008] Heimann RB. Interaction of cement and radioactive waste forms in multicomponent systems tests at 200°C Part 1: Leaching and Sorption of cesium, strontium and actinides. Cement and Concrete Research 1988;18(3) 389-400.

[1009] Ayvacıklı M, Ege A, Yerci S, Can N. Synthesis and optical properties of Er^{3+} and Eu^{3+} doped $SrAl_2O_4$ phosphor ceramic. Journal of Luminescence2011;1312432-2439.

[1010] Kacimi L, Simon-Masseron A, Salem S, Ghomari A, Derriche Z. Synthesis of belite cement clinker of high hydraulic reactivity. Cement and Concrete Research 2009;39(7) 559-565.

[1011] Ptáček P, Opravil T, Šoukal F, Havlica J, Holešinský R. Kinetics and mechanism of formation of gehlenite, Al–Si spinel and anorthite from the mixture of kaolinite and calcite. Solid State Sciences2013;2653-58.

[1012] Marocco A, Liguori B, Dell'Agli G, Pansini M.Sintering behaviour of celsian based ceramics obtained from the thermal conversion of (Ba, Sr)-exchanged zeolite A. Journal of the European Ceramic Society2011;31(11) 1965-1973.

[1013] Abe T, Sukamoto K, Sunagawa I. Nucleation, growth and stability of $CaAl_2Si_2O_8$ polymorphs. Physics and Chemistry of Minerals 1991;17(6) 473-484.

[1014] Arifov PA, Bulatova MM. Triangulation and Specific Features of the Phase Formation in Strontium Aluminosilicate Glass-Forming Systems.Glass Physics and Chemistry 2004;30(2) 198-201.

[1015] NielsenSP.The biological role of strontium. Bone 2004;35(3) 583-588.

[1016] Jensen J.-EB, Stang H, Kringsholm B. Relationship between trace element content and mechanical bone strength. Bone 1997;20(Suppl 4) 104.

[1017] Morohashi T, Sano T, Yamada S. Effects of strontium on calcium metabolism in rats: I. A distinction between pharmacologic and toxic doses. Japanese Journal of Pharmacology 1994;64(3) 155-162.

[1018] TippleBJ, ChauT, ChessonLA, FernandezDP, EhleringerJR. Isolation of strontium pools and isotope ratios in modern human hair. Analytica Chimica Acta 2013;798 64-73.

[1019] FennerJN, WrightLE. Revisiting the strontium contribution of sea salt in the human diet. Journal of Archaeological Science 2014;44 99-103.

[1020] WrightLE. Immigration to Tikal, Guatemala: Evidence from stable strontium and oxygen isotopes. Journal of Anthropological Archaeology 2012;31(3) 334-352.

[1021] BatailleCP, BowenGJ. Mapping $^{87}Sr/^{86}Sr$ variations in bedrock and water for large scale provenance studies. Chemical Geology 2012;304-305 39-52.

[1022] KremenovićA, ColombanPh, PiriouB, MassiotD, FlorianP. Structural and spectroscopic characterization of the quenched hexacelsian. Journal of Physics and Chemistry of Solids 2003;64(11) 2253-2268.

[1023] Romero-Serrano A, Cruz-Ramirez A, Zeifert B, Hallen-Lopez M, Hernandez-Ramirez A. Thermodynamic Modeling of the $BaO–SiO_2$ and $SrO–SiO_2$ Binary Melts. Glass Physics and Chemistry 2010;36(2) 171-178.

[1024] Zhou L, Guo J, Yang N, Li L. KiSolid-state magnetic resonance and infrared spectroscopy of alkali feldspars. Science in Cguba (Series D) 1997;40(2) 159-165.

[1025] Le Parc R, Champagnon B, Dianoux J, Jarry P, Martinez V. Anorthite and $CaAl_2Si_2O_8$ glass: low frequency Raman spectroscopy and neutron scattering. Journal of Non-Crystalline Solids 2003;323 155-161.

[1026] Velde B, Syono Y, Couty R, Kikuchi M. High pressure infrared spectra of diapletic anorthite glass. Physics and Chemistry of minerals 1987;14 345-349.

[1027] Dowty E. Vibrational interactions of tetrahedra in silicate glasses and crystals: II. Calculations on melilites, pyroxnes, silica polymorphs and feldspars. Physics and Chemistry of minerals 1987;14 122-138.

[1028] Yang Y, Min Y, Lococo J, Jun Y.-Sh. Effects of Al/Si ordering on feldspar dissolution: Part I. Crystallographic control on the stoichiometry. Geochimica et Cosmochimica Acta 2014;126 574-594.

[1029] Anbalagan G, Sankari G, Ponnusamy S, Thilak kumar R, Gunasekaran S.Investigation of silicate mineral sanidine by vibrational and NMR spectroscopic methods. Spectrochimica Acta Part A: Molecular and Biomolecular Spectroscopy 2009;74(2) 404-409.

[1030] Orlova LA, Popovich NV, Uvarova NE, Paleari A, Sarkisov PD.High-temperature resistant glass-ceramics based on Sr-anorthite and tialite phases. Ceramics International2012;38(8) 6629-6634.

[1031] Ibrahim DM, Mostafa AA, Khalil T. Preparation of tialite (aluminium titanate) via the urea formaldehyde polymeric route. Ceramics International 1999;25(8) 697-704.

[1032] Zabicky J, Kimmel G, Yaaran J, Zevin L.Thermal anisotropy of tialite (Al_2TiO_5) by powder XRD. Nanostructured Materials1995;6(5-8) 675-678.

[1033] WangS.-F, HsuY.-F, LuH.-Ch, LoS.-Ch, ChengCh.-S. B_2O_3-free SiO_2–Al_2O_3–SrO–La_2O_3–ZnO–TiO_2 glass sealants for intermediate temperature solid oxide fuel cell applications. International Journal of Hydrogen Energy 2012;37(7) 5901-5913.

[1034] OjhaPK, ChongdarTK, GokhaleNM, KulkarniAR. Investigation of crystallization kinetic of SrO–La_2O_3–Al_2O_3–B_2O_3–SiO_2 glass and its suitability for SOFC sealant. International Journal of Hydrogen Energy 2011;36(22) 14996-15001.

[1035] MahapatraMK, LuK, Reynolds Jr WT. Thermophysical properties and devitrification of SrO-La_2O_3-Al_2O_3-B_2O_3-SiO_2-based glass sealant for solid oxide fuel/electrolyzer cells. Journal of Power Sources 2008;179(1) 106-112.

[1036] Mallik A, Kundu P, Basumajumdar A. Nucleation, crystallization behavior and microstructure of mica glass-ceramics in the system $SrO\cdot 4MgO\cdot xAl_2O_3\cdot 6SiO_2\cdot 2MgF_2$ (x=1, 1.5 and 2). Ceramics International 2013;39(6) 6963-6969.

[1037] Duan F. The role of PbO on crystallization in PbO-SrO-TiO_2-SiO_2 glass. Journal of Non-Crystalline Solids2010;356(43) 2286-2288.

[1038] Hazeli K, Sadeghi A, Pekguleryuz MO, Kontsos A. The effect fo strontium in plasitity of magnesium alloys. Materials Science & Engineering A 2013;578 383-393.

[1039] Niu J.-x, Chen Q.-r, Xu N.-x, Wei Z.-l. Effect of combinative addition of strontium and rare earth elements oncorrosion resistance of AZ91D magnesium alloy. Transactions of Nonferrous Metals Society of China 2008;18 1058-1064.

[1040] SadeghiA, PekguleryuzM. Microstructure, mechanical properties and texture evolution of AZ31 alloy containing trace levels of strontium. Materials Characterization 2011;62(8) 742-750.

[1041] TimpelM, WanderkaN, SchlesigerR, YamamotoT, LazarevN, IsheimD, SchmitzG, MatsumuraS, BanhartJ. The role of strontium in modifying aluminium-silicon alloys. Acta Materialia 2012;60(9) 3920-3928.

[1042] ShabestariSG, KeshavarzM, HejaziMM. Effect of strontium on the kinetics of formation and segregation of intermetallic compounds in A380 aluminum alloy. Journal of Alloys and Compounds 2009;477(1-2) 892-899.